Ecological Consequences of Climate Change

Mechanisms, Conservation, and Management

Edited by
Erik A. Beever and Jerrold L. Belant

CRC Press
Taylor & Francis Group
Boca Raton London New York

CRC Press is an imprint of the
Taylor & Francis Group, an **informa** business

CRC Press
Taylor & Francis Group
6000 Broken Sound Parkway NW, Suite 300
Boca Raton, FL 33487-2742

First issued in paperback 2017

Version Date: 2011912

ISBN-13: 978-1-4200-8720-8 (hbk)
ISBN-13: 978-1-138-11469-2 (pbk)

Visit the Taylor & Francis Web site at
http://www.taylorandfrancis.com

and the CRC Press Web site at
http://www.crcpress.com

Contents

SECTION III Higher-Level Ecological Relationships to Climate Change

SECTION IV Monitoring Ecological Consequences of Climate Change

SECTION V Conservation Efforts in the Face of Rapid Climate Change

SECTION VI Conclusions and Future Research Needs

Preface

Although the ultimate impetus for the book arguably had its nascence decades earlier in the Industrial Revolution, the idea for this book came in a practical sense during the lead-up to a symposium on biological responses to contemporary climate change at the 2007 international conference of The Wildlife Society in Tucson, Arizona. Dr. Erik Beever has been researching the changing distribution of an alpine mammal (*Ochotona princeps*, the American pika) in the Great Basin of western North America since 1994, and although climate was among the original suite of hypothesized determinants of the species' distribution since the beginning of the research, continuing investigation has suggested that climatic influences have played a stronger role in shaping the pattern of local extinctions in the years after 1999 compared to during the latter half of the twentieth century. In that research, as in much of the early research on biological response to contemporary climate change, the key challenge has been trying to divine the "why" and the "how" behind the "what" and "when"—that is, the *mechanisms* behind any changes in ecological systems in response to contemporary climate change.

We have consequently asked all authors to focus part of the discussion in each of their chapters on mechanisms, for two reasons. Understanding the hows and whys of these responses not only provides a more complete understanding of the dynamics associated with ecological responses, but also it is exactly that understanding that is needed to best inform strategies of mitigation, adaptation, conservation, and management of affected ecosystems and their components. Mechanisms are just beginning to be addressed rigorously, for good reason—they require much more in-depth understanding of the relevant natural history of species and ecosystem components involved, more critical thinking before beginning the fieldwork, and often more time and money than if mechanisms were ignored (i.e., simply documenting any changes). It quickly becomes apparent that much still remains to be learned, and humility in the face of so many unknowns and sources of uncertainty seems prudent. We feel that it is this focus on mechanisms—as well as our melding of empirical research, management, and conservation—that will distinguish this book from its contemporaries and perhaps even from later books on the topic.

Both editors have spent their careers working around the nexus of research and management of ecological systems, and this book clearly reflects that applied approach. We have asked authors to not only provide case studies to exemplify their messages throughout their chapter, but also to provide the "so what?" implications of their chapter for management and conservation of the systems they discuss. We feel honored and fortunate to have been able to work with such an experienced, widely respected, and capable group of authors, who fortuitously also happen to be wonderful individuals with whom to work. They hail from six different nations around the world and work not only on diverse spatial and temporal scales, but also on a vastly divergent collection of ecosystem components. Our vision was that this heterogeneity would best exemplify the diversity of biological responses that may be expected in the coming decades (and have already begun to be observed).

Mirroring the progression of research surrounding ecological responses to contemporary climate changes, we have sought to first describe how climate is changing, then describe the basic responses of wildlife and other ecosystem components to climatic variability and change, then discuss the management strategies in response to such climatic influence, then consider implications for various scales of conservation, and then finish with a vision to the future that springs from what we know already to explore gaps in our understanding and research frontiers. We suspect that although the novelty of investigating biotic response to climate change may wane with time, climate will remain a pervasive and profound driver of ecosystem dynamics for decades to come, and likely with increasing strength. Thus, our hope is that this book initiates discussions, foments critical reviews of the ideas contained within, and informs future research, management, and conservation during this period of the worldwide natural experiment in which we currently find ourselves engaged.

Acknowledgments

An undertaking of this magnitude could never have come to fruition without the concerted effort of numerous individuals. Dr. Beever would like to thank some of the prominent biology-oriented and other mentors in his life, including Guy Malain, Sheila Ward, Tom Schoener, Frank Joyce, Peter Brussard, David Pyke, Mary Pat White, and his parents; special thanks are due to his wonderful wife Yuriko for her patience and support. Dr. Belant similarly thanks Layne Adams, Richard Dolbeer, Erich Follmann, and Bruce Leopold for their professional guidance, and Mary-Kay Belant for her support and encouragement throughout this process.

The patient staff at Taylor & Francis—Pat Roberson, Jill Jurgensen, Robin Lloyd-Starkes, and especially Randy Brehm—helped shepherd the organizational process along. Chapter authors deserve special thanks for their willingness to endure multiple rounds of editing necessary to achieve a polished final form.

Thanks are also due to the Biological Diversity Working Group of the Wildlife Society, from which a symposium sprung that fostered the germination of this book. Finally, we commend researchers seeking to "push the envelope" (not the bioclimatic type) to develop theory, algorithms, and empirical research needed to observe, understand, and interpret the multitude of biological responses to contemporary climate changes, as well as wildlife and conservation practitioners in land-managing agencies, non-governmental organizations (NGOs), and other organizations working "in the trenches" to balance livelihoods with leaving a rich biotic heritage for future generations to enjoy and take sustenance and inspiration from, and to watch climate interact with other ecosystem drivers to shape the composition and function of biological systems.

Introduction

Contemporary climate change looms as one of the most profound ecological phenomena of the 21st century. Climate change has motivated an extensive body of research across a diverse array of topics including ecological niche theory, landscape ecology, community organization, and biogeography. Climatologists, hydrologists, and physical scientists have already documented changing rainfall patterns, increased frequency of extreme climate events, decreased snow-water equivalents, earlier timing of spring melt runoff, and increased average temperatures, among numerous other climate-related phenomena (IPCC 2007, Serreze 2010). These climate changes are not within the range of typical variability of expected natural perturbations, but rather are accelerated changes considered a consequence of anthropogenic events. Not surprisingly, climate "fingerprints" have already been detected in the responses of hundreds of wildlife species, clades or functional groups of species, communities, and numerous ecosystems around the world (Root et al. 2003). Broadly speaking, four types of changes in species' traits may occur as a result of global warming. First, densities of species may change at given locations, resulting in different patterns of occupancy and altered community composition. Second, phenological changes in key natural-history traits (e.g., migration, flowering, and egg laying) that are triggered by temperature-related cues can occur. Third, changes in morphology, behavior, growth rates and body size, or demographic changes such as changes in sex ratios, can occur. Finally, genetic frequencies may shift spatially through time.

Numerous efforts to monitor contemporary climate change and its ecological effects are currently underway, including programs on local to global scales. Understanding climate change and its ecological effects across a spatial and temporal hierarchy will provide knowledge critical for effective management. Whereas historically managers have most often dealt with more proximate and spatially and temporally constrained threats, climate change represents a more ultimate, broad-scale driver of ecological change. Novel and large-scale conservation strategies in the face of this almost unprecedented rate of change are beginning to appear. However, the effectiveness of any one strategy is difficult to estimate, and forecasting the utility of a multi-faceted management approach is even more difficult to predict.

There is considerable need to better understand how recently observed rapid shifts in aspects of climate are currently altering the biotic landscape, and to contemplate how these climate shifts may be expected to drive future changes. Combining modeling with a wide range of real-world examples, this book highlights trends expected in species, communities, and overall ecological responses, and the exceptions and contingencies that may mediate those responses. Considerations for researchers and managers will include quantifying the magnitude and variability in responses to recent climate change, understanding relative vulnerability among species and community types, designing investigations and management actions, and identifying major gaps in our current knowledge.

Although many sources have offered relatively comprehensive treatments of the subject of scale (e.g., Beever et al. 2006, Wiens and Bachelet 2010), its importance to species and their ecological relationships in response to contemporary climate change remains under-appreciated by many researchers and managers. Many challenges remain to adequately address issues of scale in both ecological and jurisdictional contexts. For example, responses of individual species or communities will likely ignore administrative boundaries, unless they indicate changes in connectivity or habitat quality. Consequently, extensive collaboration will be required to address these biological responses. Furthermore, biologists and ecologists rarely know the temporal and spatial scales at which organisms experience their environment and respond to contemporary climatic change. For example, for many species it is not clear whether acute climatic exposure, prolonged climatic stress, or a combination of both causes physiological tolerances to be exceeded. Finally, we have relatively few studies that integrate physical climate models with data from either broad extents or higher trophic-level taxa (e.g., Lawler et al. 2010).

A primary intent of this book is to provide a systematic overview, from addressing species-level responses to contemporary climate change to monitoring and management strategies used to quantify and attempt to address the observed effects on species and the ecosystems where they reside. Authors of the various chapters approach these issues by demonstrating the effects of changing temperature and precipitation on diverse taxa in considerable detail. Authors address not only direct and indirect effects of climate-induced stress on these taxa, but also higher-level ecological interactions. In addition, management actions and conservation planning that may be employed in response to species and system-level responses to changing climate are outlined.

The book is divided into several sections, beginning with an overview of physical processes and meteorological mechanisms of climate change, drawing heavily on western North America as an example to understand these processes. Section II provides a diverse array of case studies addressing single- and multiple-species relationships to climatic influence change. Section III demonstrates the importance of higher-level ecological relationships and how they influence (and complicate) our understanding of climate-change effects. A series of monitoring approaches developed and implemented to track changes in biological and ecological responses to rapid change is contained in Section IV. Section V provides specific fine- and coarse-scale guidance on management strategies that may aid conservationists in addressing short- and long-term effects of climate change. We conclude with a synthesis of the state of our knowledge and provide recommendations for areas of research likely to advance our understanding of climate change and the ecological consequences of this change.

An original theme of this book was the pursuit, to the fullest extent possible, a mechanistic understanding of the effects of climate-change on organisms and ecological processes (Walther et al. 2002, Root and Schneider 2006). Although considerable progress has been made to this end, what became clear in many cases were the limits of our current knowledge of detailed mechanisms influencing biological responses and ecological processes in the face of rapid climate change. Nonetheless, the attentive reader will note that this theme is pervasively addressed throughout this book, across all levels of biological organization as well as diverse climatic, topographic, and spatio-temporal contexts. This is important both initially for understanding causation, and ultimately for refining our predictive abilities of climate-change effects on species, systems, and their interactions (Walther et al. 2002, Root and Schneider 2006, Beever et al. 2010).

Rapid changes in climate are dramatically affecting virtually every aspect of ecology and our society. Our hope is that this book will provide a foundation for greater understanding of the ecological effects of contemporary climate change and will be of interest to a broad audience, including students, conservationists, and natural resource managers. Anthropogenic rapid climate change is one of the most pervasive and critical issues facing us today; adapting to, slowing, and ultimately adapting to the effects of climate change from both ecological and societal perspectives will require continued acquisition of relevant knowledge and using that knowledge to implement effective strategies.

REFERENCES

Beever, E. A., C. Ray, P. W. Mote, and J. L. Wilkening. 2010. Testing alternative models of climate-mediated extirpations. *Ecological Applications* 20:164–178.

Beever, E. A., R. K. Swihart, and B. T. Bestelmeyer. 2006. Linking the concept of scale to studies of biological diversity: evolving approaches and tools. *Diversity and Distributions* 12:229–235.

Cole, K. L. 2010. Vegetation response to early Holocene warming as an analog for current and future changes. *Conservation Biology* 24: 29–37.

IPCC (Intergovernmental Panel on Climate Change). 2007. Climate change 2007: the physical science basis. Cambridge University Press, Cambridge, United Kingdom.

Lawler, J. J., S. L. Shafer, B. A. Bancroft, and A. R. Blaustein. 2010. Projected climate impacts for the amphibians of the western hemisphere. *Conservation Biology* 20:38–50.

Root, T. L., and S. H. Schneider. 2006. Conservation and climate change: the challenges ahead. *Conservation Biology* 20:706–708.

Root T. L., J. T. Price, K. R. Hall, S. H. Schneider, C. Rosenzweig, and A. J. Pounds. 2003. Fingerprints of global warming on wild animals and plants. *Nature* 421:57–60.

Serreze, M. 2010. Understanding recent climate change. *Conservation Biology* 24:10–17.

Walther, G., E. Post, P. Convey, A. Menzel, C. Parmesan, T. J. C. B. J. Fromentin, O. Hoegh-Guldberg, and F. Bairlein. 2002. Ecological responses to recent climate change. *Nature* 416:389–395.

Wiens, J. A., and D. Bachelet. 2010. Matching the multiple scales of conservation with the multiple scales of climate change. Conservation Biology 24:51–62.

About the Editors

Dr. Erik A. Beever received his BS in biological sciences from the University of California, Davis in 1993 and his PhD in ecology, evolution, and conservation biology from the University of Nevada, Reno, in December 1999. He has published over 50 articles in diverse scientific journals and in numerous subdisciplines of biology. He has performed field research on plants, soils, amphibians, birds, reptiles, fishes, and insects, as well as small, medium, and large mammals. His work has spanned salt-scrub, sagebrush-steppe, alpine, subalpine, subarctic, riparian, primary and secondary temperate and tropical forests, and coastal ecosystems of the western hemisphere. In addition to seeking to understand mechanisms of biotic responses to climate change, he has also focused on disturbance ecology and monitoring in conservation reserves, all at community to landscape scales, as well as other topics of conservation ecology, wildlife biology, and landscape ecology. He is a member of the IUCN Protected Areas Specialist Group, the IUCN Lagomorph Specialist Group, as well as the Wildlife Society, the Society for Conservation Biology, the American Society of Mammalogists, Sigma Xi, and the Union of Concerned Scientists.

Dr. Jerrold L. Belant is an associate professor of wildlife ecology and management, and director of the Carnivore Ecology Laboratory at Mississippi State University. He received his PhD from the University of Alaska, Fairbanks. Dr. Belant has authored over 100 publications in wildlife ecology, conservation, and management. He is the chair of the IUCN Species Survival Commission's Small Carnivore Specialist Group and a member of the International Federation of Mammalogists. Dr. Belant is also editor of *Small Carnivore Conservation* and associate editor for *Ursus*, *Natural Areas Journal*, and *Latin American Journal of Conservation*.

Contributors

Craig D. Allen
U.S. Geological Survey
Biological Resources Division
Fort Collins Science Center
Los Alamos, New Mexico

Betsy A. Bancroft
College of Forest Resources
University of Washington
Seattle, Washington

Jill S. Baron
U.S. Geological Survey
Fort Collins Science Center
Los Alamos, New Mexico

Erik A. Beever
Northern Rocky Mountain Science Center
U.S. Geological Survey
Bozeman, Montana

Jerrold L. Belant
Carnivore Ecology Laboratory and Forest and Wildlife Research Center
Mississippi State University
Starkville, Mississippi

Andrew R. Blaustein
Department of Zoology
Oregon State University
Corvallis, Oregon

Nigel Dudley
Equilibrium Research
Bristol, United Kingdom

Liesl Erb
Department of Ecology and Evolutionary Biology
University of Colorado
Boulder, Colorado

Daniel B. Fagre
U.S. Geological Survey
Northern Rocky Mountain Science Center
Glacier National Park
West Glacier, Montana

Erica Fleishman
Bren School of Environmental Science and Management
University of California, Santa Barbara
Santa Barbara, California
and
University of California, Davis
Davis, California

Andrew G. Fountain
Department of Geology
Portland State University
Portland, Oregon

Douglas G. Goodin
Department of Geography
and
Konza Prairie LTER Program
Kansas State University
Manhattan, Kansas

Robert Guralnick
Department of Ecology and Evolutionary Biology
and
University of Colorado Museum of Natural History
University of Colorado
Boulder, Colorado

David Gutiérrez
Área de Biodiversidad y Conservación
Escuela Superior de Ciencias Experimentales y Tecnología
Universidad Rey Juan Carlos, Móstoles
Madrid, Spain

Patricia J. Heglund
U.S. Fish and Wildlife Service
La Crosse, Wisconsin

Jeffrey A. Hicke
Department of Geography
University of Idaho
Moscow, Idaho

Alan R. Jones
Division of Marine Invertebrates
Australian Museum
Sydney, Australia

Melinda G. Knutson
U.S. Fish and Wildlife Service
La Crosse, Wisconsin

Linda Krueger
Wildlife Conservation Society
Bronx, New York

Joshua Lawler
College of Forest Resources
University of Washington
Seattle, Washington

Kathy MacKinnon
Haddenham, Cambridge
United Kingdom

George P. Malanson
Department of Geography
University of Iowa
Iowa City, Iowa

Enrique Martínez-Meyer
Instituto de Biología
Universidad Nacional Autónoma de México
Mexico City, Mexico

Donald McKenzie
USFS PWFS Lab
University of Washington
Seattle, Washington

Donald McLennan
Parks Canada Agency
Hull, Quebec, Canada

Philip W. Mote
Oregon Climate Change Research Institute
and
College of Oceanic and Atmospheric Sciences
Oregon State University
Corvallis, Oregon

Dennis D. Murphy
Department of Biology
University of Nevada
Reno, Nevada

Dennis S. Ojima
Natural Resource Ecology Laboratory
Colorado State University
Fort Collins, Colorado

David L. Peterson
Pacific Northwest Research Station
U.S. Forest Service
Seattle, Washington

Chris Ray
Department of Ecology and Evolutionary Biology
University of Colorado
Boulder, Colorado

Kelly T. Redmond
Western Regional Climate Center
Desert Research Institute
Reno, Nevada

Catherine Searle
Department of Zoology
Oregon State University
Corvallis, Oregon

Nathan L. Stephenson
Western Ecological Research Center
Sequoia-Kings Canyon Field Station
U.S. Geological Survey
Three Rivers, California

Sue Stolton
Equilibrium Research
Bristol, United Kingdom

Christina L. Tague
Bren School of Environmental Science and Management
University of California, Santa Barbara
Santa Barbara, California

Phillip J. van Mantgem
Western Ecological Research Center
Redwood Field Station
U.S. Geological Survey
Arcata, California

Robert J. Wilson
Centre for Ecology and Conservation
University of Exeter Cornwall Campus
Penryn, Cornwall, United Kingdom

Cover Photo Credits

Top Left. Consolidated pancake ice off the coast of Greenland. Decreases in the extent of polar sea ice have been one of the most pronounced inter-annual trends in the world, among abiotic indicators of recent climate change. Photo courtesy of Andy Mahoney, National Snow and Ice Data Center.

Top Right. Newborn Olive Ridley turtle (*Lepidochelys olivacea*), on a beach in Junquillal, on the Pacific coast of Costa Rica. Because the sex of marine turtles is determined by the incubation temperature of eggs in the sand, forecasts of increasing temperatures have spawned concern that over-abundance of females and scarcity of males may compromise population recovery. In addition, rising sea levels may erode nesting beaches in which infrastructure prevents their gradual shift landward. Photo copyright of World Wildlife Fund International/Carlos Drews.

Middle Left. Blue-spotted salamander (*Ambystoma laterale*), on Rocky Island, Apostle Islands National Lakeshore, Wisconsin, USA. Altered climatic conditions, UV radiation, and indirect effects of climate (e.g., altered suitability for pathogens and invasive species) have been suggested as contributors to recent declines in numerous amphibian species globally. Photo courtesy of Eric Ellis.

Center. Upper-elevation ecosystems of the Andes in Argentina, near Aconcagua—the highest point in the western and southern hemispheres. Montane ecosystems store drinking water for most of the world's humans, and have been demonstrated to exhibit greater vulnerability to contemporary climate changes than their lowland-habitat counterparts. Photo by Erik Beever.

Middle Right. An American pika (*Ochotona princeps*) in talus of the East Humboldt Range, northwestern Nevada, USA. This species' distribution has changed dramatically within the Great Basin since the early 20th century, apparently in association with changes in climate. Photo by Shana Weber.

Bottom Left. Shrubs, trees, and native and invasive herbaceous plants in Mojave National Preserve, southeastern California, USA. Changes in aspects of climate have been associated with the phenology of plant flowering and with dynamics of plant-species invasions. Photo by Benjamin Chemel.

Bottom Right. Hundreds of blue wildebeests (*Connochaetes taurinus*) and plains zebras (*Equus burchellii*) along their migration route through Serengeti National Park, Tanzania, Africa. This multi-species migration, the largest migration of large mammals in the world, exemplifies concerns about landscape connectivity, as community associations re-assemble and ecological niches shift spatially, under changing climate. Photo by Erik Beever.

Section I

The Basis of Recent Climate Change: Climate-Science Foundations

1 Western Climate Change

Philip W. Mote and Kelly T. Redmond

CONTENTS

INTRODUCTION

Earth's global climate is determined by a balance between absorbed solar radiation and emitted infrared radiation. The amount of absorbed solar radiation in turn is determined by the sun's emissions and the Earth's reflectivity, primarily the fraction of the planet covered by clouds and ice. Infrared emissions come predominantly from gases in the atmosphere: water vapor, carbon dioxide, methane, nitrous oxide, and many more. The atmosphere also emits energy toward Earth, keeping it warmer than it would otherwise be, and providing roughly twice the energy as is provided by absorbed solar energy (e.g., Trenberth et al. 2009). At the surface, the absorbed solar energy plus atmospheric infrared energy is balanced globally by radiation of infrared radiation plus latent and sensible heat flux (Trenberth et al. 2009), all of which are mediated by vegetation, especially moisture fluxes. In turn, the expression of global climate and of atmospheric fluctuations helps determine the distribution, health, function, reproductive rates, and much more, of organisms on the landscape.

Through technology and sheer numbers, humans have acquired the ability to modify climate in many ways. Chief among these are the production of (1) greenhouse gases, such as carbon dioxide, ozone, nitrous oxide, methane, chlorofluorocarbons; and (2) aerosols that originate from disturbed soils, soot, ash, pollution, and gases transformed through photochemistry to particles. All of these affect the flow of radiant energy through air. Deforestation, irrigation, agricultural practices, paving, and other kinds of development change land surface properties and influence the dynamics of energy exchange, heat transfer, and surface winds. Recent findings indicate that changes in atmospheric particle concentration can greatly alter cloud properties and reduce precipitation efficiency and amount (e.g., Forster et al. 2007). Anticipated changes in temperature may also affect precipitation type (rain or snow). Changes in atmospheric CO_2, ozone, and other gaseous and aerosol constituents have direct but differential physiological effects on vegetation, species competitiveness, and amount and quality of light, which in turn affect soil moisture and recharge budgets, plant species composition, and community properties.

Assessing historical and future biological responses to global climate change requires an understanding of two causal connections. The first is the connection between the local climate variables and the biological system of interest. Because so many factors are at work simultaneously, this determination is seldom straightforward. For instance, temperature may affect tree growth; however, other climatic and nonclimatic factors like competition for light or other resources might be as important. Put differently, if we knew perfectly how climate would change in the immediate vicinity of an ecological community, how well could we predict the ecological response? The second is the connection between the biologically important local climate variables and global climate drivers like greenhouse gases. In the field of climate research, this connection is called "detection and attribution." These involve whether a change has actually been observed, within measurement error (detection), and whether any such change (e.g., in global average temperature) could have occurred naturally or can confidently be attributed to human activity (attribution; Stott et al. 2000; Broccoli et al. 2003; Meehl et al. 2004). Attribution is most successful when the signal-to-noise ratio is high, that is, when the response of the variable in question to greenhouse-gas forcing is large relative to natural variability. Keeping the signal-to-noise ratio large typically requires considerable spatial averaging and a long period of record (>50 yr) for analysis. The two causal connections are in tension, owing to the inherently conflicting spatial and temporal scales. Detection and attribution are clearest at the global scale over multiple decades, but responses of ecosystems or species to climate are often clearest at the local scale and at shorter time periods.

In this chapter we highlight the meteorological and physical background of observed climate variability and change, and recent attribution efforts related to contemporary climate change. We also describe scenarios of future climate for the western United States. Climate is a principal driver of the natural and managed environmental systems of the western United States, and is such a pervasive influence that its properties and behavior in space and time must be taken into account and factored into the management of western lands and resources (Redmond 2007).

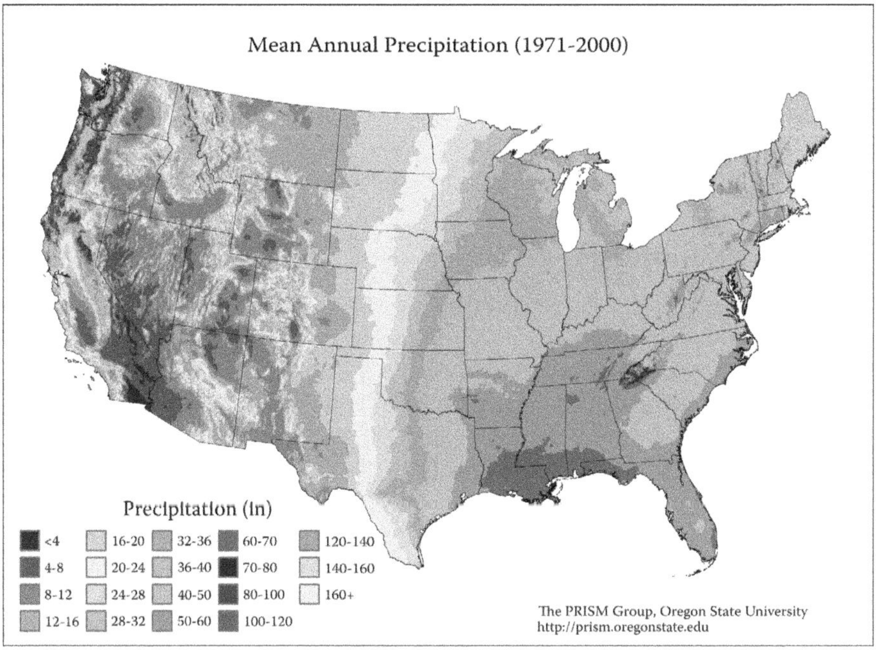

FIGURE 1.1 **(See color insert.)** Mean annual precipitation from PRISM (Daly et al. 2002, 2004, 2008) for the 1971–2000 period of record. Note the sharp gradients in much of the West.

OBSERVED CLIMATE VARIABILITY AND CHANGE IN THE WESTERN UNITED STATES

The western United States is a land of juxtapositions and sharp contrasts in physiography, climate, vegetation, and other biophysical attributes. In addition to the sharp contrasts across short distances, the West has striking seasonality in precipitation; in most of the West, summer precipitation is substantially less than winter precipitation. Fundamental attributes of average climate, notably precipitation, can change greatly over short distances (Figure 1.1), as can precipitation seasonality, annual amount, and phase (i.e., rain vs. snow). Properties of temporal variability can also vary over short distances (Redmond 2003). Elevation plays a key role in shaping the patterns of temperature and precipitation (Daly et al. 2008), and mountain ranges greatly modify and sometimes cause their own weather. Mountain time series of climatic variables can be very different from those in the adjoining valleys. Large-scale "teleconnections" with other parts of the globe lead to spatially different responses in reaction to faraway phenomena such as tropical El Niño and La Niña events. Examples of observed variability and change, and global processes affecting climate in the western United States are described in the next section.

The fundamental issue is the following: to fully understand the interrelation between ecosystems and the climate system, we must ideally first understand the properties of spatial and temporal variability (and in addition, combined spatiotemporal variability) of each of these two sets of systems, across their characteristic range of spatial and temporal scales. Ecological systems have evolved to selectively take advantage of regularities in physical environmental drivers (such as climate) across a very large range of scales, and in addition must respond to more stochastic variability in both space and time of these drivers, also across a broad range of scales, from that of a stomate or needle (1 mm) to the globe (10,000 km), a range of scales that encompasses approximately 7–10 orders of magnitude.

Temperature Variability and Change

Studies of variability and change in climate variables utilize several common gridded datasets. Station data are the basis for all of these datasets, which make use of different ways of aggregating or averaging station data over regions. Here we use gridded 0.5° (longitude) × 0.5° (latitude) annual mean temperature data from CRUv2.1 (Brohan et al. 2006), which have been widely used in global-trends analyses. We use a domain from the Pacific Ocean to 107.5° west longitude, and from 30 to 52.5° north latitude, and select the longest period for which all grid points have complete records: the 1920–2008 period of record. Figure 1.2 shows the trends in annual mean temperature from the HadCRU dataset. There are a few patches of negative trends over this time period, but for most of the western United States, the trends have been upward.

In order to better understand the time dimension of these changes, we regionally average the HadCRU data to produce a West-wide annual mean temperature each year. Only HadCRU grid points with at least some data in the first 5 and last 5 years are used. We also use the regionally averaged temperature data from WestMap (www.cefa.dri.edu/Westmap) derived from the PRISM dataset (Daly et al. 2008). These two datasets are derived from different station data and give somewhat different results that depend on how elevation–temperature relationships are treated, resulting in a systematic difference stemming from systematic station-grid elevation differences. The time series for the regionally averaged temperature (Figure 1.3) shows a strong upward trend, reflecting the warming of the West during the time period of analysis. The magnitude and frequency of negative anomalies dwindled during the 1970s and 1980s, as nearly every year since 1985 has been near average or above average in temperature. The record warmest year remains 1934 but the warmest 10- and 20-year periods are recent. The two time series differ the most in the early years and consequently have different trends (0.6°C for HadCRU and 1.0°C for WestMap). Slow variations highlighted by the (smoothed) curves are substantially the same, with a bit of warming between about 1910 and 1930, fairly level temperatures until 1970, and then warming.

Dependence of Trends on Elevation

For the mountainous West, a critical question about long-term change concerns the relative rates of warming at mountaintops, mid-slopes, and valley floors. Do these rates differ, and if so, do they vary among the seasons? Whether these rates should be similar depends ultimately on the physical mechanisms for potential variation in rates across elevations. Unfortunately, long-term climate stations in mountainous regions are fairly rare: for example, the state of Washington has no climate-quality stations above 1300 m that provide full annual measurements before 1945 and

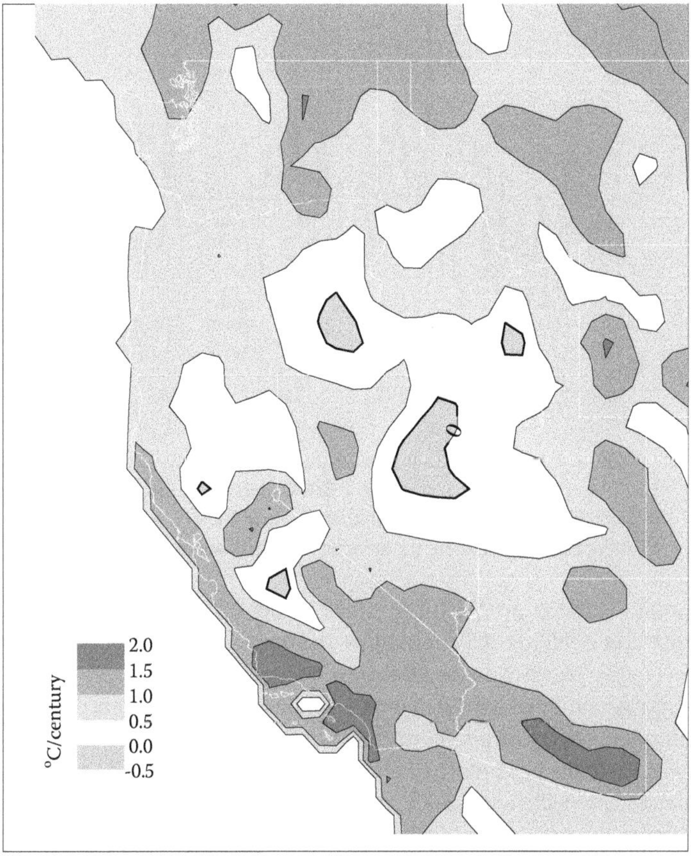

FIGURE 1.2 (See color insert.) Linear trends in temperature (°C/century) from the HadCRU 0.5° × 0.5° dataset, evaluated over the 1901–2000 period. The contour interval is 0.5°C per century.

continue today. In most western states there are at least a few stations above 2000 m, but most are in valley bottoms. In many areas the data are too sparse to draw a conclusion about whether trends in temperature depend on elevation alone.

In evaluating temperature trends in the mountains, for example, the issue of whether temperature trends depend on elevation, a critical question concerns the separate roles of "advection" (the heat carried by the wind) and local surface-energy balance, and different studies have reached different conclusions. Diaz and Bradley (1997) analyzed surface-temperature records at 116 sites and found that many high-elevation sites had warmed more than lower-elevation sites during the twentieth century, but questions have been raised about whether these findings represent true surface trends or are determined by the varying exposure to advection. Changes in free-air temperature (away from the surface, measured by balloon) have generally exceeded surface temperature changes (Karl et al. 2006), but not all studies reach this conclusion (Pepin and Losleben 2002; Vuille and Bradley 2000). Most studies found no clear relationship

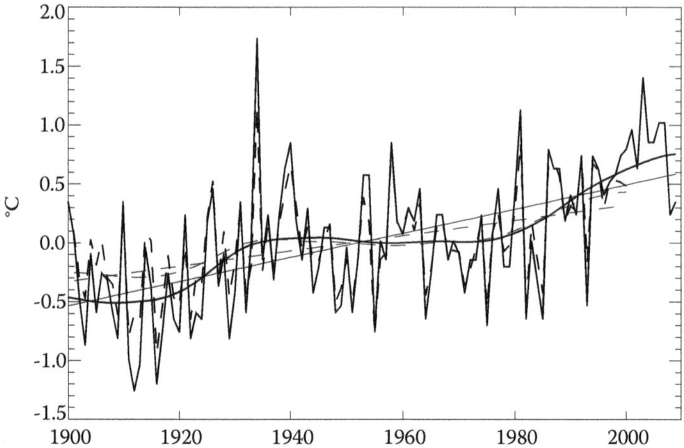

FIGURE 1.3 Values of annual mean temperature over the western United States calculated from the CRUv2.1 dataset (dashed) and the PRISM dataset (solid). Linear trends (straight lines) and slow variations (curves) were calculated using locally weighted regression (loess; Cleveland 1995). For both curves, the mean over the entire record is subtracted.

with elevation (Vuille et al. 2003; Pepin and Seidel 2005; Liu et al. 2006; You et al. 2008). Pepin and Seidel (2005) noted that the correspondence between surface and free-air variability and trends depended on the convexity (hilltop vs. valley or frost hollow) of the terrain in which the station was situated, such that variability at stations higher than surrounding terrain more closely resembled that of the free air, while stations in valley bottoms were more likely to differ from the free air.

As illustrated by Pepin and Seidel, the key question in evaluating the dependence of trends on elevation is the extent to which surface temperatures are determined by local energy balance as opposed to mere exposure to free air. Local advection (drainage winds, upslope winds) can exert significant influence as well. Some areas may remain for some time as cold-air pools and in effect serve as "climate refugia" (Ashcroft 2010; Petit et al. 2003; Bennett and Provan 2008). Another approach to evaluating trends that brought the surface-energy balance into clearer focus was the work of Pepin and Lundquist (2008). Examining trends globally, they noted that the largest warming trends were found at locations whose mean annual temperature was near 0°C, suggesting a strong role for snow-albedo feedback. This observational result was largely corroborated by the regional modeling work of Salathé et al. (2010), who noted the largest warming trends in a future scenario in montane areas presently near snowline. In short, there are ample reasons to believe that topographic complexity may produce considerable small-scale variability in change rates that could rival or exceed (positively or negatively) regionally averaged rates (Daly et al. 2009).

Variability and Changes in Precipitation

Although trends in temperature are positive almost everywhere in the western United States, trends in precipitation are far more diverse. Linear fits are a poor description

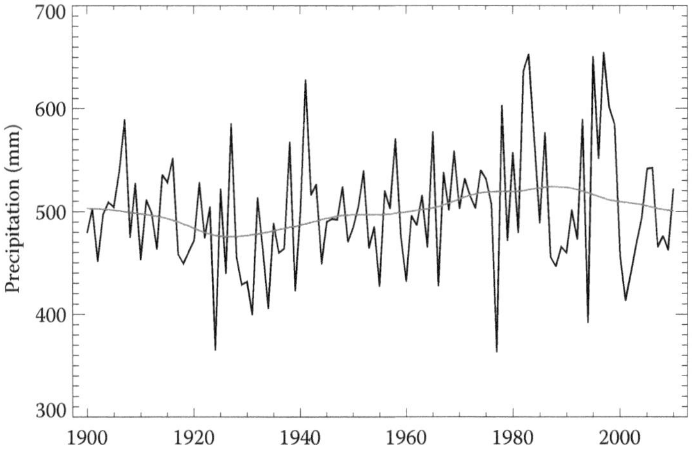

FIGURE 1.4 Water year (October–September) precipitation 1899–1900 through 2009–2010 for the 11 western United States. Data from 4-km resolution PRISM grid using the WRCC Westmap (www.cefa.dri.edu/Westmap/) application. Smooth grey curve calculated using loess as in Figure 1.3.

of the patterns of precipitation variability because the sign of the fit can change over short spatial distances and with slightly different periods of analysis. The time series of total precipitation (which includes the water equivalent in snowfall) for the October through September water year illustrate temporal patterns (Figure 1.4). The time series of average total precipitation for the 11 western states shows some interesting features, especially a pronounced increase in the mid-1970s in both the variance and the year-to-year persistence of precipitation anomalies, as has been noted by Hamlet et al. (2005) and Pagano and Garen (2005). Many of the extreme years—both dry and wet—have occurred since 1975. The increase in year-to-year persistence (i.e., multiyear episodes or regimes) is visible first as several unusually wet years in the early 1980s, which were followed by several unusually dry years in the late 1980s, then several wet years in the late 1990s, then several dry years in the early 2000s. There is little theoretical basis to expect such a shift to accompany rising greenhouse gases, and it may simply be a statistical artifact of a red noise time series.

Hydrologic Responses to Changes in Temperature

Fluctuations in streamflow are closely linked to fluctuations in precipitation, but a large body of literature emphasizes that western hydrology also responds to temperature. An analysis of fine-scale gridded meteorological data, specifically the fraction of annual precipitation falling at temperatures between 0°C and –6°C, what might be called warm snow, illustrates the West's hydrologic sensitivity to temperature fluctuations (Bales et al. 2006). Temperature increases of 2–4°C (likely to occur during the twenty-first century; see the following) during precipitation events could lead to a considerable increase in precipitation falling as rain rather than snow. The more immediate runoff has numerous hydrologic and water-management consequences. The analysis of Bales et al. (2006) shows that within certain elevation bands in the Cascade and Sierra Nevada Ranges, over half of the annual precipitation historically falls in this temperature range. The vulnerability to a change from snow to rain is next greatest in the mountains of Idaho and parts of the Great Basin, and least in the highest and thus coldest parts of Colorado and the southern High Sierra Nevada.

Consistent with the physical sensitivity analysis of Bales et al. (2006), several studies have demonstrated a statistical connection between fluctuations and trends in temperature and fluctuations and trends in various hydrologic variables. Knowles et al. (2006) used weather station data and reported that rain/snow ratios have increased in most of the West since about 1950, with spatial patterns resembling those of temperature change and the temperature sensitivity noted by Bales et al. (2006). Mote et al. (2005) and Hamlet et al. (2005), using observations and modeling, demonstrated that springtime mountain snowpack declined at roughly 75% of locations in the West since the mid-twentieth century. These changes were dependent on elevation (and thus temperature, because temperature usually decreases with elevation), with warmest locations losing the largest fraction of snow. Stewart et al. (2005) showed that streamflow in much of the West has changed in a manner consistent with the observations of declining mountain snowpack. In basins with a significant snowmelt contribution, winter and early spring flows generally increased, summer and late spring flows generally decreased, and the date of peak spring snowmelt shifted earlier by, on average, 2 weeks. Stewart et al. (2005), Hamlet et al. (2005), and Mote (2006) all evaluated the possible contributions of changes in precipitation and of changes in atmospheric circulation over the Pacific Ocean and concluded that the dominant factor in western trends in hydrology was the widespread increase in temperature unrelated to atmospheric circulation.

Increases in temperature with no change in precipitation can cause evapotranspiration (ET) to increase. Hidalgo et al. (2005) estimate an average temperature increase of +3°C could increase potential evapotranspiration (i.e., what evapotranspiration would be if not limited by water availability) by about 6% in California. However, the physiological response of plants to increased CO_2 concentration would likely act to reduce water loss.

Global Teleconnections

Spatial patterns of climate variability in the western United States are correlated with patterns of climate variability in other parts of the world. For example, winter precipitation in the West frequently exhibits a "dipole" pattern (wet in the Pacific Northwest and dry in the Southwest, or vice versa), and that this pattern is strongly related to tropical Pacific Ocean temperatures and to atmospheric pressure patterns in the Southern Hemisphere (Redmond and Koch 1991). The sense of the relationship is such that the phenomenon known as El Niño is associated with wet winters in the Southwest and dry winters in the Northwest and northern Rockies, and that La Niña is associated with dry winters in the Southwest and wet winters in the Northwest and northern Rockies.

There is much popular confusion between the El Niño phenomenon and its effects. El Niño refers to ocean warming in the top 100–200 meters in a narrow band between South America and the international date line, typically within 5° latitude of the equator. The effects of El Niño, by contrast, are global in reach. At time scales between the regular seasonal cycle and several years, El Niño is the single largest contributor to climate variability on earth. The warm area may look small on a map of the Pacific Ocean, but can easily be larger than the United States. The shape, magnitude, extent, duration, and longitudinal position of the warm-water patch can vary from one episode to the next—factors that can significantly influence the impacts of the phenomenon on the West (Hoerling and Kumar 2002). Typical events last 6–18 months and recur irregularly at 2- to 7-year intervals. La Niña refers to unusually cool temperatures in this same area.

El Niño exhibits characteristics of an oscillation in the sense that during one phase of the cycle, forces are at work that lead to the demise of that phase and often even the eventual growth of the opposite phase, like a very complicated pendulum, albeit one subject to irregular forcing by short-term weather events. The atmospheric pressure difference between stations at Tahiti and Darwin (Australia) is negatively correlated with ocean temperatures in the El Niño/La Niña area—a phenomenon known as the Southern Oscillation: it is the atmospheric counterpart of the oceanic El Niño. The magnitude of this correlation, usually strong, has varied somewhat through time (McCabe and Dettinger 1999), so the atmosphere and the ocean each carry somewhat different information. In recognition of the coupled oceanic and atmospheric nature of this vacillation, this phenomenon is called ENSO ("El Niño/Southern Oscillation").

In the Western United States, the effects of El Niño and La Niña are experienced during the cold half of the year, from approximately October through March; summer signals are very weak. The climatic effects of ENSO are also found in streamflow (Andrews et al. 2004; Barnett et al. 2004) where they are greatly accentuated with respect to precipitation (Cayan et al. 1999) in the western states. Because annual tree growth in the Southwest is strongly dependent on prior winter precipitation, these ENSO effects are clearly seen in tree ring widths (Swetnam and Betancourt 1990).

The frequency of El Niño has varied through time. During the period 1947–1976, El Niño occurred relatively infrequently and La Niña was common. A sudden and still unexplained change (the "1976–1977 shift") in the Pacific ushered in an era of much more common El Niño and a virtual dearth of La Niña. This appeared to many observers to have switched again in the late 1990s, although present evidence remains somewhat ambiguous.

In higher latitudes, this slow variation of about 50 years' duration is expressed in a pattern of ocean temperatures, atmospheric pressures, jet stream positions, and ocean currents seen from the tropics to the high latitudes in the Pacific, first described by Mantua et al. (1997) and Mantua and Hare (2002) as the Pacific Decadal Oscillation (PDO) and elaborated by others. They related the PDO to strong differences in salmon abundance between Alaska and the Pacific Northwest. There is much debate about the origin of the PDO, whether it truly is an oscillation, and even whether it really exists except as a response to ENSO (Zhang et al. 1996; Newman et al. 2003) with strong elements of chaotic behavior (Overland et al. 2000). Indeed, paleoclimatic evidence suggests that the low-frequency behavior of the PDO during the twentieth century is not a fundamental property of PDO: during most of the nineteenth century, variations had a much shorter time scale (Gedalof et al. 2002).

Climate models have performed poorly in simulating Pacific interdecadal variability, particularly transitions from one phase to the other, and in simulating the possible effects of Pacific variability on temperature or precipitation in western North America, despite tantalizing evidence that there are such effects. For now, ENSO has predictive value, but the PDO remains primarily diagnostic. Brown and Comrie (2004) saw the same western "dipole" at longer time scales in the PDO signal as seen on the shorter time scales of ENSO (e.g., Redmond and Koch 1991); McCabe and Dettinger (2002) see this dipole on both time scales in the historical western United States snow course data.

Another kind of teleconnection operates in the Pacific at intermediate time scales of approximately 40–70 days. Pairs of mostly cloudy and mostly clear regions slowly drift eastward from the Indian Ocean toward the western Pacific, a phenomenon known as the Madden-Julian Oscillation (MJO), or sometimes Intra-Seasonal Oscillation (ISO). Interactions with the eastward flowing jet stream coming off Asia can lead to multiday precipitation episodes on the West Coast 5–10 days later (Mitchell and Blier 1997; Mo 1999; Mo and Higgins 1998). These are important because much of the annual precipitation occurs in the largest 3–5 storms of the 20–25 that typically strike the coast of California in winter, and indeed there is a clear connection between MJO activity and flooding in the Northwest (Bond and Vecchi 2003).

These phenomena of Pacific climate variability exert a strong influence on western climate and complicate interpretation of long-term changes. Attempts to extend these records rely on proxy paleoclimatic data, such as tree rings (Biondi et al. 2001; D'Arrigo et al. 2001, 2005; Gedalof and Smith 2001; MacDonald and Case 2005; Wiles et al. 1998), corals (Gedalof et al. 2002), and nitrogen isotopes in salmon (Finney et al. 2002). The relevance of variations on these scales is that slow natural changes appearing in the short sample afforded by historical records (50–150 years) can masquerade (perhaps) as "climate change" if our temporal perspective for interpretation is too short. The PDO shifted abruptly in 1976, at about the time when the temperatures shown in Figure 1.2 began to rise. Our best global climate observations are unfortunately from this very same post–World War II period and thus nearly coincide with this approximately 50-year-long period of the PDO (1947 to the latter 1990s). One might interpret this in two ways: (1) some of the recent warming in western North America is a natural effect due to the mid-1970s phase change of ENSO or PDO, or (2) because the atmosphere expresses the effects of climate warming through rearrangements in circulation patterns, this pattern may in part represent such a change. However, western warming continued into 2007, despite the possible change in phase of PDO.

Another pattern of global climate variability that influences the western United States is the Arctic Oscillation (AO; Thompson and Wallace 1998, 2000, 2001; Wu and Straus 2004). It is the leading pattern of variability of sea level pressure over the Northern Hemisphere, and varies on time scales of days to years. Although it has little influence on seasonal means as far south as western North America, extremes of the AO index are associated with outbreaks of Arctic air that greatly enhance the likelihood of snow in lowland areas of Washington, Oregon, and California (Thompson and Wallace 2001).

Researchers have identified a variety of other proposed oscillations, including the Atlantic Multidecadal Oscillation (AMO), which appears to influence rainfall patterns around the Atlantic basin, Atlantic hurricane activity, and summer climate in Europe and much of North America (e.g., Enfield et al. 2001; Knight et al. 2006). There are some indications that the AMO may be related to climatic and hydrologic variability in the Upper Colorado River Basin (McCabe et al. 2004; USGS 2004; Gray et al. 2003, 2004a), even though the mechanism is unclear. Gray et al. (2004a) reconstructed the AMO back to 1567 using tree-ring data including those from North America. However, as with PDO, there are questions about the validity and independence of the AMO, the true time scale (deriving a 60- to 80-year period from 120 years of data), and the relationship between the AMO and global mean temperatures (e.g., Mann and Emanuel 2006).

The climate research community has recently begun to make a concerted effort to understand the cause of decadal-to-centennial scale climate variability. In the West, modern (e.g., Cayan et al. 1998) and paleoclimate (e.g., Cook et al. 2004; Fye et al. 2003; Gray et al. 2004a, 2004b; Hughes and Brown 1992; Stahle et al. 2000, 2001; Woodhouse 2003; Woodhouse and Overpeck 1998; Woodhouse et al. 2005) records have firmly established that long-term drought (5–20 years and more) is an inherent part of climate variability in this dry region. Recent model studies to understand the source of the 1930s Dust Bowl (Schubert et al. 2004a, 2004b; Seager et al. 2005), the 1950s Southwest drought, and the intense 2000–2004 drought (Hoerling and Kumar 2003) have shown that a major part of the answer lies in the world's oceans, in particular the Western Pacific and the Indian Oceans.

Detection and Attribution

A critical question in examining and understanding local, regional, or global variability is, what caused it? Was the source ENSO, AO, AMO, greenhouse gases, aerosols, or something else? Climate scientists use a theoretical framework (Hegerl et al. 2007) to perform detection (statistically identifying a trend) and attribution (identifying causes). A typical detection and attribution exercise fits an observed spatiotemporal pattern Y to an expected spatiotemporal pattern X: $Y = \alpha X + \varepsilon$, where α is the scaling factor and ε is the residual. The expected pattern, as well as estimates of unforced internal variability, comes from a model simulation with the posited forcing. For example, typical detection and attribution exercises examine greenhouse gases, solar, and volcanic forcings. If there is no statistical relationship between X and Y, the scaling factor is indistinguishable from zero, but if the statistical confidence range includes 1 and excludes zero, the pattern of change Y is said to be attributed to the forcing in question. Usually a few different expected patterns, generated by different climate models, are used to provide an additional indication of the confidence of the results. (For a more thorough derivation, see Hegerl et al. 2007, Appendix 9A).

Initial detection and attribution efforts focused on global mean temperature (Hasselmann 1997), but more recent studies have looked at regional temperature (Stott 2003), sea level pressure (Gillett et al. 2003), global mean precipitation (Gillett et al. 2004), zonal mean precipitation (Zhang et al. 2007), and others. For the western United States, the most relevant results are those of Stott (2003), who attributed warming to rising greenhouse gases over western North America, and Barnett et al. (2008), who attributed snowmelt-related hydrologic changes to rising greenhouse gases over the western United States. They concluded that "up to 60% of the climate-related trends of river flow, winter air temperature and snow pack between 1950–1999 are human-induced." In a set of companion papers focusing on western-US mountain regions, Barnett's colleagues attributed changes in temperature (Bonfils et al. 2008), spring snowpack (Pierce et al. 2008), and streamflow (Hidalgo et al. 2009) to rising greenhouse gases.

MODELED FUTURE CHANGE

Envisioning western-US climate in a future with much higher greenhouse gas concentrations requires the use of physically based numerical models of the ocean, atmosphere, land, and ice, often called global climate models (GCMs) or climate system models. A common set of simulations using 21 GCMs was coordinated through the Intergovernmental Panel on Climate Change (IPCC), described in the IPCC 2007 report (Randall et al. 2007), with results archived by the Program for Climate Model Diagnostics and Intercomparison (PCMDI). These models typically resolve the atmosphere with between 6,000 and 15,000 grid squares covering the surface, and with between 12 and 56 atmospheric layers. All GCMs in the PCMDI archive include a fully resolved global ocean model, usually with higher resolution than the atmospheric model, and nearly all include models of sea ice dynamics and models of the land surface. By calculating energy fluxes between the sun, atmosphere, and surface, these models compute surface temperature distributions that compare well with observations. Details of the models, as well as references, can be found in Table 8.1 of Randall et al. (2007).

Simulations of twenty-first-century climate require projections of future greenhouse gas concentrations and of sulfate aerosols, which reflect sunlight and also promote cloud droplet formation, thereby offsetting the effects of greenhouse gases locally. More than 40 such projections were produced under the auspices of the IPCC (Special Report on Emissions Scenarios [SRES]; Nakićenović and Swart 2000) after considering a wide range of future socioeconomic changes. Three of these "SRES" scenarios were commonly chosen for forcing the GCMs: B1, A1B, and A2. The climate forcing of all scenarios is similar until about 2020. A2 produces the highest climate forcing by the end of the twenty-first century, but before mid-century, none of the scenarios is consistently the highest. Because more modeling groups ran A1B than A2, and because our focus for this study was on mid-century change, we chose A1B as the higher emissions scenario and B1 as the low emissions scenario for a regional analysis of twenty-first-century Pacific Northwest (PNW) climate. Although B1 is the lowest of the IPCC illustrative scenarios, it still produces changes in global temperature in excess of 2°C. That value is increasingly viewed in the climate community as a threshold and goal to stay below (see Schellnhuber et al. 2006 for the history), and a growing number of political leaders have likewise stated their intention to avoid this outcome. At the high end, scenario A1FI (not shown) results in even higher climate forcing by 2100 than A2 or A1B. Global emissions of CO_2 from 2000 to 2006 exceeded even the A1FI scenario (Raupach et al. 2007). Whether these exceedingly high emissions will continue into the future is beyond our expertise to judge.

Spatial Patterns of Change

The IPCC report (Christensen et al. 2007) presented maps of mean changes, for the 2080–2099 average minus the 1980–1999 average, in the A1B scenario for 21 models over each continent. These results for North America are shown in Figure 1.5. Annual mean warming is greatest in the continental interior, in part, because of greater wintertime warming in the coldest climates (Figure 1.5, top row, second panel), with warming greater than 7°C in the Arctic region. Summertime warming in North America is greatest in the western states and modestly exceeds the warming in the other seasons.

Changes in precipitation are characterized by a north–south split, with Canada and Alaska getting wetter in all seasons and nearly all model runs, and Mexico and Central America getting drier in most seasons and most models. The western United States straddles an approximately east–west nodal line. In winter, a slight majority of models produce increases in the Northwest and decreases in the Southwest. Most of the models show summer drying in most of the West up to and north of the US–Canada border. The physical reasons for these changes in precipitation include (a) enhanced water-carrying capacity of warmer air, predominant in the North; (b) a poleward shift of the wintertime jet stream in many models (Yin 2005); and (c) a concomitant poleward expansion of the downward branch of the Hadley circulation, which is responsible for the existence and location of desert regions at around 20°–30° latitude globally. Patterns earlier in the twenty-first century (not shown) are generally similar but smaller in magnitude, and are projected to be already under way but have not been observed.

Temporal Change of Western Mean Temperature and Precipitation

On the PCMDI website (esg.llnl.gov), climate-modeling centers from around the world provided simulations of twentieth-century climate using observed solar, volcanic, and greenhouse gas forcing. In addition, for the twenty-first century, a total of 112 runs are available with the three scenarios B1, A1B, and A2 from the World Climate Research Program (WCRP) Coupled Model Intercomparison Project phase 3 (CMIP3) multimodel dataset. Maurer et al. (2007) describe the process of regridding the CMIP3 data to 1/8° × 1/8° (latitude × longitude), the Lawrence Livermore National Lab–Reclamation–Santa Clara University downscaled climate projections derived from the WCRP CMIP3 multimodel dataset, stored and served at the LLNL Green Data Oasis (GDO; http://gdo-dcp.ucllnl.org/downscaled_cmip3_projections/dcpInterface.html). A user can specify an averaging domain and in a few minutes

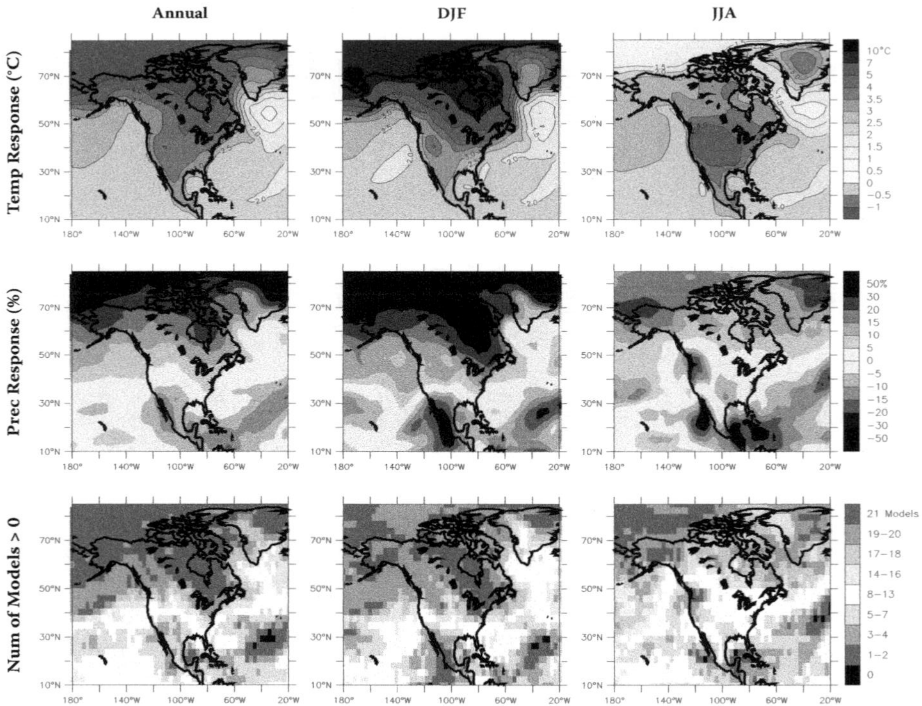

FIGURE 1.5 **(See color insert.)** Maps of changes (departures from 1971–2000 mean) in climate elements for the late-twenty-first century (2080–2099) derived from an ensemble summary of 21 global climate models (Christensen et al. 2007, their Figure 11.12) for emissions scenario A1B. Top row: Temperature departure in °C. Middle row: Precipitation departure in percent, with green indicating wetter and brown indicating drier. Bottom: Number of models agreeing on wetter than average out of 21 models; green indicates agreement on wetter and brown is agreement on drier. Columns—Left: Annual. Middle: Winter (December–February). Right: Summer (June–August).

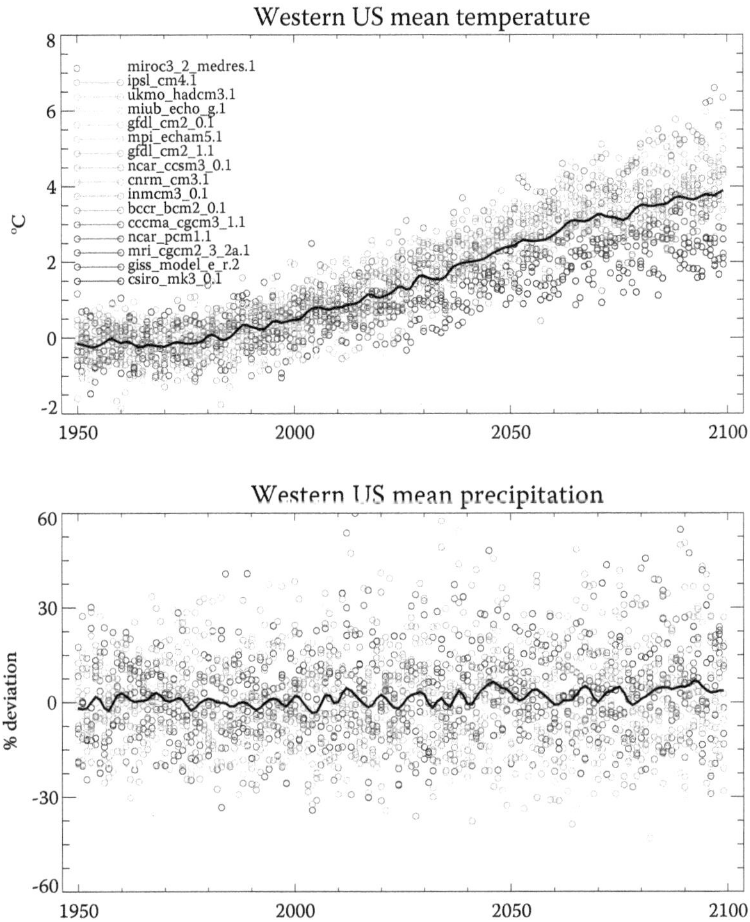

FIGURE 1.6 **(See color insert.)** Western-US annual mean temperature and precipitation for 16 climate models (colored circles, names and runs shown) along with the all-model mean (heavy black curve). Greenhouse gas and aerosol forcing follows the observed climate record for 1950–2000 and then the A1B scenario for 2000–2100.

obtain the model output for a wide range of scenarios. We obtained the first available run from each of the 16 models on the GDO averaged over the domain 30°–49°N, 107.5°W to 124.5°W. We formed annual means and compared the 2000–2100 values with the 1950–2000 reference period (Figure 1.6).

The warming rates over the twenty-first century range from 2.2°C for CSIRO3.0 to 5.3°C for MIROC3.2, and the mean warming across models is 3.4°C. By the mid-twenty-first century, the distributions have shifted considerably from those of the late-twentieth century: the coldest years in the coldest model are already as warm as the warmest years in any model for the recent historical climate. For the warmest models, the differences emerge a decade or two earlier. By contrast, the precipitation time series (Figure 1.6, bottom panel) shows no appreciable change—even in the late-twenty-first century.

We note again that this regionally averaged, annually averaged view does not mean that precipitation will not change at all. The models indicate that the character of precipitation in the West will change differently from place to place, from season to season, in type (rain versus snow), in frequency (how often), and in intensity (amount per unit time). These differences reflect the physical mechanisms at work and are internally consistent. Such changes will provide a wide variety of selection pressures to which ecological systems can and will respond in their usual complex ways. These selection pressures will exist in a context of other human-induced nonclimatic environmental changes that provide additional selection pressures. Climate is merely one of these changes, albeit a very important one.

We also note an overall finding with respect to temperature and precipitation changes. With one exception (the Parallel Climate Model [PCM]), in those locations where drying is projected, generally models that project larger temperature increases also project larger precipitation decreases. Conversely, in those locations where wetter conditions are projected, generally models that project smaller temperature increases also project larger precipitation increases.

Statistical Properties of Change

Christensen et al. (2007) also provide statistical summaries of change for regional averages. Their definition of western North America is different from ours, extending from 30° to 60° latitude and from the Pacific coast east to 100°W longitude. Table 1.1 summarizes the statistical properties of change in this region.

For this domain, summer warming exceeds that in other seasons for the minimum, 25th, 50th, and 75th percentiles but not for the hottest model. For precipitation, the median change is quite small, which is consistent with our results presented previously. Extremes are a different story: the vast majority of seasons exceed the twentieth-century 95th percentile of heat, and large shifts toward fewer extremely dry seasons and more extremely wet seasons occur in every season except summer.

Changes in Variability

So far we have described changes in the first moment of climate—that is, the mean—but one important aspect of climate, especially for biological systems, is the variability. In Figure 1.7 we compare each model's regionally averaged change in mean temperature and precipitation with its change in variance. For both variables, some models show a decrease in variance, but most indicate an increase. There is little correspondence between change in mean and change in variance for temperature, but for precipitation, the models that indicate larger increases in precipitation generally also experience an increase in variance. These clearly have implications for ecological systems.

TABLE 1.1

Changes in Temperature and Precipitation from a Set of 21 Global Models for the A1B Scenario, Averaged over Western North America (Their Definition)

	Temperature Response (°C)					Precipitation Response (%)					Extreme Seasons (%)		
Season	**Min.**	**25**	**50**	**75**	**Max.**	**Min.**	**25**	**50**	**75**	**Max.**	**Hot**	**Wet**	**Dry**
DJF	1.6	3.1	3.6	4.4	5.8	−4	2	7	11	36	80	18	3
MAM	1.5	2.4	3.1	3.4	6.0	−7	2	5	8	14	87	14	
JJA	2.3	3.2	3.8	4.7	5.7	−18	−10	−2	2	10	100	3	
SON	2.0	2.8	3.1	4.5	5.3	−3	3	6	12	18	95	17	2
Annual	2.1	2.9	3.4	4.1	5.7	−3	0	5	9	14	100	21	2

Notes: Changes in the table are means for the period 2080–2099 minus means for the period 1980–1999, expressed in degrees (temperature) or percentage (precipitation) of the 1980–2099 base period. The table shows the minimum, maximum, 25th, 50th, and 75th percentiles of the differences for the 21 models (so, roughly, the 5th, 10th, and 15th highest values). The "extreme seasons" portion of the table indicates how often late-twenty-first-century seasons exceed the twentieth-century definition of 95th percentile for hot or wet (or 5th percentile for dry). If the distribution were unchanged, the entry would be 5%. An entry appears only when at least 14 of the 21 models agree on an increase (or in a few cases where the number is less than five, a decrease).

Source: Adapted from Table 11.1 of Christensen, J.H.; Hewitson, B.; Busuioc, A.; Chen, A.; Gao, X.; Held, I.; Jones, R.; Kolli, R.K.; Kwon, W.-T.; Laprise, R.; Magaña Rueda, V.; Mearns, L.; Menéndez, C.G.; Räisänen, J.; Rinke, A.; Sarr, A.; Whetton, P. 2007. Regional climate projections. In: *Climate change 2007: The physical science basis.* Contribution of Working Group I to the Fourth Assessment Report of the Intergovernmental Panel on Climate Change [Solomon, S.; Qin, D.; Manning, M.; Chen, Z.; Marquis, M.; Averyt, K.B.; Tignor, M.; Miller, H.L. (eds.)]. Cambridge University Press, Cambridge, United Kingdom, and New York, USA.

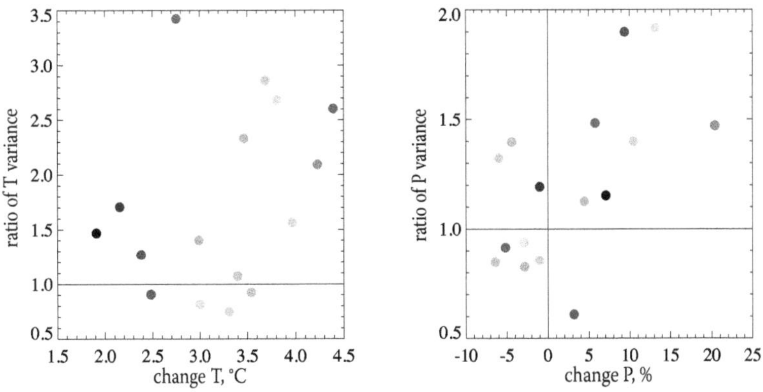

FIGURE 1.7 **(See color insert.)** Scatterplots of change in mean and ratio of change in variance, 1950–2000 versus 2050–2100, for temperature (left) and precipitation (right). Colors indicate the models and are the same as in Figure 1.6.

SUMMARY

During the past 50 years, the influence of rising greenhouse gas concentrations has emerged as an important factor in western temperature. This has directly affected snowmelt and streamflow in broad areas. Climate models are in universal agreement that rising temperatures will continue for all reasonable assumptions about future greenhouse gas emissions, but the range of warming even for a single assumption about future emissions is quite broad (roughly 2–5°C during the twenty-first century). Future changes in precipitation are even less well constrained, although most models agree on drier summers for most of the West.

It is also worth noting that the temperature increase in any particular geographic setting will quite likely not be a steady monotonic rise, but will show occasional plateaus and even decreases lasting a year to a decade or more, arising from stochastic events (e.g., volcanoes) and from the inherent internal variability that is a fundamental characteristic of the climate system. Individual runs from the IPCC climate models nearly always show such behavior, even those models that have strong warming over many decades. This facet of the way in which climate change unfolds, day by day and year by year, is usually overlooked, largely because the manner in which model results are summarized and presented (e.g., averaging as in Figure 1.6) masks what is seen in individual realizations.

Both ecology and climate have many important issues that relate to spatial scale, most especially the spatial scale of temporal variability. Important work remains to be done to understand and anticipate the fine-scale spatial structure of climate change in a topographically diverse landscape. There are intriguing hints that this structure will be more complex than is commonly conceived. Related to this, there are likewise important implications for the spatial scales at which we monitor climate over the long term.

For ecological studies, these results are important both in understanding past change and in framing the predictions of future change. The discussion here emphasizes that the limited signal-to-noise ratio of twentieth-century climate change on scales smaller than the entire West make it difficult to confidently attribute local changes to rising greenhouse gases. To state this point another way, researchers should be very cautious about attributing observed biological changes to (implied human-induced) climate change. Other patterns of variability become increasingly important as the spatial scale of analysis diminishes.

ACKNOWLEDGMENTS

We acknowledge the modeling groups, the Program for Climate Model Diagnosis and Intercomparison (PCMDI) and the WCRP's Working Group on Coupled Modeling (WGCM), for their roles in making available the WCRP CMIP3 multimodel dataset. Support of this dataset is provided by the Office of Science, US Department of Energy. KTR would like to acknowledge the support of the NOAA Western Regional Climate Center and the Regional Integrated Sciences and Assessments Program.

REFERENCES

Andrews, E.D.; Antweiler, R.C.; Nieman, P.J.; Ralph, F.M. 2004. Influence of ENSO on flood frequency along the California coast. *Journal of Climate*. 17(2): 337–348.

Applied Climate Information System (ACIS). 2011. Current anomaly maps. http://www.hprcc.unl.edu/products/current.html (20 November).

Applied Climate Information System (ACIS). 2011. Current Climate Summary Maps. http://www.hprc.unl.edu/maps (accessed April 19, 2011).

Ashcroft, M.B. 2010. Identifying refugia from climate change. *Journal of Biogeography*. 37: 1407–1413.

Bales, R.C.; Molotch, N.P.; Painter, T.H.; Dettinger, M.D.; Rice, R.; Dozier, J. 2006. Mountain hydrology of the western United States. *Water Resources Research*. 42, W08432, DOI: 10.1029/2005WR004387.

Barnett, T.; Malone, R.; Pennell, W.; Sammer, D.; Semtner, B.; Washington, W. 2004. On water resources in the West: Introduction and overview. *Climate Change*. 62: 1–11.

Barnett, T.; Pierce, D.W.; Hidalgo, H.; Bonfils, C.; Santer, B.D.; Das, T.; Bala, G.; Wood, A.W.; Nazawa, T.; Mirin, A.; Cayan, D.; Dettinger, M. 2008. Human-induced changes in the hydrology of the western United States. Science. 319: 1080–1083.

Bennett, K.D.; Provan, J. 2008. What do we mean by 'refugia'? *Quaternary Science Reviews*. 27: 2449–2455.

Biondi, F.; Gershunov, A.; Cayan, D.R. 2001. North Pacific decadal climate variability since 1661. *Journal of Climate Letters*. 14: 5–10.

Bond, N.A.; Vecchi, G.A. 2003. The influence of the Madden–Julian Oscillation on precipitation in Oregon and Washington. *Weather and Forecasting*. 18: 600–613.

Bonfils, C.; Santer, B.D.; Pierce, D.W.; Hidalgo, H.G.; Bala, G.; Das, T.; Barnett, T.P.; Dettinger, M.; Cayan, D.R.; Doutriaux, C.; Wood, A.W.; Mirin, A.; Nozawa, T. 2008. Detection and of temperature changes in the mountainous western United States. Journal of Climate. 21: 6404–6424.

Broccoli, A.J.; Dixon, K.W.; Delworth, T.L.; Knutson, T.R.; Stouffer, R.J. 2003. Twentieth-century temperature and precipitation trends in ensemble climate simulations including natural and anthropogenic forcing. *Journal of Geophysical Research*. 108(D24). DOI: 10.1029/ 2003JD003812.

Brohan, P.; Kennedy, J.J.; Harris, I.; Tett, S.F.B.; Jones, P.D. 2006. Uncertainty estimates in regional and global observed temperature changes: A new dataset from 1850. *Journal of Geophysical Research*. 111: D12106, DOI: 10.1029/2005JD006548.

Brown, D.P.; Comrie, A.C. 2004. A winter precipitation 'dipole' in the western United States associated with multidecadal ENSO variability. *Geophysics Research Letters*. 31: L09203. DOI: 10.1029/2003GL018726.

Cayan, D.R.; Dettinger, M.D.; Diaz, H.F.; Graham, N.E. 1998. Decadal variability of precipitation over western North America. *Journal of Climate*. 11: 3148–3166.

Cayan, D.R.; Redmond, K.T.; Riddle, L.G. 1999. ENSO and hydrologic extremes in the western United States. *Journal of Climate*. 12: 2881–2893.

Christensen, J.H.; Hewitson, B.; Busuioc, A.; Chen, A.; Gao, X.; Held, I.; Jones, R.; Kolli, R.K.; Kwon, W.-T.; Laprise, R.; Magaña Rueda, V.; Mearns, L.; Menéndez, C.G.; Räisänen, J.; Rinke, A.; Sarr, A.; Whetton, P. 2007. Regional climate projections. In: *Climate change 2007: The physical science basis*. Contribution of Working Group I to the Fourth Assessment Report of the Intergovernmental Panel on Climate Change [Solomon, S.; Qin, D.; Manning, M.; Chen, Z.; Marquis, M.; Averyt, K.B.; Tignor, M.; Miller, H.L. (eds.)]. Cambridge University Press, Cambridge, United Kingdom, and New York, USA.

Cleveland, W.S. 1995. *Elements of Graphing Data*. Hobart Press, Summit, NJ.

Cook, E.R., J. Esper, and R.D. D'Ariggo. 2004. Extra-tropical Northern Hemisphere land temperature variability over the past 1,000 years. *Quarternary Science Reviews.* 23: 2063–2074.

Daly, C.; Conklin, D.R.; Unsworth, M.H. 2009. Local atmospheric decoupling in complex topography alters climate change impacts. *International Journal of Climatology.* DOI: 10.1002/joc.2007.

Daly, C.; Gibson, W.; Doggett, M.; Smith, J.; Taylor, G. 2004. Up-to-date monthly climate maps for the conterminous United States. 14th AMS Conference on Applied Climatology.

Daly, C.; Gibson, W.P.; Taylor, G.H.; Johnson, G.L.; Pasteris, P. 2002. A knowledge-based approach to the statistical mapping of climate. *Climate Research.* 22: 99–113.

Daly, C.; Halbleib, M.; Smith, J.I.; Gibson, W.P.; Doggett, M.K.; Taylor, G.H.; Curtis, J.; Pasteris, P.A. 2008. Physiographically sensitive mapping of temperature and precipitation across the conterminous United States. *International Journal of Climatology.* DOI: 10.1002/joc.1688.

D'Arrigo, R.; Villalba, R.; Wiles, G. 2001. Tree-ring estimates of Pacific decadal climate variability. *Climate Dynamics.* 18: 219–224.

D'Arrigo, R.; Wilson, R.; Deser, C.; Wiles, G.; Cook, E.; Villalba, R.; Tudhope, S.; Cole, J.; Linsley, B. 2005. Tropical–North Pacific climate linkages over the past four centuries. *Journal of Climate.* 18(24): 5253–5265.

Das, T.; Hidalgo, H.G.; Dettinger, M.D.; Cayan, D.R.; Pierce, D.W.; Bonfils, C.; Barnett, T.P.; Bala, G.; Mirin, A. 2009. Structure and detectability of trends in hydrological measures over the western United States. *Journal of Hydrometeorology.* 10, 871–892.

Diaz, H.F.; Bradley, R.S. 1997. Temperature variations during the last century at high elevation sites, *Climate Change.* 36: 253–279.

Enfield, D.B.; Mestas-Nuñez, A.M.; Trimble, P.J. 2001. Rainfall and river flows in the continental US. *Geophysical Research Letters.* 28: 2077–2080.

Finney, B.P., I. Gregory-Eaves, M.S.V. Douglas, and J.P. Smol. 2002. Fisheries productivity in the northeastern Pacific Ocean over the past 2,200 years. *Nature.* 416: 729–733.

Forster, P.; Ramaswamy, V.; Artaxo, P.; Berntsen, T.; Betts, R.; Fahey, D.W.; Haywood, J.; Lean, J.; Lowe, D.C.; Myhre, G.; Nganga, J.; Prinn, R.; Raga, G.; Schu, M.; Van Dorland, R. 2007. Changes in atmospheric constituents and in radiative forcing. In: *Climate change 2007: The physical science basis.* Contribution of Working Group I to the Fourth Assessment Report of the Intergovernmental Panel on Climate Change [Solomon, S.; Qin, D.; Manning, M.; Chen, Z.; Marquis, M.; Averyt, K.B.; Tignor, M.; Miller, H.L. (eds.)]. Cambridge University Press, Cambridge, United Kingdom, and New York, USA.

Fye, F.K.; Stahle, D.W.; Cook, E.R. 2003. Paleoclimatic analogs to twentieth-century moisture regimes across the United States. *Bulletin of the American Meteorological Society.* 84(7): 901–909.

Gedalof, Z.; Mantua, N.J.; Peterson, D.L. 2002. A multi-century perspective of variability in the Pacific decadal oscillation: New insights from tree rings and coral. *Geophysical Research Letters.* 29(24): 2204. DOI: 10.1029/2002GL015824.

Gedalof, Z.; Smith, D.J. 2001. Interdecadal climate variability and regime-scale shifts in Pacific North America. *Geophysical Research Letters.* 28: 1515–1518.

Gillett, N.P.; Weaver, A.J.; Zwiers, F.W.; Wehner, M.F. 2004. Detection of volcanic influence on global precipitation. *Geophysical Research Letters.* 31(12): L12217, DOI: 10.1029/2004GL020044.

Gillett, N.P.; Zwiers, F.W.; Weaver, A.J.; Stott, P.A. 2003. Detection of human influence on sea level pressure. *Nature.* 422: 292–294.

Gray, S.T.; Betancourt, J.L.; Fastie, C.L.; Jackson, S.T. 2003. Patterns and multi-decadal oscillations in drought-sensitive tree-ring records from the central and southern Rocky Mountains. *Geophysical Research Letters.* 30: 491–494, COI: 10.1029/2002GL016154

Gray, S.T.; Fastie, C.L.; Jackson, S.T.; Betancourt, J.L. 2004b. Tree-ring-based reconstruction of precipitation in the Bighorn Basin, Wyoming, since 1260 A.D. *Journal of Climate*. 17: 3855–3865.

Gray, S.T.; Graumlich, L.J.; Betancourt, J.L.; Pederson, G.T. 2004a. A tree-ring based reconstruction of the Atlantic Multidecadal Oscillation since 1567 AD. *Geophysical Research Letters*. 31: L12205, DOI: 10.1029/2004GL019932.

Hamlet, A.F.; Mote, P.W.; Clark, M.P.; Lettenmaier, D.P. 2005. Effects of temperature and precipitation variability on snowpack trends in the western US. *Journal of Climate*. 18: 4545–4561.

Hasselman, K. 1997. Are we seeing global warming? *Science*. 276: 914–915.

Hegerl, G.C.; Zwiers, F.W.; Braconnot, P.; Gillett, N.P.; Luo, Y.; Marengo Orsini, J.A.; Nicholls, N.; Penner, J.E.; Stott, P.A. 2007. Understanding and attributing climate change. In: *Climate change 2007: The physical science basis*. Contribution of Working Group I to the Fourth Assessment Report of the Intergovernmental Panel on Climate Change [Solomon, S.; Qin, D.; Manning, M.; Chen, Z.; Marquis, M.; Averyt, K.B.; Tignor, M.; Miller, H.L. (eds.)]. Cambridge University Press, Cambridge, United Kingdom, and New York, USA.

Hidalgo, H.G.; Cayan, D.R.; Dettinger, M.D. 2005. Sources of variability of evapotranspiration in California. *Journal of Hydrometeorology*. 6: 3–19.

Hidalgo, H.G.; Das, T.; Dettinger, M.D.; Cayan, D.R.; Pierce, D.W.; Barnett, T.P.; Bala, G.; Mirin, A.; Wood, A.W.; Bonfils, C.; Santer, B.D.; Nozawa, T. 2009. Detection and attribution of streamflow timing changes to climate change in the western United States. Journal of Climate. 22: 3838–3855.

Hoerling, M.P.; Kumar, A. 2002. Atmospheric response patterns associated with tropical forcing. *Journal of Climate*. 15: 2184–2203.

———. 2003. The perfect ocean for drought. *Science*. 299: 691–694.

Hughes, M.K.; Brown, P.M. 1992. Drought frequency in central California since 101 B.B. recorded in giant sequoia tree rings. *Climate Dynamics*. 6: 161–167, DOI: 10.1007/BF00193528.

Karl, T.R.; Hassol, S.J.; Miller, C.D.; Murray, W.L. (eds.). 2006. Temperature trends in the lower atmosphere: Steps for understanding and reconciling differences. US Climate Change Science Program, Washington, D.C. (available at http://climatescience.gov/Library/sap/sap1-1/finalreport/).

Knight, J.R.; Folland, C.K.; Scaife, A.A. 2006. Climate impacts of the Atlantic Multidecadal Oscillation. *Geophysical Research Letters*. 33: L17706, DOI: 10.1029/2006GL026242.

Knowles, N.; Dettinger, M.D.; Cayan, D.R. 2006. Trends in snowfall versus rainfall in the western United States. *Journal of Climate*. 19: 4545–4559.

Liu, X.; Yin, Z.-Y.; Shao, X.; Qin, N. 2006. Temporal trends and variability of daily maximum and minimum, extreme temperature events, and growing season length over the eastern and central Tibetan Plateau during 1961–2003. *Journal of Geophysical Research*. 111: D19109, DOI: 10.1029/2005JD006915.

MacDonald, G.M.; Case, R.A. 2005. Variations in the Pacific Decadal Oscillation over the past millennium. *Geophysical Research Letters*. 32: L08703, DOI: 10.1029/2005GL022478.

Mann, M.E.; Emanuel, K.E. 2006. Atlantic hurricane trends linked to climate change. *EOS Transactions of the AGU*. 87: 233–244.

Mantua, N.J.; Hare, S.R. 2002. The Pacific Decadal Oscillation. *Journal of Oceanography*. 58(1): 35–44.

Mantua, N.J.; Hare, S.R.; Zhang, Y.; Wallace, J.M.; Francis, R.C. 1997. A Pacific interdecadal climate oscillation with impacts on salmon production. *Bulletin of the American Meteorological Society*. 78: 1069–1079.

Maurer, E.P.; Stewart, I.T.; Bonfils, C.; Duffy, P.B.; Cayan, D. 2007. Detection, attribution, and sensitivity of trends toward earlier streamflow in the Sierra Nevada. Journal of Geophysical Research. 112: D11118, DOI: 10.1029/2006JD008088.

McCabe, G.J.; Dettinger, M.D. 1999. Decadal variations in the strength of ENSO teleconnections with precipitation in the western United States. *International Journal of Climatology*. 19: 1399–1410.

———. 2002. Primary modes and predictability of year-to-year snowpack variations in the western United States from teleconnections with Pacific Ocean Climate. *Journal of Hydrometeorology*. 3: 13–25.

McCabe, G.J.; Palecki, M.A.; Betancourt, J.L. 2004. Pacific and Atlantic Ocean influences on multidecadal drought frequency in the United States. *Proceedings of the National Academy of Sciences*. 101: 4136–4141.

Meehl, G.A.; Washington, W.M.; Ammann, C.M.; Arblaster, J.M.; Wigley, T.M.L.; Tebaldi, C. 2004. Combinations of natural and anthropogenic forcings in 20th century climate. *Journal of Climate*. 17: 3721–3727.

Mitchell, T.P.; Blier, W. 1997. The variability of wintertime precipitation in the region of California. *Journal of Climate*. 10: 2261–2276.

Mo, K.C. 1999. Alternating wet and dry episodes over California and Intraseasonal Oscillations. *Monthly Weather Review*. 127: 2759–2776.

Mo, K.C.; Higgins, R.W. 1998. Tropical influences on California precipitation. *Journal of Climate*. 11: 412–430.

Mote, P.W. 2006. Climate-driven variability and trends in mountain snowpack in western North America. *Journal of Climate*. 19: 6209–6220.

Mote, P.W.; Hamlet, A.F.; Clark, M.P.; Lettenmaier, D.P. 2005. Declining mountain snowpack in western North America. *Bulletin of the American Meteorological Society*. 86: 39–49.

Nakićenović, N.; Swart, R. (eds.). 2000. Special report on emissions scenarios. A special report of working group III of the Intergovernmental Panel on Climate Change. Cambridge University Press, Cambridge, United Kingdom, and New York, USA.

Newman, M.; Compo, G.P.; Alexander, M.A. 2003. ENSO-forced variability of the Pacific Decadal Oscillation. *Journal of Climate*. 16: 3853–3857.

Overland, J.E.; Adams, J.M.; Mofjeld, H.O. 2000. Chaos in the North Pacific: Spatial modes and temporal irregularity. *Progress in Oceanography*. 47: 337–354.

Pagano, T., and D. Garen (2005), A recent increase in western US streamflow variability and persistence, *J. Hydrometeorol.*, 6, 173 – 179, DOI:10.1175/JHM410.1.

Pepin, N.C.; Losleben, M.L. 2002. Climate change in the Colorado Rocky Mountains: Free air versus surface temperature trends. *International Journal of Climatology*. 22: 311–329.

Pepin, N.C.; Lundquist, J.D. 2008. Temperature trends at high elevations: Patterns across the globe. *Geophysical Research Letters*. 35: L14701, DOI: 10.1029/2008GL034026.

Pepin, N.C.; Seidel, D.J. 2005. A global comparison of surface and free-air temperatures at high elevations, *Journal of Geophysical Research*. 110: D03104, DOI: 10.1029/2004JD005047.

Petit, R.J.; Aguinagalde, I.; De Beaulieu, J.-L.; Bitthau, C.; Breser, S.; Cheddadi, R.; Ennos, R.; Fineschi, S.; Grivet, D.; Lascoux, M.; Mohanty, A.; Muller-Starck, G.; Demesure-Musch, B.; Palme, A.; Martin, J.P.; Rendell, S.; Vendramin, G.G. 2003. Glacial refugia: Hotspots but not melting pots of genetic diversity. *Science*. 300: 1563–1565.

Pierce, D.W.; Barnett, T.P.; Hidalgo, H.G.; Das, T.; Bonfils, C.; Sander, B.; Bala, G.; Dettinger, M.; Cayan, D.; Mirin, A. 2008. Attribution of declining western US snowpack to human effects. *Journal of Climate*. 21: 6425–6444.

Randall, D.A.; Wood, R.A.; Bony, S.; Colman, R.; Fichefet, T.; Fyfe, J.; Kattsov, V.; Pitman, A.; Shukla, J.; Srinivasan, J.; Stouffer, R.J.; Sumi, A.; Taylor, K.E. 2007. Climate models and their evaluation. In: *Climate change 2007: The physical science basis*. Contribution of Working Group I to the Fourth Assessment Report of the Intergovernmental Panel on Climate Change [Solomon, S. et al., (eds)]. Cambridge University Press, Cambridge, United Kingdom, and New York, USA.

Raupach, M.R.; Marland, G.; Ciais, P.; Le Quéré, C.; Canadell, J.G.; Klepper, G.; Field, C.B. 2007. Global and regional drivers of accelerating CO_2 emissions. *Procedures of the National Academy of Science*. DOI: 10.1073/pnas.0700609104.

Redmond, K.T. 2003. Climate variability in the intermontane West: Complex spatial structure associated with topography, and observational issues. In: *Water and climate in the western United States*, Lewis W.M. Jr. (ed.), Chapter 2, pp 29–48. University Press of Colorado, Boulder.

———. 2007. Climate variability and change as a backdrop for western resource management. In: *Bringing climate change into natural resource management*, Joyce, L.; Haynes, R.; White, R.; Barbour, R.J. (eds.). Gen. Tech. Rep. PNW-GTR-706. Portland, OR: US Department of Agriculture, Forest Service, Pacific Northwest Research Station. pp. 5–40. http://www.fs.fed.us/pnw/pubs/pnw_gtr706.pdf

Redmond, K.T.; Koch, R.W. 1991. Surface climate and streamflow variability in the western United States and their relationship to large-scale circulation indices. *Water Resources Research*. 27: 2381–2399.

Rial, J.; Pielke Sr., R.A.; Beniston, M.; Claussen, M.; Canadell, J.; Cox, P.; Held, H.; de Noblet-Ducoudre, N.; Prinn, R.; Reynolds, J.; Salas, J.D. 2004. Nonlinearities, feedbacks and critical thresholds within the Earth's climate system. *Climatic Change* 65: 11–38.

Salathé, E.P.; Leung, L.R.; Qian, Y.; Zhang, Y. 2010. Regional climate model projections for the State of Washington. *Climatic Change*. 102: 51–75.

Schellnhuber H.J., et al. (eds.). 2006. *Avoiding dangerous climate change*. Cambridge University Press, New York.

Schubert, S.D.; Suarez, M.J.; Pegion, P.J.; Koster, R.D.; Bacmeister, J.T. 2004a. Causes of long-term drought in the US Great Plains. *Journal of Climate*. 17: 485–503.

———. 2004b. On the cause of the 1930s Dust Bowl. *Science*. 303: 1855–1859.

Seager, R.; Kushnir, Y.; Herweijer, C.; Naik, N.; Velez, J. 2005. Modeling of tropical forcing of persistent droughts and pluvials over Western North America: 1856–2000. *Journal of Climate*. 18: 4065ff.

Stahle, D.W.; Cook, E.R.; Cleaveland, M.K.; Therrell, M.D.; Meko, D.M.; Grissino-Mayer, H.D. 2000. Tree-ring data document 16th century megadrought over North America. *EOS, Transactions of the American Geophysical Union*. 81: 121–125.

Stahle, D.W.; Therrell, M.D.; Cleaveland, M.K.; Cayan, D.R.; Dettinger, M.D.; Knowles, N. 2001. Ancient blue oak reveal human impact on San Francisco Bay salinity. *EOS, Transactions of the American Geophysical Union*. 82(141): 144–145.

Stewart, I.T.; Cayan, D.R.; Dettinger, M.D. 2005. Changes toward earlier streamflow timing across western North America. *Journal of Climate*. 18: 1136–1155.

Stott, P.A. 2003. Attribution of regional-scale temperature changes to anthropogenic and natural causes. *Geophysical Research Letters*. 30(14): 1728-1731, DOI: 10.1029/2003GL017324.

Stott, P.A.; Tett, S.F.B.; Jones, G.S.; Allen, M.R.; Mitchell, J.F.B.; Jenkins, G.J. 2000. External control of 20th century temperature by natural andanthropogenic forcings. *Science*. 290: 2133–2137.

Swetnam, T.W.; Betancourt, J.L. 1990. Fire-Southern Oscillation relations in the southwestern US. *Science*. 249: 1017–1020.

Thompson, D.W.J.; Wallace, J.M. 1998. The Arctic Oscillation signature in the wintertime geopotential height and temperature fields. *Geophysical Research Letters*. 25: 1297–1300.

———. 2000. Annular modes in the extratropical circulation, part I: Month-to-month variability. *Journal of Climate*. 13: 1000–1016.

———. 2001. Regional climate impacts of the northern hemisphere annular mode. *Science*. 293: 85–89.

Trenberth, K.T.; Fasullo, J.T.; Kiehl, J.T. 2009. Earth's global energy budget. *Bulletin of the American Meteorological Society*. 90: 311-323, DOI: 10.1175/2008BAMS2634.1.

USGS. 2004. Climatic fluctuations, drought, and flow in the Colorado River Basin. USGS Fact Sheet 2004-3062 (available at: water.usgs.gov/pubs/fs/2004/3062/).

Vuille, M.; Bradley, R.S. 2000. Mean annual temperature trends and their vertical structure in the tropical Andes. *Geophysical Research Letters*. 27: 3885–3888.

Vuille, M.; Bradley, R.S.; Werner, M.; Keimig, F. 2003. 20th century climate change in the tropical Andes: Observations and model results. *Climate Change*. 59: 75–99.

Wiles, G.C.; D'Arrigo, R.; Jacoby, G.C. 1998. Gulf of Alaska atmosphere-ocean variability over recent centuries inferred from coastal tree-ring records. *Climate Change*. 38: 289–306.

Woodhouse, C.A. 2003. Dendrochronological evidence for long-term hydroclimatic variability. In: *Water and climate in the western United States*, W.M. Lewis Jr. (ed.). University of Colorado Press, Boulder CO, pp. 49–58.

Woodhouse, C.A.; Kunkel, K.E.; Easterling, D.R.; Cook, E.R. 2005. The 20th century pluvial in the western United States. *Geophysical Research Letters*. 32: L07701, DOI: 10.1029/2005GL022413.

Woodhouse, C.A.; Overpeck, J.T. 1998. 2000 years of drought variability in the central United States. *Bulletin of the American Meteorological Society*. 79: 2693–2714.

Wu, Q.; Straus, D.M. 2004. AO, COWL, and observed climate trends. *Journal of Climate*. 17: 2139–2156.

Yin, J.H. 2005. A consistent poleward shift of the storm tracks in simulations of 21st century climate. *Geophysical Research Letters*. 32: L18701, DOI: 10.1029/2005GL023684.

You, Q.; Kang, S.; Pepin, N.; Yan, Y. 2008. Relationship between trends in temperature extremes and elevation in the eastern and central Tibetan Plateau, 1961 – 2005, *Geophys. Res. Lett.*, 35, L04704, DOI:10.1029/2007GL032669.

Zhang, X.; Zwiers, F.W.; Hegerl, G.C.; Lambert, F.H.; Gillett, N.P.; Solomon, S.; Stott, P.A.; Nozawa, T. 2007. Detection of human influence on twentieth-century precipitation trends. *Nature*. DOI: 10.1038/nature06025.

Zhang, Y.; Wallace, J.M.; Iwasaka, N. 1996. Is climate variability over the North Pacific a linear response to ENSO? *Journal of Climate*. 9(7): 1468–1478.

Section II

Single- and Multiple-Species Relationships to Climate Change

2 Amphibian Population Declines and Climate Change

Andrew R. Blaustein, Catherine Searle, Betsy A. Bancroft, and Joshua Lawler

CONTENTS

Large losses in biodiversity are occurring around the world (Lawton and May 1995). Although the exact number of species being lost is unknown, some estimate that the rate of extinction is greater than any known in the last 100,000 years (Wilson 1992). Species loss is a major part of this "biodiversity crisis." The crisis is exemplified by population declines, range reductions, and extinctions of amphibian species around the world (Houlahan et al. 2000; Stuart et al. 2004; Lannoo 2005; Mendelson et al. 2006). In at least some regions, amphibian losses appear to be more severe than losses in other vertebrate taxa (Pounds et al. 1997, 1999; Stuart et al. 2004). From a historical perspective, amphibian losses may be representative of a sixth major extinction event (Wake and Vredenburg 2008). Moreover, declines in amphibian populations are prominent because many of them are occurring in areas that remain relatively undisturbed by humans, such as national parks, conservation sites, and rural areas some distance from urban centers.

There appears to be no single cause for amphibian population declines. Like other animals, amphibians are affected by numerous environmental stresses that often act in complex ways (Blaustein et al. 2011). The causes for the decline of a given species may be different from region to region and even in different populations of the same species. Synergistic interactions between more than one factor may be involved. There may be interspecific differences and even differences between life stages in how amphibians react to environmental changes and other stresses.

Some of the concern about amphibian population declines is due in part to the fact that amphibians are considered by many biologists to be excellent "indicators" of environmental change and contamination (Blaustein 1994; Blaustein and Wake 1995). They have permeable, exposed skin (not covered by scales, hair, or feathers) and eggs (not covered by shells) that may readily absorb substances from the environment. The complex life cycles of many species potentially expose them to both aquatic and terrestrial environmental changes. These attributes and the fact that amphibians are ectotherms make them especially sensitive to changes in temperature, precipitation, and other environmental changes such as increases in ultraviolet radiation.

KEY ISSUES

In this chapter we address several key questions concerning the effects of contemporary climate change on amphibians. Are amphibian populations affected by global climate changes? What are the major climate changes affecting amphibian populations? How are amphibian populations affected? In addressing these questions, we summarize how amphibian populations may be affected by global climate changes, and we identify some of the main agents contributing to climate change. We also attempt to incorporate several key methods/paradigms that biologists use to assess how populations may be affected by climate change.

PERSPECTIVE ON GLOBAL CLIMATE CHANGE

In the 1970s, environmental biologists and atmospheric scientists predicted that two significant human-induced environmental changes (global warming and ozone depletion) could potentially affect the biology of a wide array of plants, animals, and microorganisms (Andrady et al. 2009). Increased emissions of "greenhouse" gases resulting from burning fossil fuels and land conversion were projected to cause a significant rise in global temperatures in the coming decades. Moreover, it was shown that chlorofluorocarbons (CFCs) and other commonly used industrial gases were depleting the earth's protective ozone layer, increasing the amount of cell-damaging ultraviolet-B (UV-B; 280–315 nm) radiation that reaches the Earth's surface (van der Leun et al. 1998). Scientists projected that species might respond to these global changes by altering their behavior and shifting ranges. However, if they are unable to adapt to these environmental changes, they may experience increased mortality and significant sublethal effects. Additionally, a number of scientists suggested that global warming and ozone depletion would affect entire ecological communities (e.g., Peters and Lovejoy 1992; Reaser and Blaustein 2005; Cockell and Blaustein 2001).

An increasing body of evidence suggests a warming world accompanying other significant climate changes (Intergovernmental Panel on Climate Change [IPCC] 2001, 2007). The average global surface temperature rose by about 0.74°C from 1906 to 2005 and was more severe at higher northern latitudes (IPCC 2007). Data for the Northern Hemisphere indicate that the last half of the twentieth century experienced higher temperatures than at any other time in the past 1300 years (IPCC 2007). The global average sea level has risen (about 1.8 mm/year from 1961 to 2003) and ocean heat content has increased during the twentieth century (IPCC 2007). Changes have also occurred in other aspects of climate. For example, precipitation patterns have changed over many regions of the world (IPCC 2007). During the twentieth century, cloud cover has changed over mid and high latitudes. An increase in the number of extreme events in weather and climate are predicted for the twenty-first century. These include more heat waves, higher minimum temperatures, fewer cold days, increases in precipitation at high latitudes, decreases in precipitation on subtropical lands, and increased summer continental drying associated with drought (IPCC 2007).

Some of the predictions concerning global climate changes have been supported by recent studies. For example, global climate changes in temperature and precipitation seem to be influencing the distribution and abundance of butterflies, amphibians, reptiles, fish, and a number of other taxa (e.g., Parmesan 1996; Pounds et al. 1999; Perry et al. 2005). Warming trends may have led to the extinction of some populations (Parmesan 1996; Pounds et al. 1999; Thomas et al. 2006) and may be influencing the timing of breeding of others (Beebee 1995; Forchhammer et al. 1998; Crick and Sparks 1999; Gibbs and Breisch 2001; Blaustein et al. 2001b; Post et al. 2001; Chadwick et al. 2006; Hušek and Adamík 2008).

Increases in temperature and variability in precipitation (see Chapter 1) may especially affect amphibians that are dependent on hydrological cycles. Changes in precipitation and increasing temperatures could alter connectivity of bodies of water. This could result in either increasing or decreasing fragmentation between bodies of water and amphibian populations. Long-term changes in temperature and precipitation could affect amphibians indirectly in a variety of ways. Potential changes include impacts on terrestrial and aquatic habitats, food webs and community-level interactions (e.g., competition, predation), the spread of diseases, and the interplay among all these factors. These changes can influence the interrelationships of species occurrences on a landscape level and could lead to range shifts of entire species assemblages (Raxworthy et al. 2008; Seimon et al. 2007).

Just as climate change may have wide and varied effects on numerous organisms, so might increasing UV-B radiation (Andrady et al. 2009). And exposure of organisms to increasing UV-B radiation is often linked to fluctuations in weather and changes in climate (Ovaska 1997; Kiesecker et al. 2001; Andrady et al. 2009). In fact, the latest United Nations panel on the environment discussed the importance of the link between ozone depletion and climate (Andrady et al. 2009). As the climate changes, so does species' exposure to UV-B radiation. This can occur independently of or in concert with changes in precipitation and temperature that can affect aquatic organisms.

At the terrestrial surface, UV-B radiation is extremely important biologically. Critical biomolecules absorb light of higher wavelength (UV-A; 315–400 nm) less efficiently, and stratospheric ozone absorbs most light of lower wavelength (UV-C; 200–280 nm) (Cockell and Blaustein 2001). UV-B radiation can cause mutations and cell death (Tevini 1993; Cockell and Blaustein 2001). At the individual level, UV-B radiation can slow growth rates, cause immune dysfunction, and induce sublethal damage (Tevini 1993).

UV-B radiation has been a ubiquitous stressor on living organisms over evolutionary time (Cockell 2001). Natural events such as impacts from asteroids and comets, volcanic activity, cosmic events such as supernova explosions, and solar flares can cause large-scale ozone depletion with accompanying increases in UV radiation (Cockell 2001). However, these natural events are transitory and may only have significant effects on stratospheric ozone for a few years.

This is markedly different from human-induced production of CFCs and other chemicals that are continuously depleting stratospheric ozone, inducing long-term increases in UV-B radiation at the surface. Decreases in stratospheric ozone, climate warming, and lake acidification leading to decreases in dissolved organic carbon concentrations (e.g., Schindler et al. 1996) all result in increasing levels of UV-B radiation. Levels of UV-B radiation have risen significantly in modern times (especially since 1979) both in the tropics and in temperate regions (Kerr and McElroy 1993; Herman et al. 1996; Middleton et al. 2001).

UV-B radiation adversely affects a wide variety of organisms including amphibians (e.g., Tevini 1993; Cockell and Blaustein 2001; Bancroft et al. 2007, 2008a; Croteau et al. 2008). Moreover, UV-B radiation may interact with other stressors to affect a wide array of organisms and entire ecological communities (Tevini 1993; van der Leun et al. 1998; Cockell and Blaustein 2001; Bancroft et al. 2007).

Living organisms have had fewer than 100 years to cope with a gradual human-induced rise in UV radiation. Combined effects of UV-B radiation with pollutants, pesticides, and other agents that have been on earth for a relatively short period of time may be especially damaging to organisms that have not had time to develop mechanisms to survive their effects. Indeed, environmental change and contamination appear to be contributing to the current unprecedented loss in biodiversity (Eldridge 1998).

The diversity of locations where amphibian populations have declined has prompted consideration of global environmental changes. Climate change and other changes in the environment that are global in nature have been recognized as potential problems for amphibian populations by a number of investigators (e.g., Ovaska 1997; Donnelly and Crump 1998; Alford and Richards 1999; Alexander and Eischeid 2001; Blaustein et al. 2003; Carey and Alexander 2003; Blaustein et al. 2010).

Ovaska (1997) suggested that climate warming may result in both beneficial and deleterious effects on amphibians. For example, it was suggested that anticipated temperature increases in winter may be beneficial for amphibians in some regions of Canada, whereas anticipated decreases in precipitation in combination with elevated temperatures and evaporation in summer may be harmful to some species. Moreover, Ovaska (1997) described how changes in temperature and moisture could influence amphibian behavior physiology and life history events. For example, Ovaska (1997) suggested that changes in temperature and moisture could cause physiological stress in amphibians, affect their mobility patterns, change the timing and duration of breeding, and affect growth, hibernation, and aestivation patterns.

Changes in global temperature, precipitation, and levels of ultraviolet radiation may contribute to amphibian population declines. Moreover, contamination from airborne pollutants and other chemicals may act alone or in conjunction with global climate change to adversely affect amphibian populations. Finally, a number of investigators have proposed that emerging infectious diseases may be stimulated by climate change and changes in the levels of UV-B radiation reaching the earth's surface (e.g., Epstein 1997; Kiesecker et al. 2001).

ASSESSING THE EFFECTS OF CLIMATE CHANGE ON AMPHIBIANS

Climate Change and Amphibian Population Declines

There is some evidence that climate change may be contributing to amphibian population declines—either directly or indirectly. Several authors have suggested that unusual weather contributes to amphibian population declines. For example, McMenamin et al. (2008) described how increasing temperatures and drought have contributed to loss of amphibian habitats in parts of western North America. They observed severe reductions in the number and diversity of amphibian populations in northern Yellowstone National Park in the United States for 16 years. Moreover, they documented that amphibian declines in that region were linked to regional changes in the hydrologic landscape and overall groundwater condition, which were driven by long-term, large-scale climatic change. The disappearance of several amphibian species in southeastern Brazil in the late 1970s was attributed to unusual frost (Heyer et al. 1988). Weygoldt (1989) attributed population declines of Brazilian frogs in the Atlantic mountains to extremely dry winters. The disappearance of the golden toad (*Bufo periglenes*) in Costa Rica was originally attributed to unusual weather conditions including warmer temperatures and changes in precipitation patterns that may have affected breeding patterns (Crump et al. 1992).

D'Amen and Bombi (2009) tested the effects of global warming, habitat destruction, and ultraviolet radiation on population declines of amphibians in Italy. By analyzing spatial patterns they found that, even though multiple factors are likely involved in amphibian population declines, climate change appeared to play the greatest role. Climatic shifts were often associated with declines, even in species experiencing slow rates of population decline.

Pounds et al. (1999) illustrated the complex interrelationships among global environmental changes and amphibian population declines. They found that changes in water availability associated with changes in large-scale climate processes, such as the El Niño/Southern Oscillation (ENSO), may significantly affect amphibian, reptile, and bird populations in the Monteverde cloud forest of Costa Rica. They showed that dry periods associated with global warming are correlated with amphibian and reptile losses and changes in the bird community. In Costa Rica and potentially in other high-altitude tropical sites, global warming appears to have resulted in a decrease in the amount of mist precipitation received in the forest due to the increased altitude of the cloud bank.

Changes in ambient temperature may influence amphibian behaviors, including those related to reproduction. Potentially, changes in ambient temperature on a global scale could disrupt the timing of breeding, periods of hibernation, and the ability to find food (e.g., Donnelly and Crump 1998; Blaustein et al. 2001b).

The hypothesis that unusual weather patterns, including decreases in rainfall and continuing drought conditions in the montane rainforests of eastern Australia, caused the populations of several amphibian species to decline was not supported (Laurance 1996). Laurance suggested that other factors, including epidemic disease, may have contributed to these declines.

Remote Sensing

Several approaches have been used to examine the potential effects of climate change on amphibian populations. One strategy is the use of remote sensing information that incorporates a variety of tools and databases to examine the relationship between climate and amphibian population declines.

Alexander and Eischeid (2001) examined the relationship between amphibian declines and climate variations in Colorado, Puerto Rico, Costa Rica, Panama, and Queensland state in northeastern Australia using information gathered from airplanes, land stations, satellites, ships, and weather balloons with outputs from a weather forecast model. They showed that although declines occurred while temperature and precipitation anomalies occurred, these anomalies were not beyond the range of normal variability. They concluded that unusual climate, measured as regional estimates of temperature and precipitation, is unlikely to be a direct cause for amphibian population declines in the regions they examined.

Stallard (2001) measured time series data sets for Puerto Rico that extended back at least into the 1980s with some data sets covering the entire century. These data included forest cover; annual mean, minimum, and maximum daily temperature; annual rainfall; rain and stream chemistry; and atmospheric dust transport. Moreover, he used satellite imagery and air chemistry samples from a single aircraft flight across the Caribbean. As Alexander and Eischeid (2001) also found, none of the data sets pointed to significant changes in the parameters he measured that would directly cause amphibian population declines. He suggested that more experimental research is needed to examine the amphibian population decline problem.

Models

Predicted changes in the global climate can potentially cause shifts in the geographic ranges of plants and animals. Recent climatic changes have already resulted in species range shifts (Parmesan 2006). The more extreme changes in climate projected for the coming century (IPCC 2007) will likely produce even larger shifts in species distributions (Thomas et al. 2004; Thuiller et al. 2005). The impacts of these range shifts may have profound effects on both community structure and the functioning of ecosystems (Lovejoy and Hannah 2005). Predictions of future range shifts have relied on a variety of modeling approaches, some of which are more accurate than others.

Relatively few studies have attempted to model the potential future impacts of climate change on amphibians (e.g., Araújo et al. 2006; Thomas et al. 2004; Lawler et al. 2010). The studies that have been conducted generally make use of bioclimatic models to predict climate-driven shifts in the potential ranges of species. Bioclimatic models define the current distribution of a species as a function of current climate and then project future potential ranges based on projected future climate data (Pearson and Dawson 2003; discussed in Lawler et al. 2006, 2010). These models are generally good for projecting potential continental-scale responses of species to climate change. However, these models, in general, do not take evolution or biotic interactions into account, even though there is some empirical evidence that evolution is responsive to climate variation (Skelly et al. 2007). Although climate is likely to be the dominant driver of species distributions at broad scales, interactions with competitors, predators, prey, and other species may influence ranges even at broad scales. The ability of a bioclimatic model to capture the range of a species depends in part on whether the climate variables used in the model act as adequate surrogates for these biotic interactions and whether the relationships between climate and the biotic factors are conserved in a changing climate. Despite these limitations, bioclimatic modeling is a powerful tool that provides a first approximation of how species might respond to climate change at broad scales.

Lawler et al. (2010) used a set of bioclimatic models to assess the relative vulnerability of amphibians to climate change throughout the Western Hemisphere. The bioclimatic model projections provide a general indication of where changes in amphibian faunas might be expected to occur over the coming century. Lawler et al. (2010) supplemented these projections with two other analyses. First, they mapped the distributions of 1099 restricted-range species for which they were unable to build accurate bioclimatic models. Many of these species likely occupy very narrowly defined niches and thus may be more vulnerable to changes in climate. Second, they mapped projected climatic changes for the Western Hemisphere. They mapped areas that were consistently projected to get both warmer and drier. Such climatic changes have the potential to adversely affect amphibian populations and thus alter distributions. By combining the range-shift projections, restricted-range species distribution maps, and climate projections, Lawler et al. (2006) highlighted areas in which climate change may have large impacts on amphibians.

Using this three-part vulnerability assessment, Lawler et al. (2010) concluded that some of the greatest impacts to amphibian populations will likely occur in Central America. Several areas in this region have high concentrations of restricted-range species and are simultaneously projected to get hotter and drier and to experience high rates of species turnover. Other portions of the hemisphere were projected to experience either one or two of these factors. For example, although the Caatinga region in northeastern Brazil was projected to experience relatively little species turnover and harbors relatively few restricted-range species, it was consistently projected to experience decreases in precipitation. Although the analysis of Lawler et al (2010) included three ways in which amphibians might be vulnerable to climate change, there are several other aspects of climate vulnerability that were not addressed, including susceptibility to changes in phenology, disease dynamics, upland habitat, invasive species, and more detailed changes in hydrology.

Phenology

One direct approach for assessing whether amphibians have been affected by climate change comes from studies of amphibian breeding phenology (Table 2.1). Some examples are provided in the following.

Beebee (1995), plotting the start of breeding activities for six amphibian species in southern England over 16 years, suggested that amphibians in temperate countries may be responding to climate change by breeding earlier. Furthermore, he found that the breeding dates of two species of April–June breeding anurans were negatively correlated with average minimum temperatures in March and April and maximum temperatures in March. The spawning date of one early-breeding anuran species whose breeding date did not change from 1978 to 1994 was negatively correlated with overall winter maximum temperatures since 1978. An analysis of the most abundant newt species showed a strong negative correlation of when they arrive at the pond and average maximum temperature in the month before arrival.

Scott et al. (2008) observed the common frog (*Rana temporaria*) at various sites in the United Kingdom for 5–11 years since 1999. They found that common frogs were congregating earlier and spawning earlier but hatching was not occurring earlier. The observed changes were strongly correlated with temperature.

Chadwick et al. (2006) examined the breeding phenology of two newt species, *Triturus helveticus* and *Triturus vulgaris*, in Wales from 1981 to 1987 and from 1997 to 2005. Spring temperatures over this period explained up to 74% of among-year variability in median arrival date, with a significant advance of 2–5 days with every degree centigrade increase. Males of both species displayed greater changes than females. Moreover, changes were greater for *T. helveticus* compared with *T. vulgaris* within the sexes. Thus, climate change may have differential effects on sexes and between species.

Not all amphibians in Europe appear to be responding to climate change in the same fashion as those described above (Table 2.1). Reading's (1998) study of the common toad (*Bufo bufo*) in England from 1980 to 1998 showed that the main arrival to breeding sites was highly correlated with the mean daily temperatures over the 40 days immediately preceding the main arrival. However, a significant trend toward earlier breeding in recent years compared with previous years was not found.

Gibbs and Breisch (2001) showed that over the last century, daily temperatures increased near Ithaca, New York, and several species of anurans have shifted their breeding patterns accordingly (Table 2.1). Consequently, four species of anurans first(?) vocalized 10–13 days earlier, two were unchanged, and none called later during 1990–1999 compared with calling dates between 1900 and 1912.

In Oregon, at one site there was a nonsignificant trend for western toads (*Bufo* [=*Anaxyrus*] *boreas*) to breed earlier, and this was associated with increasing mean maximum daily temperature (Blaustein et al. 2001b). However, at four other sites

TABLE 2.1
Climate Change and Breeding Phenologies of Temperate Amphibians

Species	Breeding Earlier?	Reference
Europe		
Bufo bufo (Common toad)	No	Reading 1998
Bufo calamita (Natterjack toad)	Yes	Beebee 1995
Rana kl. esculenta (Edible frog)	Yes	Beebee 1995
R. temporaria (Common frog)	No	Beebee 1995
R. temporaria (Common frog)	Yes	Scott et al. 2008
Triturus helveticus (Palmate newt)	Yes	Beebee 1995
T. helveticus (Palmate newt)	Yes	Chadwick et al. 2006
T. vulgaris (Smooth newt)	Yes	Beebee 1995
T. vulgaris (Smooth newt)	Yes	Chadwick et al. 2006
T. cristatus (Great-crested newt)	Yes	Beebee 1995
North America		
Anaxyrus americanus (American toad; New York)	No	Gibbs and Breisch 2001
A. boreas (Western toad; Oregon)	No	Blaustein et al. 2001a
A. fowleri (Fowler's toad; Ontario, Canada)	No	Blaustein et al. 2001a
Hyla versicolor (Gray tree frog; New York)	Yes	Gibbs and Breisch 2001
Pseudacris crucifer (Spring peeper; Michigan)	No	Blaustein et al. 2001a
Pseudacris crucifer (Spring peeper; New York)	Yes	Gibbs and Breisch 2001
Rana cascadae (Cascades frog; Oregon)	No	Blaustein et al. 2001a
Lithobates catesbeianus (Bullfrog; New York)	Yes	Gibbs and Breisch 2001
L. clamitans (Green frog; New York)	No	Gibbs and Breisch 2001
L. sylvaticus (Wood frog; New York)	Yes	Gibbs and Breisch 2001
Asia		
Hynobius tokyoensis (Tokyo salamander)	Yes	Kusano and Inoue 2008
Rana ornativentris (Montane brown frog)	Yes	Kusano and Inoue 2008
Rhacophorus arboreus (Kinugasa flying frog)	Yes	Kusano and Inoue 2008

neither western toads nor Cascades frogs (*Rana cascadae*) showed significant positive trends toward earlier breeding. At three of four of these sites, breeding time was associated with warmer temperatures. The spring peeper (*Pseudacris crucifer*) in Michigan did not show a significant trend to breeding earlier but did show a significant positive relationship between breeding time and temperature. Fowler's toads (*A. fowleri*) in eastern Canada did not show a trend for breeding earlier, nor was there a positive relationship between breeding time and temperature. In Colorado, the breeding time of boreal chorus frogs (*Pseudacris maculata*) was correlated with snow levels accumulated throughout the winter (Corn and Muths 2002). Therefore, at least for this species, changes in snowpack associated with climate change could influence breeding time.

In Japan, Kusano and Inoue (2008) analyzed long-term data sets (12–31 years) for three species of amphibians and found that all populations examined for these species were breeding earlier (Table 2.1). The date of first spawning was correlated with the mean monthly temperature just before the breeding season for each population examined.

The broad pattern emerging from available breeding studies is that some temperate zone amphibian populations show a trend toward breeding earlier, whereas others do not. The reasons for this variation are not known.

ENVIRONMENTAL CHANGES AND EMERGING INFECTIOUS DISEASES

One potential consequence of global warming is the increased spread of infectious disease (Cunningham et al. 1996; Epstein 1997). This may occur, for example, if rising temperatures affect the distribution of the vectors of a pathogen, thereby making vectors more abundant, or if an environmental agent renders a host's immune system more susceptible. Changes in temperature, precipitation, acidification, pollutants, and increased UV-B radiation are some of the stressors that may affect the immune systems of amphibians. Immune system damage from multiple stressors could make amphibians more susceptible to pathogens whose ranges may change due to global warming (Blaustein et al. 1994b; Kiesecker and Blaustein 1995, 1997). Thus, rising temperatures and changes in precipitation could be stressful and might be associated with disease outbreaks in amphibian populations (Pounds et al. 1999; Kiesecker et al. 2001).

There is some evidence that climate change may influence disease emergence in amphibian populations. On a local scale, an experimental field study in Oregon by Kiesecker et al. (2001) illustrates the complex interrelationships among environmental change, UV radiation, and amphibian population declines, and it parallels a tropical study by Pounds et al. (1999). Kiesecker et al. (2001) linked ENSO events with decreased winter precipitation in the Oregon Cascade Range. They suggested that less winter snow pack resulted in lower water levels when western toads (*A. boreas*) breed in early spring. Toad embryos developing in shallower water are exposed to higher levels of UV-B radiation, which results in increased mortality from the pathogenic oomycete *Saprolegnia ferax*. *Saprolegnia* infects all stages of amphibians and has been documented to cause extreme mortality in developing embryos (e.g., Blaustein et al. 1994b; Romansic et al. 2007, 2009).

Pounds et al. (2006) identified two dynamics that are global in nature: climate change and the spread of a highly virulent pathogen (discussed in Blaustein and Dobson 2006). A pathogenic "chytrid" fungus, *Batrachochytrium dendrobatidis* (BD), is implicated as the proximate cause for *Atelopus* population crashes and species extinctions in tropical America. BD infects the mouthparts of tadpoles and the skin of adult amphibians (Longcore et al. 1999). It has been associated with amphibian population declines around the world (e.g., Berger et al. 1998; Mendelson et al. 2006). Pounds et al. (2006); (Vredenburg et al. 2010; Blaustein and Johnson 2010) presented a mechanistic explanation for how climate change influences outbreaks of BD by modifying conditions in montane areas of Central and South America where nighttime temperatures are shifting closer to the thermal optimum for BD, while increased daytime cloudiness prevents frogs from finding thermal refuges from the pathogen.

Climate change variation determines altitudinal patterns of extinction. According to Pounds et al. (2006) mid-elevation montane *Atelopus* species experience higher rates of extinction than do those restricted to the lowlands or to the highest elevations (where there are fewer extinctions), because these sites in tropical America are thermal refuges from the pathogen. In these refuges, temperatures are either too high or too low for optimal growth of BD.

Several earlier studies suggested that BD was an emergent pathogen and potentially the sole reason for amphibian population declines in the tropics (Berger et al. 1998; Daszak et al. 1999). While chytridiomycosis, the disease caused by BD, was a relatively common condition in many areas experiencing declines, the early studies presented no mechanisms explaining the emergence of the pathogen. In many cases, it was unknown if BD was responsible for causing declines directly or if it was secondary in nature and is associated with already dead and dying animals (McCallum 2005). Thus, Pounds et al. (2006) explain this "climate chytrid paradox." The climate-chytrid paradox simply states that the climatic conditions favoring chytrid growth appeared to be the opposite of those created by current climate trends. Pounds et al.'s (2006) discussion of this paradox explains how shifting temperatures may be the ultimate trigger that sets off a pathogenic fungus such as BD—a likely contributor to *Atelopus* extinctions.

Importantly, Pounds et al. (2006, 2007) suggest that the dynamics discussed above are but one of many potential scenarios and that more than one pathogen may be involved in amphibian population declines. Moreover, they suggest that more than one stressor could make amphibians susceptible to disease. Their "climate-linked epidemic hypothesis" assumed no specific disease or mechanism of outbreak (Pounds et al. 2007).

Lips et al. (2008) tested the climate-linked epidemic hypothesis by reanalyzing data on declines and extinctions of *Atelopus* species in the same region where Pounds et al. (2006) conducted their research. They found no evidence for the climate-linked epidemic hypothesis and suggested that BD is an introduced pathogen that has been spreading throughout the American tropics since the 1970s. This spatiotemporal hypothesis suggests that BD spreads independently of climate shifts.

The points made by Lips et al. (2008) have been questioned in recent papers. For example, Lampo et al. (2008) demonstrated that some of the assumptions of the spatiotemporal hypothesis are inconsistent with data on extinctions and declines. Moreover, some populations have a lag time of 7 years between pathogen arrival and amphibian population declines, while other populations crash when BD arrives. Thus, Lampo et al. (2008) reason that, at least in some populations, population crashes must not be indicative of BD arrival.

Parmesan and Singer (2008) stated that the Lips et al. (2008) analysis is statistically flawed. They conclude that Pounds et al. (2006) show spatial patterns of amphibian extinctions that are moderately consistent with climate change being the driver and that Lips et al. (2008) show spatiotemporal patterns of amphibian population declines that are moderately consistent with disease spread (BD) being the driver. They also suggest that both hypotheses are supported by numerous studies, are not mutually exclusive, and may be interactive.

Examining the climate-linked epidemic hypothesis and the spatiotemporal spread hypothesis, Rohr et al. (2008) showed that there was spatial structure to the timing of *Atelopus* extinctions, but the cause for this structure was unknown. Furthermore, Rohr et al. (2008) stated that the temporal pattern of *Atelopus* extinctions reported by Lips et al. (2006) in Panama could also be explained by spatiotemporal heterogeneities in the emergence of a novel pathogen or an abiotic stressor that caused local population declines (McCallum 2005). They also stated that in Costa Rica, BD might have been widespread long before declines were detected (Puschendorf et al. 2006).

Rohr et al. (2008) showed a positive multidecade correlation between *Atelopus* extinctions and mean air temperature in the previous year. However, the evidence that this causes population declines is weak because other variables were also good predictors of extinction events. Although Rohr et al. (2008) found no support for the "chytrid-thermal-optimum hypothesis" (Pounds et al. 2006; Blaustein and Dobson 2006), they suggest that climate change is likely to play an important role in amphibian population declines worldwide.

Climate change and outbreaks of chytridiomycosis have been reported in recent studies. Bosch et al. (2006) showed a significant association between rising temperatures and outbreaks of chytridiomycosis in Spain. Increases in chydriomycosis were correlated with low average summer maximum temperatures in Australia (Drew et al. 2006). During the decline of amphibians in New South Wales, Australia, very few moribund specimens examined were positive for BD (Alford et al. 2007). Alford et al. (2007) suggested that stress during growth and development may contribute to limb asymmetries in amphibians. They suggested that unusual climate conditions might be a significant stressor on amphibian growth and development. In Australia, Alford et al. (2007) attributed the greater frequency of limb asymmetries in "pre-declining populations" compared with nondeclining "control" populations to stress from "dramatic regional warming." They suggested that BD might not be the only factor involved in amphibian population declines in this region and that climate conditions alone may be an important factor in declines.

Di Rosa et al. (2007) showed that BD may be present without causing chytridiomycosis in frogs in Italy and that other pathogens may also contribute to frog declines in this region. In agreement with Pounds et al. (2007), they suggested that climate conditions might make amphibians more susceptible to a number of pathogens.

Evidence from studies at the Savannah River long-term-research site in South Carolina illustrates that populations of several amphibian species have been in decline (Daszak et al. 2005). At this site, the presence of BD was rare and there was no evidence of chytridiomycosis. The investigators concluded that the population declines in this region were more likely due to low rainfall and shortened hydroperiod for breeding rather than an epidemic caused by BD.

Depending upon the model, a specific region may receive more or less precipitation in the future than at present and may be colder or warmer than at present (Lawler et al. 2006, 2009). Such changes could significantly affect hydroperiods critical to amphibian life cycles. Thus, Gervasi and Foufopoulos (2008) showed that wood frog (*Lithobates sylvaticus*) tadpoles exposed to desiccation had shorter development times, weaker cellular immune system responses, and lower total leukocyte numbers than animals from control groups. Measures of immune response showed a decrease with increasing severity of the desiccation treatment. It was unclear whether the observed depression in immune response was transient or permanent. However, even temporary periods of immune system suppression shortly after metamorphosis could lead to greater susceptibility to pathogens.

AMBIENT UV-B RADIATION IS HARMFUL TO AMPHIBIANS AND MAY INTERACT WITH CLIMATE CHANGE

Increases in UV-B radiation due to ozone depletion and other dynamics have been documented in both temperate and tropical ecosystems. However, even relatively steady ambient levels of UV-B radiation can harm amphibians, especially if there are synergistic effects with other agents, such as contaminants, pathogens, or other stressors. UV-B radiation interacts with other stressors to affect a wide array of organisms and entire ecological communities (Tevini 1993; van der Leun et al. 1998; Cockell and Blaustein 2001; Bancroft et al. 2007).

Spectral measurements made at Toronto, Canada, since 1989 indicate a 35% annual increase in UV-B radiation per year in winter and a 7% increase in summer (Kerr and McElroy 1993). These increases were caused by a downward trend in ozone that was measured at Toronto during the same time. Moreover, it was suggested that increases in UV-B in late spring might have a disproportionately larger effect on some amphibian species if the increases occur at critical phases of amphibian development (Kerr and McElroy 1993). This supports the results of studies showing the adverse effects of UV-B radiation in spring-conducted field experiments on developing amphibian embryos (discussed below).

Middleton et al. (2001) assessed trends in solar UV-B radiation over 20 sites in Central and South America, derived from the Total Ozone Mapping Spectrometer satellite data. They showed that annually averaged UV-B doses, as well as the maximum values, have been increasing in both regions since 1979. The UV index was consistently higher for Central America, where many amphibian declines have been documented (Lips 1998; Pounds et al. 1999). They conclude that further investigation of the role of UV-B radiation in amphibian declines is warranted.

Using satellite data, Middleton et al. (2001) attempted to examine a potential relationship among amphibian declines in Central and South America and trends in solar radiation where declines occurred. The data collected by Middleton et al. (2001) was consistent with experimental evidence that UV-B radiation is harmful to amphibians and that increasing UV-B levels may be contributing to amphibian population declines (Bancroft et al. 2008a; Blaustein et al. 1998, 2001a). However, as discussed in detail by Middleton et al. (2001), data gathered from remote sensing have many limitations. For example, there are few long-term data sets on amphibian populations to serve as a baseline data source for remote sensing (Blaustein et al. 1994c). Moreover, there are a number of shortcomings when using data generated from satellites, including the fact that resolution of the satellite-generated data is not accurate enough to approximate ground-level interpretations (Middleton et al. 2001).

Investigators at various sites around the world have shown via experiments that ambient UV-B radiation decreases the hatching success of some amphibian species at natural oviposition sites in the field (reviewed in Bancroft et al. 2008a; Blaustein et al. 1998, 2001a). Embryos of some species are more resistant to UV-B radiation than others (reviewed in Blaustein et al. 1998; Bancroft et al. 2008a).

There are differences in how species cope with UV-B radiation. For some species, in field experiments, hatching success is lower when eggs are exposed to UV-B radiation compared with shielded controls. In other species, hatching success is not affected by UV-B exposure (e.g., Blaustein et al. 1999; Starnes et al. 2000, Häkkinen et al. 2001). This is not a contradiction. Rather, these studies illustrate clear interspecific differences in tolerance to UV-B radiation due to differences in molecular, cellular, behavioral, and ecological attributes between species (Blaustein and Belden 2003). In fact, within the same study, conducted at the same time and same site, it has been shown that the eggs of some species were sensitive to UV-B, whereas the eggs of other species were resistant (e.g., Blaustein et al. 1994a; Anzalone et al. 1998; Lizana and Pedraza 1998).

Importantly, even though hatching rates of some species may appear unaffected by ambient UV-B radiation in field experiments, an increasing number of studies illustrate a variety of sublethal effects due to UV exposure. For example, when exposed to UV-B radiation, amphibians may change their behavior (Nagl and Hofer 1997; Blaustein et al. 2000; Kats et al. 2000), growth and development may be slowed (e.g., Belden et al. 2000; Pahkala et al. 2000, 2001; Smith et al. 2000), or a number of developmental and physiological malformations may occur (e.g., Worrest and Kimeldorf 1976; Hays et al. 1996; Blaustein et al. 1997; Bruggeman et al. 1998; Fite et al. 1998) (recently reviewed by Croteau et al. 2008). Sublethal effects may become evident even in species whose embryos appeared to be resistant to UV-B radiation in field experiments. Moreover, numerous field and laboratory experiments have also shown that UV-B radiation interacts synergistically with a variety of chemicals, low pH, and pathogens (Kiesecker and Blaustein 1995; Long et al. 1995; Blaustein et al. 2001a; reviewed by Bancroft et al. 2008a).

The experimental field study in Oregon by Kiesecker et al. (2001) discussed above illustrates complex interactions among precipitation–climate change, UV radiation, and amphibian population declines, and parallels the tropical study by Pounds et al. (1999). Merilä et al. (2000) present an interesting scenario combining climate change, UV radiation, and amphibian breeding. They suggest that if amphibians are breeding earlier in northern ecosystems as suggested, as discussed earlier in this chapter, then their annual life cycle will not only start earlier relative to the calendar date but also with regard to maximum UV-B exposure. UV-B exposure would be lower than if amphibians bred later in the spring. Thus, they suggest that global warming, which may induce amphibians to breed earlier, may counteract the effects of increasing UV-B levels generated by a thinning ozone layer.

Recent studies suggest that temperature and UV-B may interact synergistically. Searle et al (2010) found reduced survival in larval amphibians exposed to UV-B radiation in colder temperatures. Van Uitregt et al. (2007) also observed reduced survival, growth, and performance when amphibian larvae were exposed to UV-B at colder temperatures. While the mechanisms for these observations are unknown, there are a number of possible explanations. For example, cold temperatures could alter larval amphibian behavior, which could expose them to higher levels of UV-B or increase their exposure time. Alternatively, cold temperatures could reduce the activity of repair enzymes, such as photolyase, that are essential in repairing DNA damage after exposure to UV-B. Because many high-elevation sites have high levels of UV-B in combination with cold temperatures, amphibians inhabiting these locations may be particularly at risk from UV-B. Global warming is likely to change temperatures in many habitats, which will alter the threat of UV-B for amphibians and make the risk of negative effects from UV-B exposure hard to predict.

Like other animals, over evolutionary time numerous selection pressures have influenced amphibians in a variety of ways. Some selection pressures have shaped life-history characteristics and behaviors that relate to how amphibians are exposed to sunlight. Amphibians seek sunlight for thermoregulation and to maximize their growth and development. Especially in temperate regions, larvae often seek shallow, warm water that ultimately results in an increase in their growth rate (Wollmuth et al. 1987).

For example, Cascades frog (*R. cascadae*) tadpoles are frequently observed in warmer, sunlit areas in the afternoon (Wollmuth et al. 1987). Mountain yellow-legged frog (*Rana muscosa*) tadpoles concentrate where water temperatures are highest (near shore during the day, deeper in the late afternoon and evening; Bradford 1984). In one study, all but the latest-stage bullfrog (*Lithobates catesbeianus*) tadpoles selected the warmest microhabitats (Wollmuth and Crawshaw 1988).

Many frog species bask in sunlight for prolonged periods (Hutchison and Dupré 1992). For example, Lillywhite (1970) found that more than 70% of bullfrogs present in a pond were basking from 1300 to 1700 hours on a sunny day, compared to less than 20% on a cloudy day. An observational study of *R. muscosa* showed that eight frogs present in a pond had from 50% to 80% of their body exposed to full sunlight for half the sunlight hours (Bradford 1984). Juvenile western toads (*A. boreas*) often bask in the thousands along the shoreline (Blaustein et al. 2005).

Many amphibian species lay their eggs in shallow open water in direct sunlight (Behler and King 1979; Ashton and Ashton 1988; Nussbaum et al. 1983; Stebbins and Cohen 1995). Eggs are often laid in shallow water or even floating on the water surface to develop in a more oxygenated environment and probably in areas with lower risk of predator or parasite attacks. Oxygenation of eggs is critical to their development (Duellman and Trueb 1986).

A recent comprehensive study incorporating field transect surveys and laboratory and field experiments illustrates that even though UV-B radiation may be harmful to amphibians, the selection pressures related to thermoregulation, growth, and development are overriding and amphibians seek sunlit microhabitats (Bancroft et al. 2008b). This was the case even in experimental-choice tests.

In summary, amphibians living today often seek sunlight because, over evolutionary time, seeking sunlight was beneficial for thermoregulation, growth, and development and perhaps for other reasons. Yet today's sunlight exposes amphibians to doses of UV-B radiation that can kill or significantly damage them.

Exposure to UV-B radiation may be especially significant for those amphibian species for which selection pressures have resulted in behaviors that expose them to relatively large doses of solar radiation. For example, as stated previously, many amphibians lay eggs in open, shallow areas where they receive maximum exposure to sunlight. This exposure can heat egg masses, which induces fast hatching and developmental rates (Stebbins and Cohen 1995). Similarly, the larvae of many amphibian species seek shallow, open regions of lakes and ponds, where it is warmest and where they can develop quickly (Hokit and Blaustein 1997; Hoff et al. 1999). Rate of development is especially critical for amphibians living in ephemeral habitats. Species living in temporary habitats must metamorphose before their habitat dries or freezes (e.g., Blaustein et al. 2001a). Thus, amphibians are often faced with conflicting selection pressures. Some species must develop quickly enough before their habitat disappears. Therefore, they seek sunlight where exposure to warmer temperatures enhances development. Yet evidence from a number of recent studies illustrates that many amphibian species, even those that seek sunlight, are sensitive to solar radiation (e.g., Fite et al. 1998; Belden et al. 2000). If exposed, amphibians may die or accrue sublethal damage that can severely affect them (discussed in the following).

EVOLUTIONARY CONSIDERATIONS

Amphibians have evolved behaviors, morphologies, and overall life styles that have allowed them to persist for millions of years. Such attributes were obviously a result of numerous selection pressures over evolutionary time. Certain behaviors and life-history attributes have been instrumental in helping amphibians survive for millions of years. However, under today's environmental conditions, these same behaviors and life-history characteristics appear to be placing amphibians in harm's way (Blaustein and Bancroft 2007; Bancroft et al. 2008b). This seems especially true in light of current conditions regarding the climate and stratospheric ozone with accompanying increases in UV-B radiation. Moreover, climate models suggest that under certain scenarios climate change will affect amphibian distribution and abundance more than those of mammals and birds.

The causes for amphibian population declines are complex and amphibians may be facing unprecedented environmental changes, some of which are intense and relatively new (e.g., increasing UV-B radiation due to ozone depletion; chemical contamination; habitat alteration). These dramatic changes in the environment may interact with more gradual changes in precipitation and temperature in unpredictable ways. Therefore, it is not easy to assess the direction a population will take in the face of both natural and an increasing array of human-induced selection pressures.

Amphibians are limited by historical constraints. Thus, adaptations in the face of environmental change may not occur before populations are impacted (Blaustein and Bancroft 2007). Limited by historical constraints, many amphibians (especially frogs and toads) must lay their eggs in water. Laying their eggs in water exposes eggs not only to aquatic predators and competitors (Wells 2007) but to a variety of abiotic parameters, both natural and human induced, that may be especially harmful to them under today's conditions. For example, extant amphibians face chemical contamination in aquatic systems that have risen dramatically in modern times due to mining, industrialization, urban sprawl, and agricultural processes (Boone and Bridges 2003). The rapid rate of contamination of aquatic systems can be measured in hundreds of years, yet amphibians have been laying their eggs in less-contaminated water for millions of years.

Many amphibians still lay their eggs in shallow water, in large communal masses, potentially subjecting them to exposure to both increasing UV-B radiation and infectious diseases. Many amphibians school in large numbers. Newly metamorphosed amphibians may form aggregations. Amphibian species mate in large communal groups. These behaviors and ecological attributes have probably persisted and were probably beneficial for millions of years (Romer 1939, 1968). But under today's conditions these very behaviors may increase the potential of acquiring an infectious disease. Amphibians exhibit such seemingly maladaptive characteristics because evolution takes time, is not perfect, can only alter existing variations, and is a compromise of numerous selection pressures (Blaustein and Bancroft 2007). Those individuals with the mechanisms, behaviors, and life-history attributes best able to cope with mounting environmental changes, emerging diseases, and the various other factors that amphibians face may produce offspring with those attributes. Perhaps these adaptations will allow amphibians to persist for millions of years more.

SUMMARY

Species loss is a major part of a "biodiversity crisis" that is exemplified by population declines, range reductions, and extinctions of amphibian species around the world. In at least some regions, amphibian losses appear to be more severe than losses in other vertebrate taxa. There appears to be no single cause for amphibian population declines. Like other animals, amphibians are affected by numerous environmental stresses that often act in complex ways. The causes for the decline of a given species may be different from region to region and even in different populations of the same species. Synergistic interactions between more than one factor may be involved. There may be interspecific differences and even differences between life stages in how amphibians react to environmental changes and other stresses. Although the major cause for amphibian population declines and extinctions is probably habitat destruction, climate change, diseases, environmental contamination, and introduced exotic species may all contribute to amphibian population declines.

The emphasis of our review is on climate change and its relationship with other factors such as increasing ultraviolet radiation and disease. In this chapter we address several key questions concerning the effects of climate change on amphibians. Are amphibian populations affected by global climate changes? What are the major climate changes affecting amphibian populations? How are amphibian populations affected? The multifaceted dynamics of climate change makes it difficult to accurately predict the outcome of global change on individual populations. A multidisciplinary approach is probably necessary for understanding the effects of climate change on amphibians. We suggest that climate modeling should be fortified with laboratory and field experiments. Experimental studies could be bolstered with molecular work to provide clues as to how amphibians are affected by climate change at the physiological level. Molecular biologists, ecologists, and modelers working together may help us understand the broad impacts of climate change on amphibians. We suggest that developing conservation strategies for species and populations should involve adaptive management where policies can change as new information is gathered.

ACKNOWLEDGMENTS

Financial support was provided by NSF (DEB-942333 and IBN-9904012) and by the Katherine Bisbee Fund of the Oregon Community Foundation. We thank Yun Soo Chung, Ah Jung, Yi Ling, Fung Sei, Gisselle Xie Yang, and Ezy Yoo for help with the manuscript.

REFERENCES

Alexander, M.A., and J.K. Eischeid. 2001. Climate variability in regions of amphibian declines. *Conservation Biology* 15:930–942.

Alford, R.A., K.S. Bradfield, and S.J. Richards. 2007. Global warming and amphibian losses. *Nature* 447:E3–E4.

Alford, R.A., and S.J. Richards. 1999. Global amphibian declines: a problem in applied ecology. *Annual Review of Ecology and Systematics* 30:133–165.

Andrady, A., P.J. Aucamp, A. Bais, C.L. Ballaré, L.O. Björn, J.F. Bornman, et al. 2009. Environmental effects of ozone depletion and its interactions with climate change: progress report, 2008. *Photochemical and Photobiological Sciences: Official Journal of the European Photochemistry Association and the European Society for Photobiology* 8:13–22.

Anzalone, C.R., L.B. Kats, and M.S. Gordon. 1998. Effects of solar UV-B radiation on embryonic development in *Hyla cadaverina, Hyla regilla* and *Taricha* torosa. *Conservation Biolology* 12:646–653.

Araújo, M.B., W. Thuiller, and R.G. Pearson. 2006. Climate warming and the decline of amphibians and reptiles in Europe. *Journal of Biogeography* 33:1712–1728.

Ashton, R.E., and P.S. Ashton. 1988. *Handbook of Reptiles and Amphibians of Florida. Part Three: The Amphibians*. Miami, FL. Windward Publishing, Inc.

Bancroft, B.A., N.J. Baker, and A.R. Blaustein. 2007. Effects of UVB radiation in marine and freshwater organisms: a synthesis through meta-analysis. *Ecology Letters* 10:332–345.

———. 2008a. A meta-analysis of the effects of ultraviolet B radiation and other stressors on survival in amphibians. *Conservation Biology* 22:987–996.

Bancroft, B.A., N.J. Baker, C.L. Searle, T.S. Garcia, and A.R. Blaustein. 2008b. Larval amphibians seek warm temperatures and do not avoid harmful UVB radiation. *Behavioral Ecology* 19:879–886.

Beebee, T.J.C. 1995. Amphibian breeding and climate. *Nature* 374:219–220.

Behler J.L., and F.W. King. 1979. *National Audubon Society Field Guide to North American Reptiles and Amphibians*. New York: Alfred A. Knopf.

Belden, L.K., E. L. Wildy, and A.R. Blaustein. 2000. Growth, survival, and behaviour of larval long-toed salamanders (*Ambystoma macrodactylum*) exposed to ambient levels of UV-B radiation. *Journal of Zoology (London)* 251:473–479.

Berger, L., R. Speare, P. Daszak, et al. 1998. Chytridiomycosis causes amphibian mortality associated with population declines in the rain forests of Australia and Central America. *Proceedings of the National Academy of Science (USA)* 95:9031–9036.

Blaustein, A.R. 1994. Chicken Little or Nero's fiddle? A perspective on amphibian population declines. *Herpetologica* 50:85–97.

Blaustein, A.R., and B.A. Bancroft. 2007. Amphibian population declines: evolutionary considerations. *Bioscience* 57:437–444.

Blaustein, A.R., B.A. Han, R.A. Relyea, P.T.J. Johnson, J.C. Buck, S. Gervasi, and L.B. Kats. 2011. The complexity of amphibian population declines: understanding the role of cofactors in driving amphibian losses. Annals of the New York Academy of Sciences. *The Year in Ecology and Conservation Biology*. 1223:108–119.

Blaustein, A.R., and L.K. Belden. 2003. Amphibian defenses against UV-B radiation. *Evolution & Development* 5:89–97.

Blaustein, A.R., L.K. Belden, A.C. Hatch, et al. 2001a. Ultraviolet radiation and amphibians. In *Ecosystems, evolution and ultraviolet radiation*. Eds. C.S. Cockell and A.R. Blaustein, 63–79. New York: Springer.

Blaustein, A.R., L.K. Belden, D.H. Olson, D.L. Green, T.L. Root, and J.M. Kiesecker. 2001b. Amphibian breeding and climate change. *Conservation Biology* 15:1804–1809.

Blaustein, A.R., D.P. Chivers, L.B. Kats, and J.M. Kiesecker. 2000. Effects of ultraviolet radiation on locomotion and orientation in roughskin newts (*Taricha granulosa*). *Ethology* 108:227–234.

Blaustein, A.R., and A. Dobson. 2006. Extinction: a message from the frogs. *Nature* 439:143–144.

Blaustein, A.R., A.C. Hatch, L.K. Belden, et al. 2003. Global change: challenges facing amphibians. In *Amphibian conservation*. Ed. R.D. Semlitsch, 187–198. Washington, D.C.: Smithsonian Press.

Blaustein, A.R., J.B. Hays, P.D. Hoffman, D.P. Chivers, J.M. Kiesecker, W.P. Leonard, A. Marco, D.H. Olson, J.K. Reaser, and R.G. Anthony. 1999. DNA repair and resistance to UV-B radiation in western spotted frogs. *Ecological Applications* 9:1100–1105.

Blaustein, A.R., P.D. Hoffman, D.G. Hokit, J.M. Kiesecker, S.C. Walls, and J.B. Hays. 1994a. UV repair and resistance to solar UV-B in amphibian eggs: a link to populations? *Proceedings of the National Academy of Science (USA)* 91:1791–1795.

Blaustein, A.R., D.G. Hokit, R.K. O'Hara, and R.A. Holt. 1994b. Pathogenic fungus contributes to amphibian losses in the Pacific Northwest. *Biological Conservation* 67:251–254.

Blaustein, A.R., and J.M. Kiesecker. 2002. Complexity in conservation: lessons from the global decline of amphibian populations. *Ecology Letters* 5:597–608.

Blaustein, A.R., J.M. Kiesecker, D.P. Chivers, and R.G. Anthony. 1997. Ambient UV-B radiation causes deformities in amphibian embryos. *Proceedings of the National Academy of Science (USA)* 94:13735–13737.

Blaustein, A.R., J.M. Kiesecker, D.P. Chivers, D.G. Hokit, A. Marco, L.K. Belden, and A. Hatch.1998. Effects of ultraviolet radiation on amphibians: field experiments. *American Zoologist* 38:799–812.

Blaustein, A.R., and P.T.J. Johnson. When an infection turns lethal. *Nature*. 465:881–882.

Blaustein A.R., J.M. Romansic, and E.A. Scheessele. 2005. Ambient levels of ultraviolet-B radiation cause mortality in juvenile western toads, *Bufo boreas*. *American Midland Naturalist* 154:375–382.

Blaustein, A.R., and D.B. Wake. 1995. The puzzle of declining amphibian populations. *Scientific American* 272:52–57.

Blaustein, A.R., D.B. Wake, and W.P. Sousa. 1994c. Amphibian declines: judging stability, persistence, and susceptibility of populations to local and global extinctions. *Conservation Biology* 8:60–71.

Blaustein, A.R., S.C. Walls, B.A. Bancroft, J.J. Lawler, C.L. Searle, and S.S. Gervasi. 2010. Direct and indirect effects of climate change on amphibian populations.

Boone, M.D., and C.M. Bridges. 2003. Effects of pesticides on amphibian populations. In *Amphibian conservation*. Ed. R.D. Semlitsch, 152–167. Washington, D.C.: Smithsonian Press.

Bosch, J., L.M. Carrascal, L. Duran, S. Walker, and M.C. Fisher. 2006. Climate change and outbreaks of amphibian chytridiomycosis in a montane area of Central Spain; is there a link? *Proceedings of the Royal Society (B)* 274:253–260.

Bradford, D.F. 1984. Temperature modulation in a high-elevation amphibian, *Rana muscosa*. *Copeia* 1984:966–976.

Bruggeman, D.J., J.A. Bantle, and C. Goad. 1998. Linking teratogenesis, growth, and DNA photodamage to artificial ultraviolet-B radiation in *Xenopus laevis* larvae. *Environmental Toxicology and Chemistry* 17:2114–2121.

Carey, C., and M.A. Alexander. 2003. Climate change and amphibian declines: is there a link? *Diversity and Distributions* 9:111–121.

Chadwick, E.A., F.M. Slater, and S.J. Ormerod. 2006. Inter- and intraspecific differences in climatically mediated phonological change in coexisting *Triturus* species. *Global Change Biology* 12:1069–1078.

Cockell, C.S. 2001. A photobiological history of earth. In *Ecosystems, evolution and ultraviolet radiation*. Eds. C.S. Cockell and A.R. Blaustein, 1–35. New York: Springer.

Cockell, C.S., and A.R. Blaustein (Eds.). 2001. *Ecosystems, evolution and ultraviolet radiation*. New York: Springer.

Corn, P.S., and E. Muths. 2002. Variable breeding phenology affects the exposure of amphibian embryos to ultraviolet radiation. *Ecology* 83:2958–2963.

Crick, H.Q.P., and T.H. Sparks. 1999. Climate change related to egg laying trends. *Nature* 399:423–424.

Croteau, M.C., M.A. Davidson, D.R.S. Lean, and V.L. Trudeau. 2008. Global increases in ultraviolet B radiation: potential impacts on amphibian development and metamorphosis. *Physiological and Biochemical Zoology* 81:743–761.

Crump, M.L., F.R. Hensley, and K.L. Clark. 1992. Apparent decline of the golden toad: Underground or extinct?. *Copeia* 1992:413–420.

Cunningham, A.A., T.E.S. Langton, P.M. Bennett, S.E.N. Drury, R.E. Gough, and S.K. MacGregor. 1996. Pathological and microbiological findings from incidents of unusual mortality of the common frog (*Rana temporaria*). *Philosophical Transactions of the Royal Society, London B* 351:1529–1557.

D'Amen, M., and P. Bombi. 2009. Global warming and biodiversity: evidence of climate-linked amphibian declines in Italy. *Biological Conservation*. 42:3060–3067.

Daszak, P., L. Berger, A.A. Cunningham, A.D. Hyatt, D.E. Green, and R. Speare. 1999. Emerging infectious diseases and amphibian population declines. *Emerging Infectious Diseases* 5:735–748.

Daszak, P., D.E. Scott, A.M. Kilpatrick, C. Faggioni, J.W. Gibbons, and D. Porter. 2005. Amphibian population declines at Savannah River site are linked to climate, not chytridiomycosis. *Ecology* 86:3232–3237.

Di Rosa, I., F. Simoncelli, A. Fagotti, and R. Pascolini. 2007. The proximate cause of frog declines? *Nature* 447:E4–E5.

Donnelly, M.A., and M.L. Crump. 1998. Potential effects of climate change on two neotropical amphibian assemblages. *Climate Change* 39:541–561.

Drew, A., E.J. Allen, and L.J.S. Allen. 2006. Analysis of climatic and geographic factors affecting the presence of chytridiomycosis in Australia. *Diseases of Aquatic Organisms* 68:245–250.

Duellman, E.D., and L. Trueb. 1986. *Biology of amphibians*. New York: McGraw-Hill.

Eldridge, N. 1998. *Life in the balance: humanity and the biodiversity crisis*. New Jersey: Princeton University Press.

Epstein, P.R. 1997. Climate, ecology, and human health. *Consequences* 3:3–19.

Fite, K.V., A.R. Blaustein, L. Bengston, and H.E. Hewitt. 1998. Evidence of retinal light damage in *Rana cascadae*: a declining amphibian species. *Copeia* 1998:906–914.

Forchhammer, M.C., E. Post, and N.C. Stenseth. 1998. Breeding phenology and climate. *Nature* 391:29–30.

Gervasi, S., and J. Foufopoulos. 2008. Pond desiccation rate affects immune system responsiveness in a temporary pond breeding amphibian, *Rana sylvatica* (wood frog). *Functional Ecology* 22:100–108.

Gibbs, J.P., and A.R. Breisch. 2001. Climate warming and calling phenology of frogs near Ithaca, New York, 1900–1999. *Conservation Biology* 15:1175–1178.

Häkkinen, J., S. Pasanen, and J.V.K. Kukkonen. 2001. The effects of solar UV-B radiation on embryonic mortality and development in three boreal anurans (*Rana temporaria, Rana arvalis,* and *Bufo bufo*). *Chemosphere* 44:441–446.

Hays, J.B., A.R. Blaustein, J.M. Kiesecker, P.D. Hoffman, I. Pandelova, I.C. Coyle, and T. Richardson. 1996. Developmental responses of amphibians to solar and artificial UV-B sources: a comparative study. *Photochemistry and Photobiology* 64:449–456.

Herman, J.R., P.K. Bhartia, J. Ziemke, Z. Ahmad, and D. Larko. 1996. UV-B increases (1979–1992) from decreases in total ozone. *Geophysical Research Letters* 23:2117–2120.

Heyer, W.R., A.S. Rand, C.A. Goncalvez da Cruz, and O.L. Peixoto. 1988. Decimations, extinctions, and colonizations of frog populations in southeast Brazil and their evolutionary implications. *Biotropica* 20:230–235.

Hoff, K.S., A.R. Blaustein, R.W. McDiarmid, and R. Altig. 1999. Behavior: interactions and their consequences. In *Tadpoles: the biology of anuran larvae.* Eds. R.W. McDiarmid and R. Altig, 125–239. Illinois: University of Chicago Press.

Hokit, D.G., and A.R. Blaustein. 1997. The effects of kinship on interactions between tadpoles of *Rana cascadae. Ecology* 78:1722–1735.

Houlahan, J.E., C.S. Findlay, B.R. Schmidt, A.H. Myer, and S.L. Kuzmin. 2000. Quantitative evidence for global amphibian population declines. *Nature* 404:752–755.

Hušek, J., and P. Adamik. 2008. Long-term trends in the timing of breeding and brood size in the red-backed shrike *Lanius collurio* in the Czech Republic, 1964–2004. *Journal of Ornithology* 149:97–103.

Hutchison, V.H., and R.K. Dupré. 1992. Thermoregulation. In *Environmental physiology of the amphibians.* Eds. M.E. Feder and W.W. Burggren, 206–249. Illinois: University of Chicago Press.

Intergovernmental Panel on Climate Change (IPCC). 2001. Climate change 2001: the scientific basis. Contribution of Working Group I to the Third Assessment Report of the Intergovernmental Panel on Climate Change [Houghton, J.T., Y. Ding, D.J. Griggs, M. Noguer, P.J. van der Linden, X. Dai, K. Maskell, and C.A. Johnson (Eds).]. Cambridge, UK, and New York: Cambridge University Press.

———. 2007. Climate change 2007: the physical science basis. Contribution of Working Group I to the Fourth Assessment Report of the Intergovernmental Panel on Climate Change [Solomon, S., D. Qin, M. Manning, Z. Chen, M. Marquis, K.B. Averyt, M. Tignor, and H.L. Miller (Eds.)]. Cambridge, UK, and New York: Cambridge University Press.

Kats, L.B., J.M. Kiesecker, D.P. Chivers, and A.R. Blaustein. 2000. Effects of UV-B on antipredator behavior in three species of amphibians. *Ethology* 106:921–932.

Kerr, J.B., and C.T. McElroy. 1993. Evidence for large upward trends of ultraviolet-B radiation linked to ozone depletion. *Science* 262:1032–1034.

Kiesecker, J.M., and A.R. Blaustein. 1995. Synergism between UV-B radiation and a pathogen magnifies amphibian embryo mortality in nature. *Proceedings of the National Academy of Science (USA)* 92:11049–11052.

———. 1997. Egg laying behavior influences pathogenic infection of amphibian embryos. *Conservation Biology* 11:214–220.

Kiesecker, J.M., A.R. Blaustein, and L.K. Belden. 2001. Complex causes of amphibian population declines. *Nature* 410:681–684.

Kusano, T., and M. Inoue. 2008. Long-term trends toward earlier breeding of Japanese amphibians. *Journal of Herpetology* 42:608–614.

Lampo, M., D. Sánchez, F. Nava-González, C.Z. García, and A. Acevedo. 2008. Lips et al. 2008 reply: Wavelike epidemics in Venezuela?. *Plos Biology* 1 December.

Lannoo, M. (Ed). 2005. *Amphibian declines. The conservation status of United States species.* Berkeley: University of California Press.

Laurance, W.F. 1996. Catastrophic declines of Australian rainforest frogs: is unusual weather responsible? *Biological Conservation* 77:203–212.

Lawler, J.J., S.L. Shafer, B.A. Bancroft, and A.R. Blaustein. 2010. Projected climate impacts for the amphibians of the Western Hemisphere. *Conservation Biology*. 24:38–50.

Lawler, J.J., S.L. Shafer, D. White, P. Kareiva, E.P. Maurer, A.R. Blaustein, and P.J. Bartlein. 2009. Projected climate-induced faunal change in the Western Hemisphere. *Ecology* 90:588–597.

Lawler, J.J., D. White, R.P. Neilson, and A.R. Blaustein. 2006. Predicting climate-induced range shifts: model differences and model reliability. *Global Change Biology* 12:1568–1584.

Lawton, J.H., and R.M. May (Eds.). 1995. *Extinction rates.* United Kingdom: Oxford University Press.

Lillywhite, H. 1970. Behavioral temperature regulation in the bullfrog, *Rana catesbeiana. Copeia* 1970:158–168.

Lips, K.R. 1998. Decline of a tropical montane amphibian fauna. *Conservation Biology* 12106–12117.

Lips, K.R., F. Brem, R. Brenes, et al. 2006. Emerging infectious disease and the loss of biodiversity in a neotropical amphibian community. *Proceedings of the National Academy of Science (USA)* 103:3165–3170.

Lips, K.R., J. Diffendorfer, J.R. Mendelson III, and M.W. Sears. 2008. Riding the wave: reconciling the roles of disease and climate change in amphibian declines. *Plos Biology* 6:441–454.

Lizana, M., and E.M. Pedraza. 1998. Different mortality of toad embryos (*Bufo bufo* and *Bufo calamita*) caused by UV-B radiation in high mountain areas of the Spanish Central System. *Conservation Biology* 12:703–707.

Long, L.E., L.S. Saylor, and M.E. Soule. 1995. A pH/UV-B synergism in amphibians. *Conservation Biology* 9:1301–1303.

Longcore, J.E., A.P. Pessier, and D.K. Nichols. 1999. *Batrachochytrium dendrobatidis* gen. et sp. nov., a chytrid pathogenic to amphibians. *Mycologia* 91:219–227.

Lovejoy, T.E., and L. Hannah. 2005. *Climate change and biodiversity.* New Haven, CT: Yale University Press.

McCallum, H. 2005. Inconclusiveness of chytridiomycosis as the agent in widespread frog declines. *Conservation Biology* 19:1421–1430.

McMenamin, S.K., E.A. Hadley, and C.K. Wright. 2008. Climatic change and wetland desiccation cause amphibian decline in Yellowstone National Park. *Proceedings of the National Academy of Science (USA)* 105:16988–16993.

Mendelson, J.R. III, K.R. Lips, R.W. Gagliardo, et al. 2006. Biodiversity: mitigating amphibian extinctions. *Science* 313:48.

Merilä, J., M. Pahkala, and U. Johanson. 2000. Increased ultraviolet-B radiation, climate change and latitudinal adaptation—a frog perspective. *Annales Zoologici Fennici* 37:129–134.

Middleton, E.M., J.R. Herman, E.A. Celarier, J.W. Wilkinson, C. Carey, and R.J. Rusin. 2001. Evaluating ultraviolet radiation exposure with satellite data at sites of amphibian declines in Central and South America. *Conservation Biology* 15:914–929.

Mote, P.W., and K.T. Redmond. Western climate change. In *Ecological consequences of climate change: mechanisms, conservation and management*. Eds. J. Belant and E. Beever. New York: Taylor & Francis Publishing 2012.

Nagl, A.M., and R. Hofer. 1997. Effects of ultraviolet radiation on early larval stages of the alpine newt, *Triturus alpestris*, under natural and laboratory conditions. *Oecologia* 110:514–519.

Nussbaum, R.A., E.D. Brodie Jr., and R.M. Storm. 1983. *Amphibians and reptiles of the Pacific Northwest*. Moscow, ID: University Press of Idaho.

Ovaska, K. 1997. Vulnerability of amphibians in Canada to global warming and increased ultraviolet radiation. In *Amphibians in decline: Canadian studies of a global problem*. Herpetological Conservation, Number 1. Ed. D.M. Green, 206–225. St. Louis, MO: Society for the Study of Amphibians and Reptiles.

Pahkala, M., A. Laurila, and J. Merilä. 2000. Ambient ultraviolet-B radiation reduces hatchling size in the common frog *Rana temporaria*. *Ecography* 23:531–538.

———. 2001. Carry-over effects of ultraviolet-B radiation on larval fitness in *Rana temporaria*. *Proceedings of the Royal Society of London B* 268:1699–1706.

Parmesan, C. 1996. Climate and species' range. *Nature* 383:765–766.

———. 2006. Ecological and evolutionary responses to recent climate change. *Annual Review of Ecology and Systematics* 37:637–669.

Parmesan, C., and M.C. Singer. 2008. Amphibian extinctions: disease not the whole story. *Plos Biology* 28 March.

Pearson, R.G., and T.P. Dawson. 2003. Predicting the impacts of climate change on the distribution of species: are climate envelope models useful? *Global Ecology and Biogeography* 12:361–371.

Perry, A.L., P.J. Low, J.R. Ellis, and J.D. Reynolds. 2005. Climate change and distribution shifts in marine fishes. *Science* 308:1912–1915.

Peters, R.L., and T.E. Lovejoy (Eds.). 1992. *Global warming and biological diversity*. New Haven, CT: Yale University Press.

Post, E., M.C. Forchhammer, N.C. Stenseth, and T.V. Callaghan. 2001. The timing of life-history events in a changing climate. *Proceedings of the Royal Society of London Series B-Biological Sciences* 268:15–23.

Pounds, J.A., M.R. Bustamante, L.A. Coloma, et al. 2006. Widespread amphibian extinctions from epidemic disease driven by global warming. *Nature* 439:161–167.

Pounds, J.A., M.R. Bustamante, L.A. Coloma, et al. 2007. Pounds et al. reply. *Nature* 447:E5–E6.

Pounds, J.A., M.P.L. Fogden, and J.H. Campbell. 1999. Biological responses to climate change on a tropical mountain. *Nature* 398:611–615.

Pounds, J.A., M.P.L. Fogden, J.M. Savage, and G.C. Gorman. 1997. Test of null models for amphibian declines on a tropical mountain. *Conservation Biology* 11:1307–1322.

Puschendorf, R., F. Bolanos, and G. Chaves. 2006. The amphibian chytrid fungus along an altitudinal transect before the first reported declines in Costa Rica. *Biological Conservation* 132:136–142.

Raxworthy, C.J., R.G. Pearson, N. Rabibisoa, A.M. Rakotondrazafy, J. Ramanamanjato, A.P. Raselimanan, S. Wu, R.A. Nussbaum, and D.A. Stone. 2008. Extinction vulnerability of tropical montane endemism from warming and upslope displacement: a preliminary appraisal for the highest massif in Madagascar. *Global Change Biology* 14:1703–1720.

Reading, C.J. 1998. The effect of winter temperatures on the timing of breeding activity in the common toad *Bufo bufo*. *Oecologia* 117:469–475.

Reaser, J.K., and A.R. Blaustein. 2005. Repercussions of global change in aquatic systems. In *Status and conservation of North American amphibians*. Ed. M. Lanoo. Berkeley: University of California Press 60–63.

Rohr, J.R., T.R. Raffel, J.M. Romansic, H. McCallum, and P.J. Hudson. 2008. Evaluating the links between climate, disease spread, and amphibian declines. *Proceedings of the National Academy of Science (USA)* 105:17436–17441.

Romansic, J.M., K.A. Diez, E.M. Higashi, and A.R. Blaustein. 2007. Susceptibility of newly metamorphosed frogs to a pathogenic water mold (*Saprolegnia sp.*). *Herpetological Journal* 17:161–166.

Romansic, J.M., K.A. Diez, E.M. Higashi, J.E. Johnson, and A.R. Blaustein. 2009. Effects of the pathogenic water mold *Saprolegnia* on survival of amphibian larvae. *Diseases of Aquatic Organisms* 83:187–193.

Romer, A.S. 1939. An amphibian graveyard. *Scientific Monthly* 49:337–339.

———. 1968. *The procession of life*. New York: The World Publishing Company.

Schindler, D.W., P.J. Curtis, B.R. Parker, and M.P. Stainton. 1996. Consequences of climate warming and lake acidification for UV-B penetration in North American boreal lakes. *Nature* 379:705–708.

Scott, W.A., D. Pithart, and J.K. Adamson. 2008. Long-term United Kingdom trends in the breeding phenology of the common frog, *Rana temporaria*. *Journal of Herpetology* 42:89–96.

Searle, C.L., L.K. Belden, B.A. Bancroft, B.A. Han, L.M. Biga, and A.R. Blaustein. 2010. Experimental examination of the effects of ultraviolet-B radiation in combination with other stressors on frog larvae. *Oecologia* 162:237–245.

Seimon, T.A., A. Seimon, P. Daszak, S.R.P. Halloy, L.M. Schloegel, C.A. Aguilar, P. Sowell, A.D. Hyatt, B. Konecky, and J.E. Simmons. 2007. Upward range extension of Andean anurans and chytridiomycosis to extreme elevations in response to tropical deglaciation. *Global Change Biology* 13:288–299.

Skelly, D.S., L.N. Joseph, H.P. Possingham, L.K. Freidenburg, T.J. Farrugia, M.T. Kinnison, and A.P. Hendry. 2007. Evolutionary responses to climate change. *Conservation Biology* 21:1353–1355.

Smith, G.R., M.A. Waters, and J.E. Rettig. 2000. Consequences of embryonic UV-B exposure for embryos and tadpoles of the plains leopard frog. *Conservation Biology* 14:1903–1907.

Stallard, R.F. 2001. Possible environmental factors underlying amphibian decline in eastern Puerto Rico: analysis of U.S. government data archives. *Conservation Biology* 15:943–953.

Starnes, S.M., C.A. Kennedy, and J.W. Petranka. 2000. Sensitivity of southern Appalachian amphibians to ambient solar UV-B radiation. *Conservation Biology* 14:277–282.

Stebbins, R.C., and N.W. Cohen. 1995. *A natural history of amphibians*. New Jersey: Princeton University Press.

Stuart, S.N., J.S. Chanson, N.A. Cox, B.E. Young, A.S. L. Rodrigues, D.L. Fishman, and R.W. Waller. 2004. Status and trends of amphibian declines and extinctions worldwide. *Science* 306:1783–1786.

Tevini, M. (Ed.). 1993. *UV-B radiation and ozone depletion: effects on humans, animals, plants, microorganisms, and materials*. Boca Raton, FL: Lewis Publishers.

Thomas, C.D., M.A. Aldina, and J.K. Hill. 2006. Range retractions and extinction in the face of climate warming. *Trends in Ecology and Evolution* 21:415–416.

Thomas, C.D., A. Cameron, R.E. Green, et al. 2004. Extinction risk from climate change. *Nature* 427:145–148.

Thuiller, W., S. Lavorel, M.B. Araújo, M.T. Sykes, and I.C. Prentice. 2005. Climate change threats to plant diversity in Europe. *Proceedings of the National Academy of Science (USA)* 102:8245–8250.

Van der Leun, J.C., X. Tang, and M. Tevini (Eds.). 1998. *Environmental effects of oz*one *depletion 1998 Assessment*. Lausanne, Switzerland: Elsevier.

Van Uitregt, V.O., R.S. Wilson, and C.E. Franklin. 2007. Cooler temperatures increase sensitivity to ultraviolet B radiation in embryos and larvae of the frog *Limnodynastes peronii*. *Global Change Biology* 13:1114–1121.

Vredenburg, V.T., R.A. Knapp, T.S. Tunstall, and C.J. Briggs. 2010. Dynamics of an emerging disease drive large-scale amphibian population extinctions. *Proceedings of the National Academy of Science (USA)* 107:9689–9694.

Wake, D.B., and V.D. Vredenburg. 2008. Are we in the midst of the sixth mass extinction? A view from the world of amphibians. *Proceedings of the National Academy of Science (USA)* 105:11466–11473.

Wells, K.D. 2007. *The ecology and behavior of amphibians*. Illinois: University of Chicago Press.

Wilson, E.O. 1992. *The diversity of life*. Cambridge, MA: Harvard University Press.

Wollmuth, L.P., and L.I. Crawshaw. 1988. The effect of development and season on temperature selection in bullfrog tadpoles. *Physiological Zoology* 61:461–469.

Wollmuth, L.P., L.I. Crawshaw, R.B. Forbes, and D.A. Grahn. 1987. Temperature selection during development in a montane anuran species, *Rana cascadae*. *Physiological Zoology* 60:472–480.

Worrest, R.D., and D.J. Kimeldorf. 1976. Distortions in amphibian development induced by ultraviolet-B enhancement (290–310 nm) of a simulated solar spectrum. *Photochemistry and Photobiology* 24:377–382.

Weygoldt, P. 1989. Changes in the composition of mountain stream frog communities in the Atlantic Mountains of Brazil: frogs as indicators of environmental deterioration? *Studies on Neotropical Fauna & Environment* 24:249–255.

3 Minimizing Uncertainty in Interpreting Responses of Butterflies to Climate Change

Erica Fleishman and Dennis D. Murphy

CONTENTS

INTRODUCTION

Butterflies are highly responsive to atmospheric conditions at all stages of their life cycle. Over short periods of time, butterflies are affected strongly by weather. Over years, butterflies respond to climatic variability, such as El Niño events. Over decades or longer, population dynamics of butterflies respond to climate change. Paradoxically, the responsiveness of butterfly populations to short-term weather patterns may confound interpretation of their responses to climate over the long term. As a result, it can be remarkably difficult to definitively attribute variability in occurrences or population dynamics of butterflies that may be caused by temperature, precipitation, and other meteorological variables (McLaughlin et al. 2002; Fleishman and Mac Nally 2003).

When studying populations of adult butterflies, it is common to obsess about the weather. During the flight season, on a day of unseasonably cool temperatures, precipitation, or high wind, relatively few butterflies can be detected, and few if any reliable data can be collected on that day. Intense or prolonged inclement weather can depress butterfly activities for longer periods, to the point of changing birth, death, and dispersal rates for the season (e.g., Kingsolver 1989). Within a given biogeographic region, it is fairly certain that individuals of many species will be active during a particular block of time each year. It also is possible to anticipate a core set of species that are likely to be present during that period. Yet, even with real-time data on weather, it can be challenging to predict accurately from year to year the dates of first and last appearance of butterflies of certain species, the number of individuals that will eclose and where, or when an abundance of a given species will peak (Weiss and Weiss 1998). Changes in local butterfly abundances over orders of magnitude are observed regularly in the absence of obvious deterministic environmental changes, whether natural or anthropogenic in origin. Moreover, the degree of synchrony among multiple populations of the same species in a given year often cannot be forecast.

Whereas butterflies may be among the first taxonomic groups to show the effects of climate change, shifts in the distribution and abundance of butterflies are not necessarily a direct reflection of climate change. It may be possible to explain much of the variance in a trend in abundance on the basis of precipitation. However, an observed trend in abundance of a butterfly species is unlikely to signal that a specific pattern of precipitation has occurred. Furthermore, changes in population dynamics of any organism rarely are triggered by a change in one environmental attribute. Instead, those changes tend to reflect suites of interacting or cascading changes in the abiotic and biotic environment. Mechanistic pathways by which atmospheric conditions affect butterflies often involve larval host plants, sources of nectar for adults, and rates of parasitism and disease rather than thresholds of physical tolerance (Singer 1972; Weiss et al. 1988). When human land uses affect abiotic and biotic attributes of butterfly habitat, interpreting the role of weather and climate becomes substantively more difficult.

Previous authors, especially Dennis (1993) and Hellmann (2002), have explained in detail the physiological responses of butterflies to weather and climate. Here, we review recent evidence that butterflies are responding to climate change. We reexamine previous projections of changes in light of improved understanding of mechanisms. We also provide guidance on future efforts to track responses of butterflies to climate change. Our discussion focuses on butterflies in temperate regions. Long-term data on tropical butterflies are comparatively rare, and the composition, let alone the dynamics, of many tropical butterfly faunas are comparatively poorly known.

EVIDENCE OF DIRECTIONAL RESPONSES OF BUTTERFLIES TO CLIMATE

The recent literature provides evidence that climate change is affecting butterflies via their phenologies, the location and size of their geographic ranges, and their abundances. Most published evidence for virtually all taxonomic groups focuses on apparent responses (it is generally unknown whether responses are direct or indirect) of population dynamics to temperature, for which records generally are more complete than for precipitation (see Parmesan et al. 1999; Thomas et al. 2001). Much of the evidence is derived from long-established monitoring programs in the United Kingdom. Data and inferences from these monitoring programs largely are responsible for the conventional wisdom that measures of butterfly occurrence and demography are highly correlated with current or incipient responses of other taxonomic groups to climate change. Most attention has concentrated on species-level shifts in geographic range rather than patch-level extirpation events. Patchy extirpations are important even if they do not have a clear latitudinal or elevational signal because they affect probabilities of persistence and long-term evolutionary potential, especially on human-dominated landscapes with barriers to movement.

Both historically (1880s to 1940s) and more recently (1970s to 1990s), mean dates of appearance of British butterflies were associated with mean air temperatures (Sparks and Yates 1997). Dates of appearance of more than 150 species of moths and butterflies in southern England in the late 1800s were positively correlated with temperature in October of the previous year and negatively correlated with temperature in spring of the current year (Sparks et al. 2006). However, even when phenology is linked with temperature across a butterfly fauna, phenological patterns of some species may not track spatial gradients in climate (Roy and Asher 2003).

Temperature interacts with precipitation to determine the timing of snowmelt, with which emergence of arthropods in northeastern Greenland was strongly associated (Hoye and Forchhammer 2008). In that region, butterflies emerged during a short time within the warmest period of the year (Hoye and Forchhammer 2008). Patterns of protandry also may be affected by apparent responses of phenology to shifts in climate. Decadal increases of 0.7–1.8°C in spring temperature in Switzerland were associated with earlier emergence—9 days per decade for males, and 12 days for females—of *Apatura iris* (Dell et al. 2005). As emergence dates of males and females converge, less food may be available to adult males (Dell et al. 2005).

Many studies have documented expansions in the geographic ranges of individual species of butterflies in direct and indirect response to climate change. An increase in mean annual temperature of 0.8–1.8°C during the past 100 years was associated with colonization of the coastal northwestern United States by *Atalopedes campestris* (Crozier 2004). Among migratory lepidoptera, annual species richness at a site in the southern United Kingdom was correlated positively with increases in temperature at butterfly source locations in southwestern Europe (Sparks et al. 2007). Menendez et al. (2008) attributed successful northward expansion of *Aricia agestis* in Britain during the last 30 years to lower rates of parasitism in the newly colonized locations.

The ability of a species to expand its geographical range often is related to its capacity for adaptation through either evolution or phenotypic plasticity. *Pararge aegeria* in the United Kingdom has expanded its range northward, but increases in dispersal rate are associated with decreases in reproductive rate (Hughes et al. 2003). Northward expansion of *Polygonia c-album* in the United Kingdom appears to be supported by its use of two novel host plants in addition to a traditional host (Braschler and Hill 2007). *A. agestis* has become more flexible in its use of multiple host plants, and *Hesperia comma* in its use of microhabitats; both species appear to have dispersed across vegetation types previously thought to impede movement (Thomas et al. 2001).

Adaptation in butterflies reflects responses to vegetation structure as well as vegetation composition. Davies et al. (2006) documented a shift in understory attributes of oviposition neighborhoods of *H. comma* that may be related to regional increases in temperature. The temperature of host plants on which females oviposited also varied as a function of ambient temperature. In this case, the range of slope exposures and other conditions suitable for oviposition increased as regional temperature increased (Davies et al. 2006). The frequency and magnitude of extreme climatic events is likely to increase as climate changes, and extreme events can precipitate sudden shifts in vegetation structure. Singer and Thomas (1996) demonstrated evolutionary responses by a metapopulation of *Euphydryas editha* to changes in relative abundance of different larval hosts that were driven by human activity and intense climatic events (Singer and Thomas 1996).

In some cases, geographic ranges of butterflies have shifted rather than expanded. Parmesan et al. (1999) found that the ranges of 63% of 35 nonmigratory European butterflies shifted to the north by 35–240 km during the 1900s. Not only latitudinal but elevational range shifts have been observed. As mean annual temperature in central Spain rose by 1.3°C over 30 years, the mean lower elevational limit of 16 species of butterflies increased by 212 m (Wilson et al. 2005). Hill et al. (2002) did not detect a systematic northward shift in the ranges of 51 species of butterflies in Britain during the 1900s. However, the distribution of species that occupied northern or montane areas tended to shift to higher elevations (Franco et al. 2006).

A small number of studies have emphasized relationships between population dynamics of butterflies and patterns of precipitation. Fluctuations in abundance of two populations of *E. editha*, the mechanism that ultimately led to extirpations, appear to have been exacerbated by increasing variability in precipitation (McLauglin et al. 2002). Since 1978, abundances of 28 of 31 species of British butterflies have been correlated positively with low precipitation in the current year and high precipitation in the previous year, as well as with warm temperatures during summer of the current year (Roy et al. 2001). In general, since the mid-1800s, abundance of British butterflies tended to increase when summers were warm and dry and to decrease when summers and winters were wet and winters were warm (Dennis and Sparks 2007). These observations are not necessarily linked to climate change, although changes in the amount, timing, and variability of precipitation are consistent with projections of climate change.

DRAWING RELIABLE INFERENCE FROM BUTTERFLIES

It is reasonable to assume that occurrence and population dynamics of butterflies reflect changes in climate, but tracking that response is confounded by noise introduced by responsiveness to weather. In many instances, the population dynamics of butterflies actually may be so responsive to minor variations in ecological conditions that they may not serve as signals of meaningful environmental trends. Interpretation of responses of population dynamics of butterflies also is confounded by a general misconception that butterflies can provide more-general information on myriad environmental variables and trends (Fleishman and Murphy 2009).

Many apparent associations between butterflies and weather or climate are mediated through the structure or composition of vegetation, rather than directly by climate. There is concern that differences in the phenological responses of butterflies and their larval host plants to changes in climate may lead to declines in abundance or distribution. However, evidence suggests that these relationships may be more plastic than widely understood. The assumption that butterflies have predictable associations with the composition and structure of vegetation communities (Hermy and Cornelis 2000) can be quite tenuous. At a local level, the larvae of many species of butterflies are restricted to one or a few closely related species of larval host plants, and adults of some species are linked closely with particular species of plants from which they draw nectar. Nevertheless, the breadth of host-plant use and preferences for individual species sometimes differs dramatically in space, time, and even among individuals in the same population (Singer 1983; Boughton 1999). Furthermore, in temperate regions, let alone the tropics, the identities of larval host plants used by many butterflies still are not known at the species level, and the geographic distribution of a given butterfly species typically is less extensive than the geographic distribution of its larval host plants and nectar sources (Scott 1986; Ehrlich and Hanski 2004).

The feasibility of monitoring a biological response to climate change depends in part on the ease with which the target of monitoring can be measured. Butterflies generally meet reasonable standards of measurement tractability—they are diurnal, colorful, and typically are readily detected—but there can be substantive caveats. As noted previously, phenologies of butterfly populations can vary dramatically among locations and years and are affected by short-term weather conditions. These attributes are a challenge to conducting, and analyzing data from, annual surveys of butterfly populations. Estimation of population size and emergence curves can require intensive surveys or mark-recapture efforts by trained personnel (Weiss et al. 1993; Haddad et al. 2008). It is generally appreciated that targets of monitoring also should be easy to identify. Certainly many species of butterflies can be identified on the wing or in the net by an experienced observer, but morphological similarities among species in some species-rich genera (e.g., *Speyeria*, *Euphydryas*, and *Euphilotes*) confound reliable identification, even by seasoned lepidopterists. In addition, many small-bodied species of butterflies, especially hesperiids and lycaenids, are difficult to capture and handle without injuring the animals.

Past assumptions about the elevational distributions of butterfly habitats and about habitat–area relationships may have led to overestimates of the probability of climate-driven extirpations and underestimates of the ability of butterflies to adapt to climate change. It often is assumed implicitly, but incorrectly (because evolution occurs at the level of the organism), that multiple plant species comprising a vegetation association have synchronous responses to climate. As a result, projections about shifting distributions of host plants, nectar sources, or other vegetational attributes of the habitat of butterflies may be erroneous. For example, woodlands dominated by singleleaf pinyon (*Pinus monophylla*) and Utah and western juniper (*Juniperus osteosperma*, *Juniperus occidentalis*) are widespread across the Great Basin. Researchers, managers, and the public often refer to pinyon–juniper as a single unit. Nevertheless, the two conifers have different climatic tolerances and resiliencies to disturbances such as fire. Thus the compositional stability of vegetation across the region is uncertain.

In addition, for computational simplicity, it is common to assume that elevational boundaries of vegetation associations are homogeneous in space. Previous work cited 2300–2800 m (Murphy and Weiss 1992) or 2300–2700 m (Boggs and Murphy 1997) as the elevation within which climate was appropriate for pinyon–juniper woodland and therefore was appropriate for montane butterflies (taxa believed to complete their entire life cycle in mountain ranges and not found in nearby valleys). More recent data, derived from remote sensing and field measurements, demonstrates that the elevational boundaries of pinyon–juniper woodland are highly variable both within and among mountain ranges (Bradley and Fleishman 2008; Fleishman and Dobkin 2009).

Assumptions similar to those for the Great Basin (Murphy and Weiss 1992; Boggs and Murphy 1997) have been made elsewhere. As mean annual temperature in central Spain rose by 1.3°C over 30 years, the mean lower elevational limit of 16 species of butterflies increased by 212 m (Wilson et al. 2005). Wilson et al. (2005) estimated that the mean geographic range of the species decreased by approximately 33% as a result. But many montane species inhabit lower elevations, and each species is not distributed uniformly across higher elevations (Fleishman et al. 1998). Vegetation and thermal zones are patchy, reflecting topographic variation in microclimate and land-use history, and are highly unlikely to simply move upward across entire mountain ranges. Not only erroneous assumptions about vegetation, but empirical data on the elevational ranges of individual species of butterflies (Fleishman et al. 1998) suggest the inference that 20–50% of the species in a given mountain range were likely to be extirpated (Murphy and Weiss 1992; Boggs and Murphy 1997) likely is too high.

Regional extirpations of butterflies for which robust data exist often have been driven by disruptions to metapopulation dynamics. The distributions of most species are patchy, reflecting disjunct distributions of abiotic and biotic attributes of their habitat as well as barriers to dispersal and stochastic factors. Accordingly, simple latitudinal or elevational shifts in range are unlikely. *E. editha* is cited widely as evidence of range shifts in response to climate change. Parmesan (1996) resurveyed 151 sites in western North America from which the species previously had been recorded. Sites where necessary resources for the species (e.g., larval host plants) no longer existed were excluded from analysis. Consistent with expectations of climate-driven shifts in the distribution of the species, the incidence of extirpations fell as both latitude and elevation increased. Disappearances of populations were more frequent at lower latitudes and elevations. But the analysis did not account for geographically correlated life-history variation, or loss or conversion of habitats. Low-elevation populations of *E. editha* generally feed on annual plants, the availability of which is relatively unpredictable, whereas populations at higher elevations feed on perennials. The low-elevation populations inhabit areas that are more likely to have been invaded by non-native annual grasses or urbanized, both of which may or may not be driven by climate change. A number of the lowest latitude sites were in northern Mexico, where land use was especially intense.

Changes in climate across extensive areas vary in their effects on different populations of the same species. In the southern part of its distributional range, *E. editha* is believed to have been historically abundant and widely distributed. These populations oviposit on a native annual *Plantago* that varies in abundance as a function of precipitation. A series of drought years that both decrease abundance and accelerate senescence of *Plantago* increases the probability of local extirpation of the butterfly. A series of years with relatively high precipitation, by contrast, typically are associated with large populations of larvae that locally defoliate their host plants (Murphy and White 1984). The frequency with which *E. editha* recolonize patches of habitat after extirpation events has decreased because of the establishment of non-native annual plants in former patches of habitat after fire events and local decreases in abundance of *Plantago*. The remaining patches of habitat are increasingly isolated and often are fragmented. Given these conditions, drought increases the probability of regional as well as local extirpation of the butterfly.

A second set of *Plantago*-feeding populations of *E. editha* that once likely was widespread across much of central California now is restricted to a relatively small number of patches of grasslands on serpentine-based soils. These soils impede colonization by non-native annual grasses. However, because serpentine-based soil is porous, *Plantago* senesces relatively early, especially on south-facing slopes. When temperature increases and precipitation decreases even modestly, recolonization of these locations following extirpation becomes less frequent (Murphy and Weiss 1989).

Females from a population of *E. editha* just east of the Sierra Nevada once oviposited nearly exclusively on *Collinsia parviflora*, a native annual plant that grows in the understory beneath sagebrush (*Artemisia* spp.) and mixed coniferous woodland. In response to a decrease in abundance and earlier senescence of *C. parviflora* over a decade, the butterfly began to oviposit almost exclusively on non-native *Plantago lanceolata* in a nearby patch of grassland (Thomas et al. 1987). Recent reduction in the intensity of local grazing by domestic livestock reduced the abundance of *Plantago* and facilitated colonization by other grasses, which shaded surviving *Plantago*. As a result, habitat quality for *E. editha* decreased. Abundance of the butterflies decreased considerably, and no individuals have been detected for several years. It appears that a shift by ovipositing females to the use of a novel species of host plant prevented a potential weather-induced extirpation event, yet another change in habitat quality resulted in extirpation (Singer et al. 2008).

AN IDEAL SCHEME FOR TRACKING THE EFFECTS OF CLIMATE CHANGE ON BUTTERFLIES

Inference on the responses of butterflies to climate change has been based on retrospective analysis of time-series data rather than studies designed explicitly to test competing hypotheses about the potential effects of shifts in climate versus other environmental stressors. Two criteria must be satisfied for one to infer that changes in climate are driving changes in the distribution or population dynamics of butterflies. First, data on the butterflies and climatic variables must be available across extensive areas and many years. Second, one must understand mechanistically how climate changes can affect individuals and populations.

If one is focused on potential assemblage-level signals of climate change, it may be most valuable to examine multiple butterfly species with different predicted responses to climate change. This criterion was largely met by Forister et al. (2010), who reported annual occurrence data from multiple sites across an elevational gradient, differentiating between specialized, less dispersive species and disturbance-associated species. One might analyze data for several species whose ranges are expected to expand northward and several whose ranges are expected to contract rather than concentrating only on species for which northward expansion is anticipated (Wilson et al. 2007).

At the species level, it is most straightforward to track species that are not highly constrained by patch dynamics. Metapopulation dynamics across habitat patches that vary in resource availability and quality obfuscate signals of climate change. Patches supporting populations with a relatively high probability of extirpation likely will disappear long before manifestation of incremental climate change. For example, wide-ranging Papilionids and Pierids might be more appropriate for study than well-known Nymphalids, such as *Euphydryas* and other Melitaea.

The demographic, spatial, and temporal resolution of each set of data must be appropriate for the given question. Data on abundance or population dynamics usually are more informative than data on presence or absence, but more sensitive to potential confounding factors such as short-term weather and differences in observers. If one is evaluating whether appearance dates in a particular location over many years have changed, then an estimate of regional means and variance in temperature and precipitation derived from just one or two weather stations may be sufficient. If one instead is examining asynchronous emergences within a region in one or more years, then fine-resolution data on climatic variables likely are necessary.

Known mechanisms by which the abiotic environment affects butterflies can allow the generation of specific, testable predictions related to climate change (Hellmann 2002; Parmesan 2003). The ideal species would have fairly well-resolved and highly specific requirements for survival, so relatively few stressors would confound interpretation of responses to climate. Most extant species in developed countries are resilient to moderate levels of human activity, which reduces the number of likely drivers of a population response (Parmesan 2003). Especially when species already are rare, short-term weather events that may not be causally related to deterministic changes in climate may lead to extirpation or extinction. One must separate the effects of deterministic changes in climate on, for example, the availability of larval host plants from stochastic, but not necessarily surprising, anomalies in weather. Furthermore, many alternative hypotheses for changes in distributional range or demographic parameters, such as parasitoids or invasion of non-native species of plants, may themselves be affected by changes in climate.

Data from long-term butterfly monitoring programs in the United Kingdom and, to a lesser extent, other European countries sparked a belief among scientists and the public that changes in climate will be reflected in the distribution and population dynamics of butterflies sooner than in many other popular taxonomic groups (Fleishman and Murphy 2009). Butterflies indeed respond to atmospheric conditions both physiologically and indirectly, mediated by changes in sources of food for larvae and adults, interactions with parasites and other predators, and habitat structure. However, robust data and biological understanding are critical to discriminating between signal and noise and to validating mechanistic hypotheses. Projected range shifts and regional extirpations of butterflies that are based on simple assumptions about habitat quality and relationships among land cover, latitude, and elevation well may overestimate the magnitude of climate change as a stressor or an ultimate driver of extinction. Thoughtful tracking of butterfly demography and trends and patterns of resource use that specifically is designed to examine potential effects of multiple drivers of faunal change is most likely to yield strong inference.

SUMMARY

Activity patterns and vital rates of butterflies respond to both weather and climate. Responses to the noise of weather confound the interpretation of responses to the signal of climate change, albeit some fluctuations in weather patterns are consistent with projected effects of increases in emissions of greenhouse gases. Moreover, ongoing changes in land use and other stressors affect butterflies both directly and indirectly. Accordingly, shifts in the population dynamics of butterflies cannot necessarily be interpreted as responses to climate change. Some projections of effects of climate change on butterflies have reflected assumptions about the dependence of butterflies on particular larval host plants or other elements of habitat. Evidence suggests that in fact, many species are relatively plastic in their resource requirements. Two criteria must be satisfied to infer that changes in climate are driving a change in the distribution or population dynamics of butterflies. First, data on the butterflies and climatic variables must be available across extensive areas and many years. Second, mechanisms by which not only climate but different types of land use may affect individuals and populations must be understood. Climate change occurs in the context of multiple existing and interacting stressors. Changes in the quality, quantity, and spatial and temporal diversity of resources resulting from land use increase the probability that short-term changes in weather and long-term changes in climate will lead to changes in the population dynamics of butterflies.

REFERENCES

Boggs, C.L., and D.D. Murphy. 1997. Community composition in mountain ecosystems: climatic determinants of montane butterfly distributions. *Global Ecology and Biogeography Letters* 6:39–48.

Boughton, D.A. 1999. Empirical evidence for complex source–sink dynamics with alternative states in a butterfly metapopulation. *Ecology* 80:2727–2739.

Bradley, B., and E. Fleishman. 2008. Relationships between expanding pinyon–juniper cover and topography in the central Great Basin, Nevada. *Journal of Biogeography* 35:951–964.

Braschler, B., and J.K. Hill. 2007. Role of larval host plants in the climate-driven range expansion of the butterfly *Polygonia c-album*. *Journal of Animal Ecology* 76:415–423.

Crozier, L. 2004. Warmer winters drive butterfly range expansion by increasing survivorship. *Ecology* 85:231–241.

Davies, Z.G., R.J. Wilson, S. Coles, and C.D. Thomas. 2006. Changing habitat associations of a thermally constrained species, the silver-spotted skipper butterfly, in response to climate warming. *Journal of Animal Ecology* 75:247–256.

Dell, D., T.H. Sparks, and R.L.H. Dennis. 2005. Climate change and the effect of increasing spring temperatures on emergence dates of the butterfly *Apatura iris* (Lepidoptera: Nymphalidae). *European Journal of Entomology* 102:161–167.

Dennis, R.L.H. 1993. *Butterflies and climate change*. Manchester University Press, United Kingdom.

Dennis, R.L.H., and T.H. Sparks. 2007. Climate signals are reflected in an 89-year series of British Lepidoptera records. *European Journal of Entomology* 104:763–767.

Ehrlich, P.R., and I. Hanski. 2004. *On the wings of checkerspots*. Oxford University Press, New York.

Fleishman, E., G.T. Austin, and A.D. Weiss. 1998. An empirical test of Rapoport's rule: elevational gradients in montane butterfly communities. *Ecology* 79:2482–2493.

Fleishman, E., and D.S. Dobkin. 2009. Current and potential future elevational distributions of birds associated with pinyon–juniper woodlands in the central Great Basin, U.S.A. *Restoration Ecology* 17:731–739.

Fleishman, E., and R. Mac Nally. 2003. Distinguishing between signal and noise in faunal responses to environmental change. *Global Ecology and Biogeography* 12:395–402.

Fleishman, E., and D.D. Murphy. 2009. A realistic assessment of the indicator potential of butterflies and other charismatic taxonomic groups. *Conservation Biology* 23:1109–1116.

Forister, M.L., A.C. McCall, N.J. Sanders, J.A. Fordyce, J.H. Thorne, J. O'Brien, D.P. Waetjen, and A.M. Shapiro. 2010. Compounded effects of climate change and habitat alteration shift patterns of butterfly diversity. *Proceedings of the National Academy of Sciences* 107:2088–2092.

Franco, A.M.A., J.K. Hill, C. Kitschke, Y.C. Collingham, D.B. Roy, R. Fox, B. Huntley, and C.D. Thomas. 2006. Impacts of climate warming and habitat loss on extinctions at species' low-latitude range boundaries. *Global Change Biology* 12:1545–1553.

Haddad, N.M., B. Hudgens, C. Damiani, K. Gross, D. Kuefler, and K. Pollock. 2008. Determining optimal population modeling for rare butterflies. *Conservation Biology* 22:929–940.

Hellmann, J.J. 2002. Butterflies as model systems for understanding and predicting climate change. Pages 93–126 in S. H. Schneider and T. L. Root, Editors. *Wildlife responses to climate change*. Island Press, Washington, D.C.

Hermy, M., and J. Cornelis. 2000. Toward a monitoring method and a number of multifaceted and hierarchical biodiversity indicators for urban and suburban parks. *Landscape and Urban Planning* 49:149–162.

Hill, J.K., C.D. Thomas, R. Fox, M.G. Telfer, S.G. Willis, J. Asher, and B. Huntley. 2002. Responses of butterflies to twentieth century climate warming: implications for future ranges. *Proceedings of the Royal Society Biological Sciences Series B* 269:2163–2171.

Hoye, T.T., and M.C. Forchhammer. 2008. Phenology of high-arctic arthropods: effects of climate on spatial, seasonal, and inter-annual variation. *Advances in Ecological Research* 40:299–324.

Hughes, C.L., J.K. Hill, and C. Dytham. 2003. Evolutionary trade-offs between reproduction and dispersal in populations at expanding range boundaries. *Proceedings of the Royal Society Biological Sciences Series B* 270 (Supplement 2): S147–S150.

Kingsolver, J.G. 1989. Weather and the population dynamics of insects: integrating physiological and population ecology. *Physiological Zoology* 62:314–334.

McLaughlin, J.F., J.J. Hellmann, C.L. Boggs, and P.R. Ehrlich. 2002. Climate change hastens population extinctions. *Proceedings of the National Academy of Sciences (U.S.A.)* 99:6070–6074.

Menendez, R., A. Gonzalez-Megias, O.T. Lewis, M.R. Shaw, and C.D. Thomas. 2008. Escape from natural enemies during climate-driven range expansion: a case study. *Ecological Entomology* 33:413–421.

Murphy, D.D., and S.B. Weiss. 1989. Ecological studies and the conservation of the endangered bay checkerspot butterfly, *Euphydryas editha bayensis*. *Biological Conservation* 46:183–200.

———. 1992. Effects of climate change on biological diversity in western North America: species losses and mechanisms. Pages 355-368 in R.L. Peters and T.E. Lovejoy, Editors. *Global warming and biological diversity*. Yale University Press, New Haven, Connecticut.

Murphy, D.D., and R.R. White. 1984. Rainfall, resources, and dispersal in southern populations of *Euphydryas editha* (Lepidoptera: Nymphalidae). *Pan-Pacific Entomologist* 60:350–354.

Parmesan, C. 2003. Butterflies as bioindicators for climate change effects. Pages 541–560 in C.L. Boggs, W.B. Watt, and P.R. Ehrlich. *Butterflies: ecology and evolution taking flight*. University of Chicago Press, Illinois.

Parmesan, C., N. Ryrholm, C. Stefanescu, J.K. Hill, C.D. Thomas, H. Descimon, B. Huntley, L. Kaila, J. Kullberg, T. Tammaru, W.J. Tennent, J.A. Thomas, and M. Warren. 1999. Poleward shifts in geographical ranges of butterfly species associated with regional warming. *Nature* 399:579–583.

Parmesan, C. 1996. Climate and species' range. *Nature*. 382:765–766.

Roy, D.B., and J. Asher. 2003. Spatial trends in the sighting dates of British butterflies. *International Journal of Biometeorology* 47:188–192.

Roy, D.B., P. Rothery, D. Moss, E. Pollard, and J.A. Thomas. 2001. Butterfly numbers and weather: predicting historical trends in abundance and the future effects of climate change. *Journal of Animal Ecology* 70:201–217.

Scott, J.A. 1996. *The butterflies of North America*. Stanford University Press, California.

Singer, M.C. 1972. Complex components of habitat suitability within a butterfly colony. *Science* 176:75–77.

———. 1983. Determinants of multiple host use by a phytophagous insect population. *Evolution* 37:389–403.

Singer, M.C., and C.D. Thomas. 1996. Evolutionary responses of a butterfly metapopulation to human- and climate-caused environmental variation. *American Naturalist* 148:S9–S39.

Singer, M.C., B. Wee, S. Hawkins, and M. Butcher. 2008. Rapid natural and anthropogenic diet evolution: three examples from checkerspot butterflies. Pages 311–324 in K.J. Tilmon, editor. *Specialization, speciation and radiation: the evolutionary biology of herbivorous insects*. University of California Press, Berkeley.

Sparks, T.H., R.L.H. Dennis, P.J. Croxton, and M. Cade. 2007. Increased migration of Lepidoptera linked to climate change. *European Journal of Entomology* 104:139–143.

Sparks, T.H., K. Huber, and R.L.H. Dennis. 2006. Complex phenological responses to climate warming trends? Lessons from history. *European Journal of Entomology* 103:379–386.

Sparks, T.H., and T.J. Yates. 1997. The effect of spring temperature on the appearance dates of British butterflies 1883–1993. *Ecography* 20:368–374.

Thomas, C.D., E.J. Bodsworth, R.J. Wilson, A.D. Simmons, Z.G. Davies, M. Musche, and L. Conradt. 2001. Ecological and evolutionary processes at expanding range margins. *Nature* 411:577–581.

Thomas, C.D., D. Ng, M.C. Singer, J.L.B. Mallet, and C. Parmesan. 1987. Incorporation of a European wedd into the diet of a North American herbivore. *Evolution* 41:892–901.

Weiss, S.B., D.D. Murphy, P.R. Ehrlich, and C.F. Metzler. 1993. Adult emergency phenology in checkerspot butterflies: the effects of macroclimate, topography, and population history. *Oecologia* 96:261–270.

Weiss, S.B., D.D. Murphy, and R.R. White. 1988. Sun, slope, and butterflies: topographic determinants of habitat quality for *Euphydryas editha*. *Ecology* 69:1486–1496.

Weiss, S.B., and A.D. Weiss. 1998. Landscape-level phenology of a threatened butterfly: a GIS-based modeling approach. *Ecosystems* 1:299–309.

Wilson, R.J., D. Gutierrez, J. Gutierrez, D. Martinez, R. Agudo, and V.J. Monserrat. 2005. Changes to the elevational limits and extent of species ranges associated with climate change. *Ecology Letters* 8:1138–1146.

Wilson, R.J., D. Gutierrez, J. Gutierrez, and V.J. Monserrat. 2007. An elevational shift in butterfly species richness and composition accompanying recent climate change. *Global Change Biology* 13:1873–1887.

4 Advances, Limitations, and Synergies in Predicting Changes in Species' Distribution and Abundance under Contemporary Climate Change

Enrique Martínez-Meyer

CONTENTS

One of the most widely documented ecological consequences of the current climatic-change episode is that of geographic range shifts of species (Parmesan and Yohe 2003; Root et al. 2003; Parmesan 2006). The geographic range is the spatial manifestation of underlying population processes, such as local colonization and extinction, which are naturally dynamic but become even more lively when sudden or abrupt changes in the environment (biotic or abiotic) occur (Brown et al. 1996; Holt 2003).

At the most fundamental level, when environmental conditions are favorable for a species, its populations grow and more individuals disperse to other suitable areas, thus expanding the geographic range. In contrast, when conditions become unsuitable, populations may respond in one of three ways: (1) if they are preadapted or their genetic/phenotypic plasticity is sufficient, they can survive under novel conditions and the geographic range of the species will remain more or less constant; (2) if adaptation does not occur but the dispersal capacity of individuals allows them to track suitable environments across space, then the species' range will shift; or (3) when neither adaptation nor movement is possible, populations experience local reductions in density or extirpations and the species' range shrinks—in the extreme case, the range will collapse ubiquitously and the species will go extinct globally (Holt 1990). In real life, however, intrinsic (e.g., dispersal potential, population growth rate, life history) and extrinsic (e.g., interacting species' dynamics, human facilitation or hindrance, disease, environmental catastrophes) factors and processes complicate the fundamental-level dynamics, making prediction of consequences of contemporary climate change on species' distribution and abundance a challenging endeavor (Thuiller 2004).

PREDICTING SPECIES' DISTRIBUTIONS UNDER CLIMATE CHANGE

Ecological Niche Modeling

Currently, the most popular approach for forecasting species' rangewide distributional shifts due to climate change is termed Ecological Niche Modeling (ENM), also known in the literature as Species Distribution Modeling (SDM), Habitat Suitability Modeling (HSM), and Climate Envelope Modeling (CEM), among others. Ecological niche models are logically elegant because they simply look for nonrandom relationships between species occurrences and a set of environmental variables that are known (or assumed) to constrain species' distributions. These relationships are then used to produce potential distribution maps. Next, those environmental combinations suitable for the species are incorporated in a spatially explicit manner with ≥1 climate scenarios for the future (or the past) to produce a new map representing the species' potential distribution under such altered conditions (Pearson and Dawson 2003; Guisan and Thuiller 2005). Further Geographic Information Systems (GIS) manipulations and analyses allow spatially explicit quantification of the amount of area expected to be added or lost through time from the species' range (Figure 4.1). Several works have addressed the assumptions, strengths, and shortcomings of this modeling approach (e.g., Pearson and Dawson 2003; Guisan and Thuiller 2005; Araújo and Guisan 2006; Guisan et al. 2006; Heikkinen et al. 2006; Botkin et al. 2007; Soberón and Nakamura 2009). Here, I will review some of the most important ones and present the trend of current research and perspectives in the field.

The conceptual simplicity of ENMs dictates that they must rely on a number of assumptions, the tenability of which has been questioned by numerous authors (Araújo and Guisan 2006). First, species are assumed to be at equilibrium with climate (i.e., they occupy all climatically suitable areas available for them) when niche models are projected onto future scenarios (Araújo and Pearson 2005). In practice,

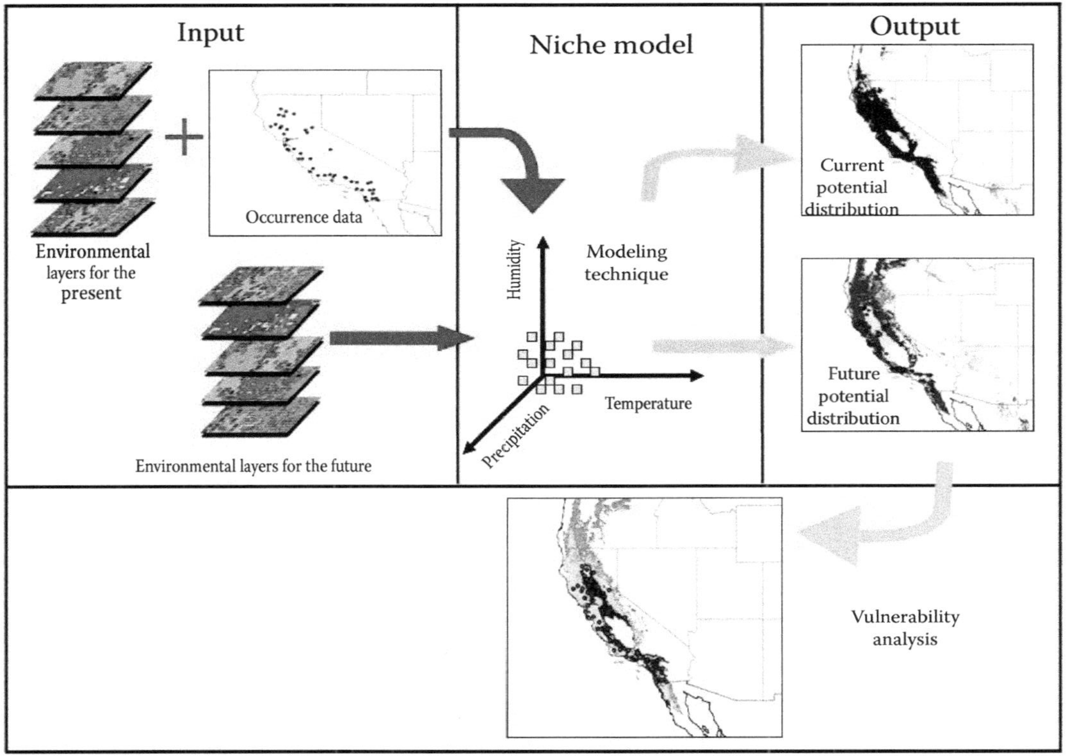

FIGURE 4.1 **(See color insert.)** Diagrammatic representation of the ecological niche modeling process.

however, this assumption does not often hold, due to biological reasons or artifacts of the analytical process. For example, species with limited dispersal capabilities or that are in the process of range expansion do not fill all their potential distributions; for instance, herptiles consistently show lower degrees of equilibrium with climate compared to other vertebrates such as birds (Araújo and Pearson 2005). Furthermore, failing to include critical environmental variables during the modeling process may also produce models falsely appearing to be at disequilibrium (Dormann 2007). Failing to meet the equilibrium assumption overestimates the distributional potential of species in altered scenarios (e.g., due to contemporary climate change). However, implementation of distance constraints obtained from calculating distances among neighboring occurrences and spatial autocorrelation data typically improves model performance (Allouche et al. 2008).

Another assumption used when projecting ENMs through time is that of niche stability (Nogués-Bravo 2009), also referred to as niche conservatism. Projected models assume that species will not adapt to novel conditions present in changed climates; rather, species will maintain evolutionary stability and thus they will track the environmental conditions suitable for them across space. In other words, the quantitative relationships to climatic variables that a species' distribution has exhibited recently are assumed to remain unchanged through time (and through altered climates). This seems a reasonable assumption for most species with relatively low evolutionary rates (e.g., higher plants and vertebrates), given that the duration of climate change in this warming episode covers only a few generations. However, for fast-evolving (e.g., insects and microorganisms) or genetically plastic species this may not be the case, and adaptation can occur (Rodríguez-Trelles and Rodríguez 1998; Rodríguez-Trelles 2010). In the latter cases, ENMs will underestimate species' potential distributions.

Strengths and Limitations of ENM

The conceptual simplicity of ENMs plus the growing availability of source data (i.e., species records and climate time series) and software for analysis have made this approach very popular in recent years, and the literature is replete with the use of such models for addressing a wide variety of questions, for many taxonomic groups, and in virtually all regions of the world. Evidence of this can be found by reviewing the contents of recent issues of *Ecography*, the *Journal of Biogeography*, and *Global Ecology and Biogeography*. Ironically, the features that make ENM appealing constitute, at the same time, both strengths and weaknesses.

Owing to their need for only relatively basic input data, ENMs have been used to analyze the response of large numbers of species to climate change in different regions of the world, allowing detection of regional and global patterns. For example, Peterson et al. (2002) modeled the potential impacts of two climate-change scenarios out to the year 2055—a conservative-emission and a liberal-emission scenario—for 1870 species across México, including butterflies, birds, and mammals. The authors found that levels of species turnover, rather than extinctions or massive range reductions, may be the most important ecological consequence of contemporary climate change for this region. In another study, Thuiller et al. (2005) modeled the spatial responses of 1350 species of plants across Europe by using an ensemble of seven future scenarios. They found that a large but variable proportion of pixels (up to 42% across the continent) could potentially lose species, and that more than half of the species could become vulnerable or threatened by 2080, mainly those inhabiting mountains. In an analogous investigation that modeled species responses across an area covering about 20% of the Earth's surface, a contentious meta-analysis estimated that 15–37% of species will be "committed to extinction" by 2050 (Thomas et al. 2004).

On the other hand, the simplicity and versatility of this analytical tool frequently drives users to misuse or abuse it when no clear understanding of its assumptions and capabilities exists. In effect, this converts what should be testing of scientific hypotheses into mere menial tasks. Below, I present the main conceptual and technical limitations and associated uncertainties of ENMs when they are used for predicting species distributions through time; I also offer advances to overcome, or at least evaluate, these limitations. One such limitation in anticipating the effects of climate change on the geographic distribution of species relates to species' intrinsic dispersal ability (Best et al. 2007). Equilibrium with climate at the present time indicates that dispersal capability of a species has been sufficient to occupy the available suitable habitat; however, this does not ensure that it will be able to keep pace with climate change, which (at least for temperature) has been more rapid than it has been during the last tens of millennia, across most of the world (Rahmstorf et al. 2007; Randall et al. 2007). In fact, ability to keep pace would seem to not be the norm. There is a variable amount of lag between the onset of suitable climatic conditions and the time of arrival of new immigrants (or populations), and also between the time when conditions shift from suitable to unsuitable and when local population extinctions occur, and these two lags are asymmetrical (Devictor et al. 2008).

These "leading-edge" and "trailing-edge" dynamics have started to be incorporated in modeling exercises to better understand and represent distributional shifts (Thomas et al. 2001; Thuiller et al. 2008). For example, Anderson et al. (2009) coupled a niche model with a dynamic metapopulation model into which dispersal information was incorporated to analyze the dynamics of range margins of two lagomorphs at risk of extinction (*Romerolagus diazi* and *Lepus timidus*). Under this approach, the authors were able to detect differences in migration rates between the leading and trailing edges, providing insights into the processes behind changes in distributional patterns and thus improving upon performance of traditional, static ENMs. In another study, Thuiller et al. (2008) present some alternatives for including ancillary information into ENMs in order to produce projections for the future that are more realistic, in the face of climate change. They propose incorporating field data on migration rates, as well as demographic information, into ENMs to capture leading- and trailing-edge dynamics.

A second line of research for improving ENMs involves integration of biotic interactions into the modeling process. Evidence is mounting that interspecific interactions have a significant effect on the distribution of species not only at the local scale, but also at the macroecological scale (Araújo and Luoto 2007; Gotelli et al. 2010). Some forms of interactions are quite clearly determinants of species' distributions, such as (1) parasite species that depend on hosts and vectors to complete their life cycles, and (2) in strict coevolutionary relationships, such as the butterfly *Baronia brevicornis* and its exclusive food plant *Acacia cochliacantha* (Soberón and Peterson 2005). In these cases, information regarding geographic distributions of all species involved is necessary to achieve accurate prediction (Nakazawa et al. 2007; Peterson 2008).

However, recent work suggests that even in biotic interactions that are more relaxed (e.g., interspecific competition, mutualism), the inclusion of distributional information on interacting species improves distribution models of target species (Anderson et al. 2002). For instance, Araújo and Luoto (2007) tested whether the distribution of three food-plant species of the clouded Apollo butterfly (*Parnassius mnemosyne*) influenced the butterfly's current distribution and predictions of the species' shifted range under climate change. Their results indicate a significant role of biotic interactions in all species' distributions both in the present and in the future, suggesting that climatic variables are not the only drivers of species' ranges at macroecological scales, as has often been assumed. Analogously, Gotelli et al. (2010) found that competitive interactions among ecologically similar bird species in Denmark can be detected at spatial resolutions up to four orders of magnitude larger than that of individual territories. Accordingly, the authors concluded that biotic interactions should be incorporated into the modeling process, in order to refine distributional predictions.

Sources of Error and Uncertainty in ENM

In addition to the limitations and biases produced by complexities inherent in biological systems, ENMs are also affected by problems in source data and analytical implementations. As mentioned previously, ENMs require two sources of primary data: species occurrences (presence or presence and absence data), and environmental variables in the form of GIS layers. The latter variables also require a parallel set of environmental layers for hindcasts to the past or projections to the future. In addition to these data sources, ENMs also require modeling techniques (Figure 4.1). Each of these three elements contains its own biases that affect the accuracy and uncertainty of distributional models to varying degrees.

Occurrence data come from a variety of sources, such as scientific collections (museums, herbaria), observational surveys (e.g., Breeding Bird Surveys in different countries), and field studies. From these, the most taxonomically comprehensive and widely available data comes from scientific collections (e.g., see the GBIF portal: http://www.gbif.org/). Data drawn from collections have both taxonomic and positional inaccuracies that affect niche modeling; thus, a filtering process is necessary before using them (Newbold 2010). Nonetheless, positional uncertainty remains in georeferenced collecting locations even after dubious records have been removed (Guo et al. 2008). For example, a typical locality description in a specimen voucher could be "20 mi W Lawrence, Kansas, USA." Because Lawrence, Kansas, is a town (polygon), uncertainty exists whether distance from the town has to be taken from the centroid of the polygon or from its margin. Furthermore, ecologists seldom have additional information to decide whether the 20 miles described in the tag were measured on a map or in the field, and whether the distance was simply rounded to the nearest 10-mile increment. Similarly, we typically have no information to know whether direction was straight West or "West-ish." All of these inaccuracies produce a degree of uncertainty in occurrence data that has been demonstrated to impact distributional patterns of species (Rowe 2005) and also niche models (Graham et al. 2008). Georeferencing techniques that incorporate a measure of uncertainty have been developed (Wieczorek et al. 2004; Guo et al. 2008), and formal tests are under development to assess the degree to which positional uncertainty biases model results (J. Wieczorek, pers. comm.), but these are topics that clearly demand further research.

Environmental layers constitute the other fundamental source of primary data for building ecological niche models. Transfer of niche models through time requires parallel sets of climatic layers for the present and at least one period in the past or the future. Climatic surfaces in the present are interpolations from climatic stations that vary in their degree of accuracy, depending on the spatial dispersion of stations and the length of time measuring climatic variables (Hijmans et al. 2005; Mitchell and Jones 2005). For past and future modeling scenarios, the most widely used data sets are derived from General or Regional Circulation Models (Randall et al. 2007). Several works have evaluated different aspects of environmental layers in the performance of niche models, such as the type and number of layers used (Thuiller et al. 2004; Luoto et al. 2007; Randin et al. 2009) and the spatial resolution of the environmental data (Guisan et al. 2007a, 2007b; Trivedi et al. 2008). In the context of climate change, important efforts have focused on analyzing uncertainty generated by climatic scenarios in model output (Peterson et al. 2004; Araújo et al. 2005a, 2005b). Similarly, numerous works have evaluated uncertainty associated with different modeling techniques for transferring species distributions across space and time (Pearson et al. 2006; Peterson et al. 2007; Barbosa et al. 2009). Results indicate very high variability among scenarios and methods, to the degree of considering ENM of little help for guiding conservation actions (Dormann 2007).

As a result of this situation, new analytical tools have been implemented to analyze and reduce uncertainty in model projections. Particularly, ensemble approaches have been developed to capture variability among various modeling techniques and scenarios, assess uncertainty, and obtain results that are more stable and robust (Araújo et al. 2005b; Araújo and New 2007; Coetzee et al. 2009; Diniz et al. 2009; Buisson et al. 2010). For example, Araújo and New (2007) proposed a framework for generating and analyzing model ensembles, demonstrating that the ensemble approach produces more reliable results compared to single-model approaches. Without doubt, ensemble forecasting is an approach that will develop more fully in the field of niche modeling in the coming years; in fact, computerized tools for manipulating and analyzing ensembles have already been developed (Thuiller et al. 2009) and others are about to appear (M. Araújo, pers. comm.).

Model Validation

In addition to the variability in forecasts among niche models due to input-data and analytical-technique idiosyncrasies, another issue of concern is the evaluation of model accuracy. For example, if we produce a series of models for current and future scenarios using different techniques but use the same input data (i.e., same training and validation data sets), we will end up with a number of distributional maps that often vary in their similarity both for the present and future. The logical question that arises in this situation is, which one of the maps is more accurate? The answer is by no means an easy one. Several validation methods have been adopted to assess model accuracy, including the chi-square/binomial test, Kappa index, True Skills Statistics (TSS), Area Under the Curve (AUC) of the Receiver-Operating Characteristics (ROC) Plots, and others that have been less popular (Fielding and Bell 1997). All of these validation methods are based on the calculation of omission (false absence) and commission (false presence) errors derived from comparing model predictions against a validation data set in a contingency table.

All of these tests provide some information regarding model quality, but none of them are fully adequate for evaluating these types of geographic models, for several reasons. First, tests that incorporate both types of errors (i.e., Kappa index, TSS, AUC) do not take into account the fact that omission and commission errors are not equivalent because the former is estimated with presence data and the latter with absence data. Although presence data may contain positional and identification errors (Hortal et al. 2008), these are relatively reliable data compared to absence data, because even with exhaustive sampling there is always the possibility that the target species went undetected (Kéry 2002; MacKenzie et al. 2006). Furthermore, even if we are sure that the species is not present, it is not always possible to determine whether its absence reflects historical or dispersal reasons, or instead indicates that the site lacks suitable conditions. This is critical because only the latter is within the scope of niche modeling efforts (Jiménez-Valverde and Lobo 2007).

Second, all these tests strongly depend on the proportion of the predicted area relative to the extent of the whole area (i.e., species' prevalence) (Manel et al. 2001). If the predicted area is minimal with respect to the study area (e.g., 10% occupancy), then our expectation of predicted likelihood of presences under a null model would be so low that even a relatively large amount of omission error (e.g., 50%) will still produce statistical results significantly higher than random, and this may lead investigators to conclude that the model is acceptably predictive, when failing to predict half of the validation points is clearly a poor performance (Lobo et al. 2008; Peterson et al. 2008). Alternatively, if prevalence is too great, the power of statistical tests to detect significant differences against random expectations is dramatically reduced, potentially confounding investigators on the biological meaning of those conclusions (Manel et al. 2001).

Third, none of the above-mentioned validation tests are designed for geographic analyses. It is not uncommon that two prediction maps produced from randomizations of the same data set have exactly the same values of sensitivity (i.e., presences correctly predicted), specificity (i.e., absences correctly predicted), omission, and commission errors, but different spatial configuration. If these maps were compared with such statistical tests, both would obtain the same scores because they do not account for geographic differences and only pay attention to the numbers in the success/error matrix (i.e., confusion matrix).

Finally, the legitimacy of these tests relies on the independence of the data set used for validation; in other words, data used for testing the models should be independent from the data used to evaluate them. However, the geographic distribution of a species is of an autocorrelative nature in that the presence of a species in a specific site in one way or another depends on the presence of that species somewhere else nearby in space and time (Brown and Lomolino 1998; Araújo et al. 2005a). Thus, in order to have truly robust validation tests, geostatistical techniques must be developed.

If validating niche models in the present time seems complex, doing so for events that have not yet occurred—as do the models that use climatic conditions in the future—is even more challenging. Strictly speaking, distribution models for the future cannot be validated. Probably the only way to test the trustworthiness of such distribution models is by assessing whether the modeling exercise is reliable for sets of data from the recent past (decades) to the present. For example, Araújo et al. (2005a) modeled 116 species of birds in the United Kingdom for two time periods (1968–1972 and 1988–1991) in which both species' occurrences and climatologies were chronologically aligned. Then they generated models with four methods (Generalized Linear Models [GLM], Generalized Additive Models [GAM], Classification Tree Analysis [CTA], and an Artificial Neural Network [ANN]) using the data from 1968–1972 and projected onto climatic conditions for 1988–1991 and validated with occurrences associated with the second period. Their results showed that ANN and GAM produced projections that are more reliable than did the other methods. In another study, Hijmans and Graham (2006) produced distribution predictions for the past, present, and future with four niche modeling techniques (BIOCLIM, DOMAIN, GAM, and Maxent) and compared their performance against that of a mechanistic model. They found that Maxent produced models more similar to those obtained with the mechanistic models.

In summary, model projections for future scenarios are, in fact, highly variable and may or may not be reliable. Production of highly reliable and statistically significant models that describe present relationships is not enough to guarantee that models for the future will also be reliable. Although that is certainly a mandatory first step, whenever possible it is greatly valuable to cross-validate models with recent-past data to establish which method works better for a particular case.

Alternative Approaches for Predicting Species' Distributions under Climate Change

Ecological Niche Modeling is to date by far the most popular approach for modeling species' ranges and distributional shifts due to climate change. However, it is not the only approach; others have been developed under a more knowledge-oriented paradigm. Mechanistic or process-based models (Figure 4.2) aim to (1) understand the genetic, physiological, demographic, or behavioral mechanisms that underlie the way(s) in which environmental factors affect the performance and survival of individuals and populations, and (2) use that knowledge to infer species' geographic distributions and shifts that may result from climatic changes (Crozier and Dwyer 2006; Helaouet and Beaugrand 2009).

Mechanistic models are certainly more robust than ENMs, but have the disadvantage that they require greater amounts of data. In addition to spatial and

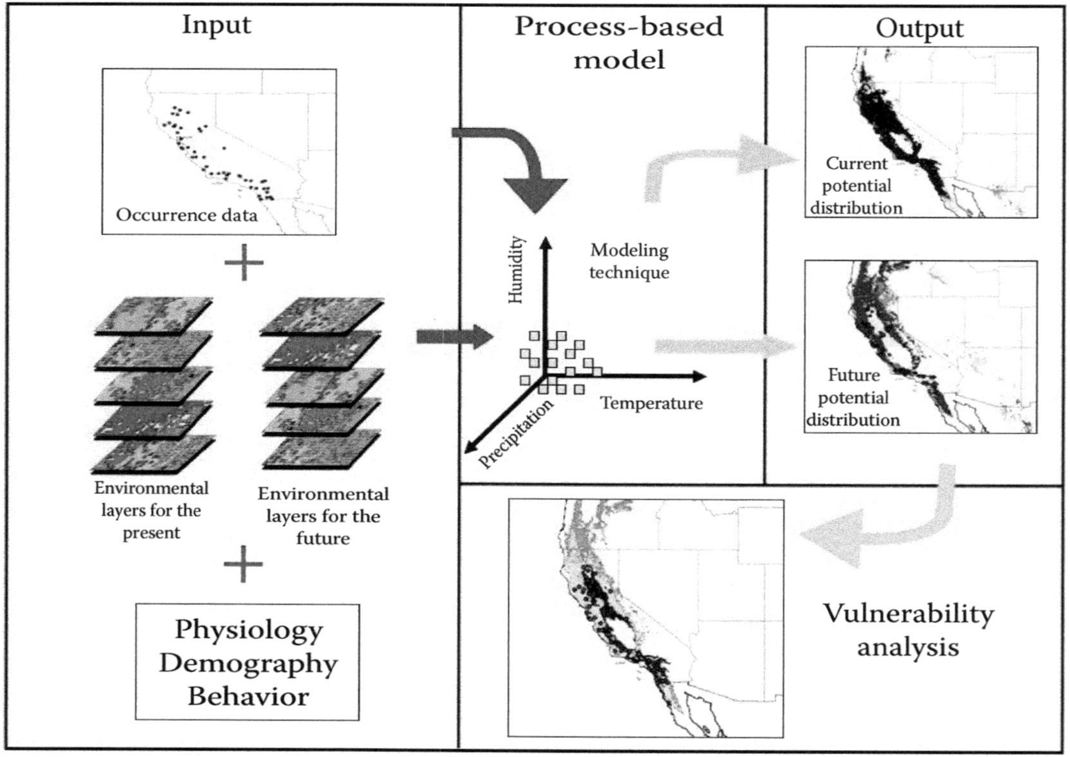

FIGURE 4.2 Diagrammatic representation of a generic mechanistic modeling process.

environmental data, mechanistic models also require information regarding physiological tolerances to critical variables that are specific for each species (Helmuth 2009; Kearney and Porter 2009). Their main advantage over niche models is that mechanistic models unveil the processes underneath distributional patterns. That is, they provide an explanation not only of *what* environmental variables are critical for delimiting a species' distribution, but also *how* and *why* the species will respond to changes in environmental conditions (Morin et al. 2007; Kearney et al. 2009).

Both ENMs and mechanistic models have advantages and limitations. Consequently, attempts to take the best from each of the two modeling approaches and produce hybrid models are starting to emerge, with promising results. For example, Crozier and Dwyer (2006) developed a population-dynamic model to analyze potential range shifts of the skipper butterfly (*Atalopedes campestris*) in which ambient temperature determines population growth asymmetrically across seasons. Their results indicate differential response to climate change during summer and winter, and an overall reduction in spatial distribution. In another study, Morin and Thuiller (2009) compared range-shift predictions from an ENM and a mechanistic model for 15 tree species in North America. They observed that niche models predicted higher extinction and colonization rates than did the mechanistic model, but found that the two approaches identified generally similar trends of shifts and areas of local extinction. This comparison allows the quantification of differences between the two approaches and thus the assessment (and ultimately reduction of) uncertainty. This is possible because the two modeling processes are based upon different assumptions; hence, identifying the same sites that are prone to species extinction with the two approaches is complementary and provides more confidence in the results (Morin and Thuiller 2009).

Spatially explicit mechanistic and hybrid models are in the early stages of development, but their potential is crystal clear. It is desirable to have stronger interdisciplinary collaboration among physiologists, ecologists, biogeographers, and computer scientists. Such collaboration could be used first to harvest all valuable physiological and demographic information from the literature generated from centuries of empirical field and experimental research, and then to develop tools that allow the integration and analysis of spatial, ecological, and physiological data to derive reliable predictive distributional models.

PREDICTING SPECIES' ABUNDANCE AT THE RANGEWIDE SCALE

A general biogeographic pattern in nature is that population numbers vary drastically across species' ranges, with the majority of places holding relatively few individuals and only some sites containing orders of magnitude more individuals (Brown et al. 1995). Because ENMs use either presence or presence and absence data, associated predictive maps are typically not expected to contain any information regarding the geographic variation in abundance. However, a few attempts have been made to predict the distribution of abundance from occurrence data, with quite interesting results. He and Gaston (2000) proposed a model based on the relationship between the area of the sampling unit and the area of occupancy

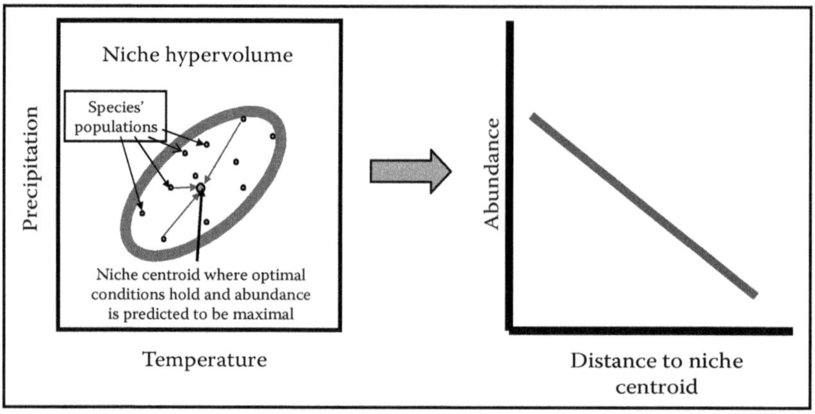

FIGURE 4.3 Relationship between abundance and distance to niche centroid. This is the conceptual basis for predicting spatial patterns of abundance.

(i.e., an "area-area curve"). They derived a series of functions that successfully predicted abundance from occurrence data at local scales, but the model was not accurate at large spatial scales. As an improvement of previous attempts to model abundance patterns from occupancy data, Hui and McGeoch (2007) developed the "droopy-tail" model (DTM), which captures the logistic relationship between occupancy and abundance more accurately than alternative models, across all spatial scales tested.

However, looking for an explanation of biogeographic patterns of abundance in the geographic domain (i.e., as a consequence of landscape configuration) may be misleading, because despite the fact that spatial autocorrelation exerts an important effect on the distribution of abundance (Brown et al. 1995), the well-known pattern of species' higher abundance at the center of the geographic range compared to at the periphery ("the abundant-center hypothesis") seems to be incidental and not a general rule (Sagarin et al. 2006). Local population abundance is largely explained by the extent to which local niche conditions are met (Brown 1984). If this is true, models that evaluate the suitability of sites conferred by niche conditions should reflect abundance patterns to some degree. VanDerWal et al. (2009) tested the abundance-suitability relationship using a probabilistic niche model (Maxent), interpreted as a suitability surface, and local abundance for 69 vertebrate species. They found a positive relationship between suitability and observed abundance for most species, and concluded that such a modeling approach provides reliable information on spatial patterns of the upper limit of abundance (VanDerWal et al. 2009).

The possibility of modeling the spatial patterns of abundance is exciting. The key seems to be in the analysis of the internal structure of the ecological niche. Maguire (1973) demonstrated that optimal conditions for a species are found near the niche centroid in environmental space, where birth rate is maximized and death rate is minimized. Thus, abundance should be highest in sites where niche-centroid conditions are met (Figure 4.3). Preliminary tests of this idea have yielded very promising results (Díaz-Porras 2006), but further testing is necessary.

CONCLUSIONS

Climate change is affecting natural systems in many ways. To the degree that we are able to understand biotic responses to these changes, we will be able to anticipate and possibly avoid undesirable future consequences. Considering that the distribution of a species is the ultimate spatial expression of population processes, distributional shifts resulting from climatic changes are among the last signals able to be detected. However, irrefutable evidence exists that distributional shifts are already occurring in many taxonomic groups worldwide. Hence, the future is here, and management and conservation strategies must account for these effects from this day forward. Predictive biogeography has seen an explosive growth during the last decade in important conceptual and operational progress; nonetheless, much work remains ahead before distribution models can be reliably used for guiding policy decisions. Uncertainty in model predictions remains the main issue that demands further work, but results of recent approaches are promising. The next few years will be a time for maturity of the field, during which the integration of correlative and mechanistic models and the development of new approaches for predicting abundance patterns will observe important progress.

SUMMARY

Geographic distributional shifts are a widely documented response of biodiversity to past and current climatic changes. These shifts mainly reflect the capacity of species to track their suitable environments across space when environmental conditions become inappropriate and they are unable to adapt. Therefore, if we know the environmental conditions that determine species' geographic ranges, prediction of the direction and magnitude of range shifts due to climatic changes becomes possible. Two main approaches have been developed for this task: Ecological Niche Models (ENMs) and mechanistic models. In this chapter, I briefly explain the general principles of each, and discuss the main assumptions and conceptual and implementation complexities that have limited the usefulness of ENM for predicting future distributional consequences of climate change more accurately. I address the topics of data and methodological biases, uncertainty, and model validation, and offer some strategies for improving the modeling practice. Finally, I discuss the future trends for modeling species' distribution and abundance under climate change.

REFERENCES

Allouche, O., O. Steinitz, D. Rotem, A. Rosenfeld, and R. Kadmon. 2008. Incorporating distance constraints into species distribution models. *Journal of Applied Ecology* 45 (2):599–609.

Anderson, B. J., H. R. Akcakaya, M. B. Araújo, D. A. Fordham, E. Martinez-Meyer, W. Thuiller, and B. W. Brook. 2009. Dynamics of range margins for metapopulations under climate change. *Proceedings of the Royal Society B-Biological Sciences* 276 (1661):1415–1420.

Anderson, R. P., A. T. Peterson, and M. Gómez-Laverde. 2002. Using niche-based GIS modeling to test geographic predictions of competitive exclusion and competitive release in South American pocket mice. *Oikos* 98 (1):3–16.

Araújo, M. B., and A. Guisan. 2006. Five (or so) challenges for species distribution modelling. *Journal of Biogeography* 33 (10):1677–1688.

Araújo, M. B., and M. Luoto. 2007. The importance of biotic interactions for modelling species distributions under climate change. *Global Ecology and Biogeography* 16 (6):743–753.

Araujo, M. B., and M. New. 2007. Ensemble forecasting of species distributions. *Trends in Ecology & Evolution* 22 (1):42–47.

Araújo, M. B., and R. G. Pearson. 2005. Equilibrium of species' distributions with climate. *Ecography* 28 (5):693–695.

Araújo, M. B., R. G. Pearson, W. Thuiller, and M. Erhard. 2005a. Validation of species-climate impact models under climate change. *Global Change Biology* 11 (9):1504–1513.

Araújo, M. B., R. J. Whittaker, R. J. Ladle, and M. Erhard. 2005b. Reducing uncertainty in projections of extinction risk from climate change. *Global Ecology and Biogeography* 14 (6):529–538.

Barbosa, A. M., R. Real, and J. M. Vargas. 2009. Transferability of environmental favourability models in geographic space: The case of the Iberian desman (*Galemys pyrenaicus*) in Portugal and Spain. *Ecological Modelling* 220 (5):747–754.

Best, A. S., K. Johst, T. Munkemuller, and J. M. J. Travis. 2007. Which species will succesfully track climate change? The influence of intraspecific competition and density dependent dispersal on range shifting dynamics. *Oikos* 116 (9):1531–1539.

Botkin, D. B., H. Saxe, M. B. Araújo, R. Betts, R. H. W. Bradshaw, T. Cedhagen, P. Chesson, T. P. Dawson, J. R. Etterson, D. P. Faith, S. Ferrier, A. Guisan, A. S. Hansen, D. W. Hilbert, C. Loehle, C. Margules, M. New, M. J. Sobel, and D. R. B. Stockwell. 2007. Forecasting the effects of global warming on biodiversity. *Bioscience* 57 (3):227–236.

Brown, J. H. 1984. On the relationship between abundance and distribution of species. *American Naturalist* 124:255–279.

Brown, J. H., and M. V. Lomolino. 1998. *Biogeography*. 2nd ed. Sinauer Associates, Sunderland, Massachusetts.

Brown, J. H., D. W. Mehlman, and G. C. Stevens. 1995. Spatial variation in abundance. *Ecology* 76 (7):2028–2043.

Brown, J. H., G. C. Stevens, and D. M. Kaufman. 1996. The geographic range: Size, shape, boundaries, and internal structure. *Annual Review of Ecology and Systematics* 27:597–623.

Buisson, L., W. Thuiller, N. Casajus, S. Lek, and G. Grenouillet. 2010. Uncertainty in ensemble forecasting of species distribution. *Global Change Biology* 16 (4):1145–1157.

Coetzee, B. W. T., M. P. Robertson, B. F. N. Erasmus, B. J. van Rensburg, and W. Thuiller. 2009. Ensemble models predict important bird areas in southern Africa will become less effective for conserving endemic birds under climate change. *Global Ecology and Biogeography* 18 (6):701–710.

Crozier, L., and G. Dwyer. 2006. Combining population-dynamic and ecophysiological models to predict climate-induced insect range shifts. *American Naturalist* 167 (6):853–866.

Devictor, V., R. Julliard, D. Couvet, and F. Jiguet. 2008. Birds are tracking climate warming, but not fast enough. *Proceedings of the Royal Society B-Biological Sciences* 275 (1652):2743–2748.

Díaz-Porras, D. 2006. Relación entre la abundancia y la distancia al centroide del nicho ecológico de las especies. MsC thesis. Biological Sciences. Universidad Nacional Autónoma de México.

Diniz, J. A. F., L. M. Bini, T. F. Rangel, R. D. Loyola, C. Hof, D. Nogues-Bravo, and M. B. Araujo. 2009. Partitioning and mapping uncertainties in ensembles of forecasts of species turnover under climate change. *Ecography* 32 (6):897–906.

Dormann, C. F. 2007. Promising the future? Global change projections of species distributions. *Basic and Applied Ecology* 8 (5):387–397.

Fielding, A. H., and J. F. Bell. 1997. A review of methods for the assessment of prediction errors in conservation presence/absence models. *Environmental Conservation* 24:38–49.

Gotelli, N. J., G. R. Graves, and C. Rahbek. 2010. Macroecological signals of species interactions in the Danish avifauna. *Proceedings of the National Academy of Sciences of the United States of America* 107 (11):5030–5035.

Graham, C. H., J. Elith, R. J. Hijmans, A. Guisan, A. T. Peterson, B. A. Loiselle, and Gro Nceas Predect Species Working. 2008. The influence of spatial errors in species occurrence data used in distribution models. *Journal of Applied Ecology* 45 (1):239–247.

Guisan, A., C. H. Graham, J. Elith, F. Huettmann, and Nceas Species Distri. 2007a. Sensitivity of predictive species distribution models to change in grain size. *Diversity and Distributions* 13 (3):332–340.

Guisan, A., A. Lehmann, S. Ferrier, M. Austin, J. M. C. Overton, R. Aspinall, and T. Hastie. 2006. Making better biogeographical predictions of species' distributions. *Journal of Applied Ecology* 43 (3):386–392.

Guisan, A., and W. Thuiller. 2005. Predicting species distribution: Offering more than simple habitat models. *Ecology Letters* 8 (9):993–1009.

Guisan, A., N. E. Zimmermann, J. Elith, C. H. Graham, S. Phillips, and A. T. Peterson. 2007b. What matters for predicting the occurrences of trees: Techniques, data, or species' characteristics? *Ecological Monographs* 77 (4):615–630.

Guo, Q., Y. Liu, and J. Wieczorek. 2008. Georeferencing locality descriptions and computing associated uncertainty using a probabilistic approach. *International Journal of Geographical Information Science* 22 (10):1067–1090.

He, F. L., and K. J. Gaston. 2000. Estimating species abundance from occurrence. *American Naturalist* 156 (5):553–559.

Heikkinen, R. K., M. Luoto, M. B. Araújo, R. Virkkala, W. Thuiller, and M. T. Sykes. 2006. Methods and uncertainties in bioclimatic envelope modelling under climate change. *Progress in Physical Geography* 30 (6):751–777.

Helaouet, P., and G. Beaugrand. 2009. Physiology, ecological niches and species distribution. *Ecosystems* 12 (8):1235–1245.

Helmuth, B. 2009. From cells to coastlines: How can we use physiology to forecast the impacts of climate change? *Journal of Experimental Biology* 212 (6):753–760.

Hijmans, R. J., S. E. Cameron, J. L. Parra, P. G. Jones, and A. Jarvis. 2005. Very high resolution interpolated climate surfaces for global land areas. *International Journal of Climatology* 25 (15):1965–1978.

Hijmans, R. J., and C. H. Graham. 2006. The ability of climate envelope models to predict the effect of climate change on species distributions. *Global Change Biology* 12:2272–2281.

Holt, R. D. 1990. The microevolutionary consequences of climate change. *Trends in Ecology & Evolution* 5 (9):311–315.

———. 2003. On the evolutionary ecology of species' ranges. *Evolutionary Ecology Research* 5 (2):159–178.

Hortal, J., A. Jimenez-Valverde, J. F. Gomez, J. M. Lobo, and A. Baselga. 2008. Historical bias in biodiversity inventories affects the observed environmental niche of the species. *Oikos* 117:847–858.

Hui, C., and M. A. McGeoch. 2007. Capturing the 'droopy-tail' in the occupancy-abundance relationship. *Ecoscience* 14:103–108.

Jiménez-Valverde, A. and J. M. Lobo. 2007. Threshold criteria for conversion of probability of species presence to either-or presence-absence. *Acta Oecologica* 31:361–369.

Kearney, M., and W. Porter. 2009. Mechanistic niche modelling: Combining physiological and spatial data to predict species' ranges. *Ecology Letters* 12 (4):334–350.

Kearney, M., W. P. Porter, C. Williams, S. Ritchie, and A. A. Hoffmann. 2009. Integrating biophysical models and evolutionary theory to predict climatic impacts on species' ranges: The dengue mosquito *Aedes aegypti* in Australia. *Functional Ecology* 23 (3):528–538.

Kéry, M. 2002. Inferring the absence of a species—A case study of snakes. *Journal of Wildlife Management* 66:330–338.

Lobo, J. M., A. Jimenez-Valverde, and R. Real. 2008. AUC: a Misleading measure of the performance of predictive distribution models. *Global Ecology and Biogeography* 17:145–151.

Luoto, M., R. Virkkala, and R. K. Heikkinen. 2007. The role of land cover in bioclimatic models depends on spatial resolution. *Global Ecology and Biogeography* 16 (1):34–42.

MacKenzie, D. I., J. D. Nichols, J. A. Royle, K. H. Pollock, L. L. Bailey, and J. E. Hines. 2006. *Occupancy Estimation and Modeling: Inferring Patterns and Dynamics of Species Occurrence*. Elsevier, San Diego, California.

Maguire, B. Jr. 1973. Niche response structure and the analytical potentials of its relationship to the habitat. *American Naturalist* 107:213–246.

Manel, S., H. C. Williams, and S. J. Ormerod. 2001. Evaluating presence-absence models in ecology: The need to account for prevalence. *Journal of Applied Ecology* 38:921–931.

Mitchell, T. D., and P. D. Jones. 2005. An improved method of constructing a database of monthly climate observations and associated high-resolution grids. *International Journal of Climatology* 25: 693–712.

Morin, X., C. Augspurger, and I. Chuine. 2007. Process-based modeling of species' distributions: What limits temperate tree species' range boundaries? *Ecology* 88 (9):2280–2291.

Morin, X., and W. Thuiller. 2009. Comparing niche- and process-based models to reduce prediction uncertainty in species range shifts under climate change. *Ecology* 90 (5):1301–1313.

Nakazawa, Y., R. Williams, A. T. Peterson, P. Mead, E. Staples, and K. L. Gage. 2007. Climate change effects on plague and tularemia in the United States. *Vector-Borne and Zoonotic Diseases* 7 (4):529–540.

Newbold, T. 2010. Applications and limitations of museum data for conservation and ecology, with particular attention to species distribution models. *Progress in Physical Geography* 34 (1):3–22.

Nogués-Bravo, D. 2009. Predicting the past distribution of species climatic niches. *Global Ecology and Biogeography* 18 (5):521–531.

Parmesan, C. 2006. Ecological and evolutionary responses to recent climate change. *Annual Review of Ecology Evolution and Systematics* 37:637–669.

Parmesan, C., and G. Yohe. 2003. A globally coherent fingerprint of climate change impacts across natural systems. *Nature* 421 (6918):37–42.

Pearson, R. G., and T. P. Dawson. 2003. Predicting the impacts of climate change on the distribution of species: Are bioclimate envelope models useful? *Global Ecology and Biogeography* 12 (5):361–371.

Pearson, R. G., W. Thuiller, M. B. Araújo, E. Martinez-Meyer, L. Brotons, C. McClean, L. Miles, P. Segurado, T. P. Dawson, and D. C. Lees. 2006. Model-based uncertainty in species range prediction. *Journal of Biogeography* 33 (10):1704–1711.

Peterson, A. T. 2008. Biogeography of diseases: A framework for analysis. *Naturwissenschaften* 95 (6):483–491.

Peterson, A. T., E. Martinez-Meyer, C. Gonzalez-Salazar, and P. W. Hall. 2004. Modeled climate change effects on distributions of Canadian butterfly species. *Canadian Journal of Zoology-Revue Canadienne De Zoologie* 82 (6):851–858.

Peterson, A. T., M. A. Ortega-Huerta, J. Bartley, V. Sanchez-Cordero, J. Soberon, R. H. Buddemeier, and D. R. B. Stockwell. 2002. Future projections for Mexican faunas under global climate change scenarios. *Nature* 416 (6881):626–629.

Peterson, A. T., M. Papes, and M. Eaton. 2007. Transferability and model evaluation in ecological niche modeling: A comparison of GARP and Maxent. *Ecography* 30 (4):550–560.

Peterson, A. T., M. Papeş, and J. Soberón. 2008. Rethinking receiver operating characteristic analysis applications in ecological niche modelling. *Ecological Modelling* 213 (1):63–72.

Rahmstorf, S., A. Cazenave, J. A. Church, J. E. Hansen, R. F. Keeling, D. E. Parker, and R. C. J. Somerville. 2007. Recent climate observations compared to projections. *Science* 316 (5825):709.
Randall, D. A., R. A. Wood, S. Bony, R. Colman, T. Fichefet, J. Fyfe, V. Kattsov, A. Pitman, J. Shukla, J. Srinivasan, R. J. Stouffer, A. Sumi and K. E. Taylor. 2007. Climate models and their evaluation. In: *Climate Change 2007: The Physical Science Basis.* Contribution of Working Group I to the Fourth Assessment Report of the Intergovernmental Panel on Climate Change. Solomon, S., D. Qin, M. Manning, Z. Chen, M. Marquis, K. B. Averyt, M. Tignor and H. L. Miller (Eds.). Cambridge University Press, Cambridge, United Kingdom, and New York.
Randin, C. F., H. Jaccard, P. Vittoz, N. G. Yoccoz, and A. Guisan. 2009. Land use improves spatial predictions of mountain plant abundance but not presence-absence. *Journal of Vegetation Science* 20 (6):996–1008.
Rodríguez-Trelles, F. 2010. Measuring evolutionary responses to global warming: Cautionary lessons from Drosophila. *Insect Conservation and Diversity* 3 (1):44–50.
Rodríguez-Trelles, F., and M. A. Rodriguez. 1998. Rapid micro-evolution and loss of chromosomal diversity in Drosophila in response to climate warming. *Evolutionary Ecology* 12 (7):829–838.
Root, T. L., J. T. Price, K. R. Hall, S. H. Schneider, C. Rosenzweig, and J. A. Pounds. 2003. Fingerprints of global warming on wild animals and plants. *Nature* 421 (6918):57–60.
Rowe, R. J. 2005. Elevational gradient analyses and the use of historical museum specimens: a cautionary tale. *Journal of Biogeography* 32 (11):1883–1897.
Sagarin, R. D., S. D. Gaines, and B. Gaylord. 2006. Moving beyond assumptions to understand abundance distributions across the ranges of species. *Trends in Ecology & Evolution* 21 (9):524–530.
Soberón, J., and M. Nakamura. 2009. Niches and distributional areas: Concepts, methods, and assumptions. *Proceedings of the National Academy of Sciences of the United States of America* 106:19644–19650.
Soberón, J., and A. T. Peterson. 2005. Interpretation of models of fundamental ecological niches and species' distributional areas. *Biodiversity Informatics* 2:1–10.
Thomas, C. D., E. J. Bodsworth, R. J. Wilson, A. D. Simmons, Z. G. Davies, M. Musche, and L. Conradt. 2001. Ecological and evolutionary processes at expanding range margins. *Nature* 411 (6837):577–581.
Thomas, C. D., A. Cameron, R. E. Green, M. Bakkenes, L. J. Beaumont, Y. C. Collingham, B. F. N. Erasmus, M. F. de Siqueira, A. Grainger, L. Hannah, L. Hughes, B. Huntley, A. S. van Jaarsveld, G. F. Midgley, L. Miles, M. A. Ortega-Huerta, A. T. Peterson, O. L. Phillips, and S. E. Williams. 2004. Extinction risk from climate change. *Nature* 427 (6970):145–148.
Thuiller, W. 2004. Patterns and uncertainties of species' range shifts under climate change. *Global Change Biology* 10 (12):2020–2027.
Thuiller, W., C. Albert, M. B. Araújo, P. M. Berry, M. Cabeza, A. Guisan, T. Hickler, G. F. Midgely, J. Paterson, F. M. Schurr, M. T. Sykes, and N. E. Zimmermann. 2008. Predicting global change impacts on plant species' distributions: Future challenges. *Perspectives in Plant Ecology Evolution and Systematics* 9 (3–4):137–152.
Thuiller, W., M. B. Araujo, and S. Lavorel. 2004. Do we need land-cover data to model species distributions in Europe? *Journal of Biogeography* 31 (3):353–361.
Thuiller, W., B. Lafourcade, R. Engler, and M. B. Araujo. 2009. BIOMOD—A platform for ensemble forecasting of species distributions. *Ecography* 32 (3):369–373.
Thuiller, W., S. Lavorel, M. B. Araujo, M. T. Sykes, and I. C. Prentice. 2005. Climate change threats to plant diversity in Europe. *Proceedings of the National Academy of Sciences of the United States of America* 102 (23):8245–8250.

Trivedi, M. R., P. M. Berry, M. D. Morecroft, and T. P. Dawson. 2008. Spatial scale affects bioclimate model projections of climate change impacts on mountain plants. *Global Change Biology* 14 (5):1089–1103.

VanDerWal, J., L. P. Shoo, C. N. Johnson, and S. E. Williams. 2009. Abundance and the environmental niche: Environmental suitability estimated from niche models predicts the upper limit of local abundance. *American Naturalist* 174 (2):282–291.

Wieczorek, J., Q. G. Guo, and R. J. Hijmans. 2004. The point-radius method for georeferencing locality descriptions and calculating associated uncertainty. *International Journal of Geographical Information Science* 18 (8):745–767.

5 Mammalian Distributional Responses to Climatic Changes

A Review and Research Prospectus

Robert Guralnick, Liesl Erb, and Chris Ray

CONTENTS

PART I: GENERAL PRINCIPLES OF MAMMALIAN SPECIES RESPONSE TO CLIMATIC CHANGES

MAGNITUDE AND RATE OF CURRENT CLIMATE CHANGE

We live in a warming world. During the twentieth century, increases in the concentrations of atmospheric greenhouse gases, along with other natural and man-made alterations to the environment, led to rising land and ocean temperatures of approximately 0.6°C. The Intergovernmental Panel on Climate Change (IPCC) has produced emission scenarios that suggest, across the range of possible scenarios, a temperature increase of 1.4°C to 5.8°C by 2100 (Houghton et al. 2001). Warming is likely to be more pronounced on land and at higher latitudes compared to oceans and equatorial regions. Precipitation is expected to increase in higher-latitude temperate and polar regions and in the tropics, with decreases in the subtropics. These long-term trends are accompanied by an increased likelihood of extreme weather events. These extreme events may prove to be as important as or more important than changing averages in terms of direct impacts on species distribution and diversity (Easterling et al. 2000).

The climatic conditions of the recent past and near future should be considered within the broader context of climatic changes over longer time scales, ranging from millenia to millions of years. Barnosky et al. (2003) used paleoclimatic reconstructions and the mammalian fossil record to examine the rate of climatic changes and the nature of species responses to those changes under different rates of warming and over three different timescales (350 years, 10,000–20,000 years, and 4 million years). We focus only on salient climatic data presented by the authors. Barnosky et al. (2003) conclude that the current rate of warming is within the range of warming seen in the past—for example during the rapid warming that followed the last ice age. However, if the IPCC predictions are accurate, at some point during this century the rate of warming is likely to exceed that of any period within the past 65 million years.

Mammalian Response Processes

What can organisms in general and terrestrial mammals in particular do in the face of such unprecedented changes in climatic conditions? The answer depends both on the severity of direct and indirect effects of climatic changes and on the physiology and behavior of the organisms in question. For some mammals, climatic conditions are rarely directly limiting factors, due to either broad climatic tolerances or the ability to behaviorally regulate the climate they experience. For these mammals, biotic factors such as food availability, competition, and predation may control species diversity and distribution. Although climate may not have direct effects on such species, there might still be indirect effects due to changes in the diversity and distribution of food sources, predators, or competitors. For other mammal species with narrower physiological tolerances to temperature or precipitation, there may be a direct response to climate change. In such cases, individuals and populations can either disperse in order to find more favorable conditions, adapt in situ physiologically or behaviorally, or go locally extinct. Both local extinction and dispersal away from unfavorable habitats will result in range alteration. However, the genetic consequences are quite different. Species that are capable of dispersing may maintain their genetic diversity. In the case of local extinction, overall population size declines and genetic diversity may be lost (Dalén et al. 2007; but see Ray 2001).

It is presumed that the likelihood of adaptation or dispersal as opposed to extinction depends both on the tolerances and plasticity of the species in question and on the rapidity and magnitude of climate change. The evolution of novel adaptations to climatic change, such that a population can persist in a newly unsuitable area, has long been thought to be a slow process. Therefore, the very rapid (in terms of geologic time) changes we are now experiencing will be unlikely to allow for novel adaptive response (Etterson and Shaw 2001; Jump and Peñuelas, 2005; Parmesan, 2006). It is possible that strong selection may favor existing genotypes that are more plastic or tolerant to an altered climate, but asymmetric gene flow from range core to range edges may inhibit the rate and amount of local adaptation. Because many mammals, especially larger ones, have relatively long generation times and are capable of significant dispersal, it is unlikely that adaptive evolutionary responses will be common.

If adaptation is indeed unlikely, then the probability of dispersal versus extinction is dependent on a set of factors: the dispersal capability and habitat specializations of the mammal in question (Warren et al. 2001), its physiological tolerance for climatic conditions (Walther et al. 2002; Root et al. 2003), and the rate and magnitude of climate change. Broadly tolerant and highly vagile species may be the least likely to show range collapse in response to rapid and significant climate change, while narrowly tolerant and poorly dispersing species are likely prone to rapid local extinctions and potential range collapse (Warren et al. 2001; Walther et al. 2002). In addition to the obvious impacts of species tolerance and climatic trends on mammalian range response, a fourth main factor controlling species response to climate change less discussed in the literature is topographic habitat heterogeneity, a topic to which we turn in the following. While we do recognize the importance of landscape changes and the interacting effects of landscape change and climate change in relation to colonization–extinction dynamics, we instead focus here primarily on climatic changes. The major factors affecting mammal response to climate change are summarized in Table 5.1 for easy reference.

The Importance of Topographic Heterogeneity

Guralnick (2006a, 2006b) discussed the importance of topographic heterogeneity as a factor in mammal distributional response to climatic warming. He argued that animals living in mountainous regions might respond to warming by preferentially moving along elevational gradients as opposed to latitudinal gradients. Climate can quickly change along steep elevational gradients, thus providing mammals with much shorter dispersal distances to favorable climatic conditions than along

TABLE 5.1
Factors That Affect Mammalian Response to Climatic Changes

Key Abiotic Factors	Key Biotic Factors
• Magnitude of local climatic change	• Thermal physiology of species
• Rate of local climatic change	• Ability to behaviorally thermoregulate
• Habitat availability and topographic heterogeneity locally and across the range of the species	• Dispersal capability of species
	• Habitat specialization of species

latitudinal gradients. This means that a species' distribution might more quickly reach equilibrium within a new climate by using elevational gradients rather than latitudinal ones, all else being equal. In addition, Peterson (2003) and Guralnick (2006a) argued that mountain and flatland species would show different patterns of range expansion and contraction under warming scenarios. In particular, wide-ranging flatland species likely shift their ranges farther than mountain species under warming. During warming episodes, as favorable habitats expand northward, North American mountain species follow, extending their northern range boundaries. Southern geographic-range boundaries, on the other hand, remain relatively static due to topographic heterogeneity and the resulting ability of mountain species to track favorable conditions by simply moving up elevational gradients. Flatland species are not able to select habitats along elevational gradients, leading to latitudinal range shifts and potential range contractions, depending on the dynamics of northern and southern range boundaries.

Distributional dynamics and population connectivity are greatly affected by topography. For cold-adapted species living in mountainous areas, continued warming may cause populations to retreat upward in elevation. Upward range shifts have two important consequences. First, the total area occupied by the species decreases because mountains typically narrow from base to peak, which can lead to fewer available resources and lower population sizes and ultimately local extinction (Figure 5.1, especially B; see also Sekercioglu et al. 2008). Second, previously contiguous populations are divided when lowland routes are no longer suitable for dispersal (Figure 5.1—note the loss of lowland connections in A and B during warming). Loss of lowland habitats cuts off gene flow between populations (Ditto and Frey 2007) and can lead to more stochastic extinctions of small populations that can no longer be recolonized. Alternatively, surviving populations may become better adapted to local conditions. Evidence for local adaptation in North American mammals comes from Guralnick (2006b), who showed that there is more variation in experienced climate in the southern portions of ranges occupied by mountain species than in the northern portions. The persistence of populations in these southern areas likely reflects long-term adaptation to these variable local conditions. In contrast, northern populations in the Holarctic are recent immigrants that colonized newly suitable habitat as continental ice sheets receded during warming in the latest Late Pleistocene (18kya–10kya) and Holocene (10kya–present). These populations most likely do not represent the whole of the gene pool found in the South and thus may be tracking a subset of climatic conditions as they disperse. In flatland areas, such as eastern North America, warming is more likely to cause a whole-cloth shift of distributions toward the North, at both range edges. At the same time, warming is not as likely to cause vicariance, and gene flow from the center of ranges to the edges may limit the potential for local adaptations.

Mountain orientation may also be an important factor affecting mountain species response to warming. In the Americas, most mountain chains are oriented approximately North-South, which means that elevation and concomitant temperature gradients are generally oriented East-West, perpendicular to latitudinal gradients. Therefore, mammal species tracking climatic changes along elevational gradients move perpendicularly to those tracking climatic changes along latitude. In Europe,

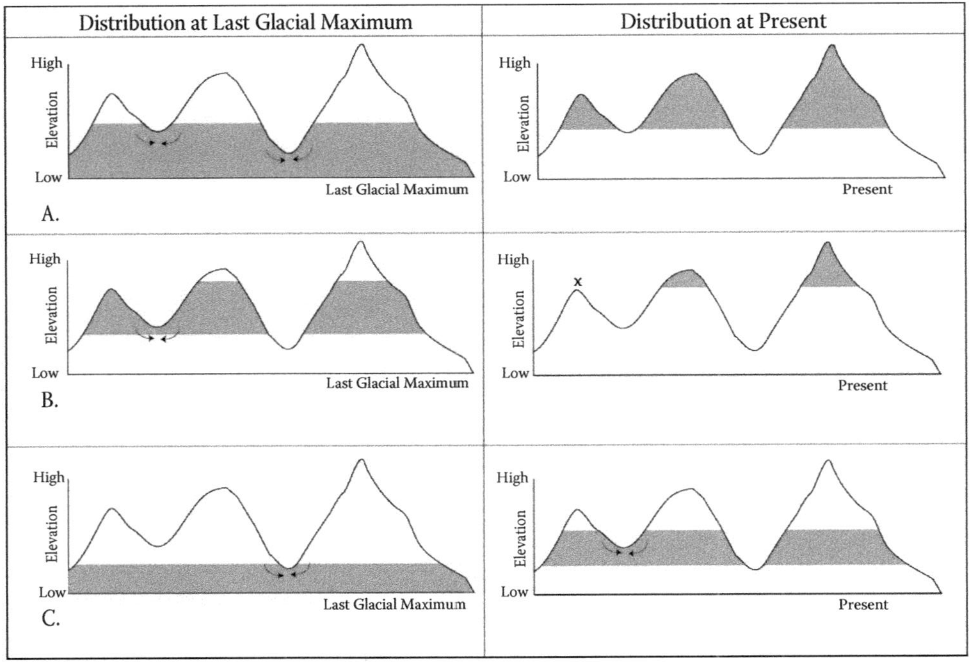

FIGURE 5.1 Distributions along mountaintops for three different mountain-dwelling taxa, labeled A, B, and C, at different time periods—the Last Glacial Maximum (LGM, left-hand panels) and the present (right-hand panels). The figure shows different possible responses to warming from the LGM to the present along elevational gradients depending on initial range position and range breadth. The taxon labeled A is widely distributed at lower elevations at the LGM and is still found across all three peaks today. However, gene flow across intervening valleys is limited as lowland habitat has disappeared. Taxon B is found at higher elevations than in the past and shows an even stronger pattern of isolation on peaks. In addition, the taxon has gone locally extinct on a low-elevation peak. Taxon C is more narrowly distributed but is found at the lowest elevation and there is still gene flow between at least two of the peaks. Arrows represent locations where gene flow across mountain peaks is possible, given suitable lowland routes.

however, mountain chains are often oriented East-West. The Alps, Pyrenees, and Carpathian mountain ranges are among the East-West–oriented chains. For such ranges, elevational temperature gradients run parallel to latitudinal gradients. This allows mammals to quickly come into equilibrium with climatic changes as they move northward and upward into cooler climes, but potentially limits the possibility for farther northward colonizations, as such range shifts would require animals to move down elevation gradients into warmer valleys between the mountain ranges.

PART 2: EVIDENCE FOR CONTEMPORARY RANGE SHIFTS IN RESPONSE TO CLIMATIC CHANGE

We focus in this chapter on the evidence for changing mammal distributions in the very recent past and present. The fossil record is essential for documenting mammalian species responses to climate changes over multiple time scales. The importance of the fossil record cannot be overestimated, but the focus here is recent evidence of response over contemporary time frames likely caused by anthropogenic warming. Blois and Hadly (2009) provide an excellent summary of the contributions of the paleontological record to our understanding of mammalian species response to climatic changes.

Contemporary mammalogists have utilized the work of their nineteenth- and twentieth-century predecessors, new surveys, and climate records from the recent past and present to gain a better understanding of mammal response to recent climate changes (e.g., those occurring over the past 100 years). The typical climatic changes and concomitant mammal distributional changes that can occur over a century may seem trivial when compared to those documented over much longer stretches of time in the paleontological record. However, the last 100 years on Earth have been anything but average, demonstrating rapid and heterogeneous changes in climate and landscape. As discussed previously, it is likely that the twenty-first century will see climate warming occurring at a faster rate than during any previous time in the past 65 million years.

Such unique climatic changes provide an ideal opportunity for studying the relative importance of the deterministic factors for mammal response to climate change highlighted previously in this chapter (e.g., species characteristics, magnitude and rate of climate change, topographic heterogeneity). Much of the research up to this point has followed in the natural progression of scientific understanding—first documenting the trends in mammalian distributional changes, and then developing more in-depth studies of the potential drivers of these observed trends. While these latter studies are under way, much of our current understanding is limited to inferences based on correlational trends between distributional and climatic shifts.

Studies of modern mammal distributional shifts due to climate change generally fall into one of two categories: (1) studies focusing on multiple species in a geographic area smaller than the species' distributional range, or (2) research focusing on a single species over a broad geographic distribution. At the heart of these studies lies a similar question: what determines the geographic range of a species, and how will a change in climate affect those range boundaries? Studies that investigate range shifts for multiple species over a relatively small geographic area have the benefit of comparing species with different natural- and life-history characteristics as they respond to a relatively homogeneous change. Such studies allow us to understand which traits may make certain species more sensitive than others to changing climatic conditions. The other approach is to focus on single species that are known to be thermally sensitive and thus likely impacted by climate change more rapidly than others. By studying single, sensitive species over a broad geographic range, we can gain a better understanding of the landscape or climatic factors determining range edges and how changes in such factors could affect other species' ranges over longer time scales. Single-species and multispecies studies have been informative and laid the foundation for gaining a mechanistic understanding of distributional shifts. We elaborate on a few key studies in the remainder of this section.

Multiple Species within a Limited Geographic Area

Multispecies mammal distribution studies often focus on either elevational or latitudinal range shifts driven by temperature change. Moritz et al. (2008) undertook an elevation-based study spanning a central portion of the Sierra Nevada mountain range of California. Revisiting the "Yosemite transect" studied by Joseph Grinnell and colleagues from 1914 to 1920, Moritz et al. assessed changes in the elevational ranges of 43 small mammals over the past century. The transect spans elevations 60- to 3300-m above sea level. This region has experienced substantial climatic change in the past century, with average minimum monthly temperature increasing 3.7°C over that period. Using occupancy modeling techniques, the authors found that most species' ranges shifted upward an average of ~500 m, but results were quite variable among species. Elevation shifts occurred more frequently for high-elevation species than those typically dwelling at low elevations; however, certain life-history traits appear to increase the probability of upward range shift for some low- to mid-elevation species. In a post-hoc analysis of 10 low- to mid-elevation species, Moritz et al. found that species with shorter life spans and larger litter sizes were more likely to exhibit upper range limit expansion. Lower limits shifted upward more frequently than upper limits, leading to range contraction for many high-elevation species. Range collapse (simultaneous increase in low limits and decrease in high limits) occurred for only two species. Despite some clear trends, there was a great deal of variation in species response, even among closely related species (e.g., those in the genus *Peromyscus*).

Overall, the results of Moritz et al. (2008) provide some of the most informative data to date on mammal elevational shifts due to climate change. The conclusions remain correlational, primarily due to the nature of this study as a one-time resurvey of the Yosemite region, rather than a continuous monitoring program of species and climate within this study area over the last century. Despite these limitations, this study has the advantage of land use stability, in that the majority of the study area is in Yosemite National Park, which has been under land use protection since before the Grinnell surveys were conducted. This scenario allows for the isolation of climate change impacts from those resulting from other, typical human disturbances such as agricultural and urban development. While most places are not similarly isolated from human land use impacts, studies such as Moritz et al.'s allow us to isolate climate impacts and develop hypotheses for further study in more complex landscapes. Their work clearly demonstrated that mammals are shifting their elevational ranges in this region, and climate change is the most probable cause of these shifts. They also highlight the importance of life history traits and species' tolerances in predicting mammal reaction to climate change. The range contraction and upward shifts seen in this study support the predictions summarized in Table 5.1.

Although the Yosemite resurvey focused on elevational shifts, multispecies studies from the northern Great Lakes region of the United States have documented rapid shifts in latitudinal range boundaries (Myers et al. 2009; Jannett et al. 2007). The northern Great Lakes region is an ideal location for studying such shifts, as it represents a transition between mammal assemblages commonly associated with southern versus northern North American habitats. Myers et al. (2009) focused their study on nine common mammal species reaching either their southern or northern limit in the area. Using museum and other collection records from 1883 to 1980 and museum and field survey results since 1980, the authors documented distributional shifts for these species. As predicted, species whose northern boundary falls in this region are experiencing northward range boundary expansion. Over the last 15 years, species such as the white-footed mouse (*Peromyscus leucopus*) and southern flying squirrel (*Glaucomys volans*) extended their northern range edge by as much as 225 km (15 km/year) based on available sampling. Although differential sampling effort can lead to biased estimates of movement rates (Shoo et al. 2006), these rates of range expansion would be fast at half the estimated speed. Also as predicted, species for which the northern Great Lakes region represents a southern geographic boundary are experiencing contraction of this southern range edge and are being replaced in community assemblages by southern species. These distributional changes correspond to increases in temperature in the region of 2.1°C for daily minima and 0.46°C for daily maxima since 1970, though local trends were highly heterogeneous. Although additional factors could affect these species' distributions, the authors argue that the overwhelming unidirectional nature of mammal distribution shifts, despite heterogeneity in forest management regimes and human land-use patterns, indicate a single dominant causal factor—climate change.

The work of Myers et al. (2009) and Moritz et al. (2008) shows an obvious change in mammal community compositions and range limits in northern Michigan and Yosemite National Park. These studies clearly demonstrate that species at temperate latitudes and in mountainous areas are responsive to climatic changes; what is less clear is whether any of the species used in these studies are showing overall range collapses. In the following section, we highlight studies on single species over larger spatial scales, focusing on single species that have suspected vulnerabilities to climatic changes. These studies more directly address questions about overall distributional changes.

Single Species in a Broad Geographic Area

Species with narrow physiological tolerances, such as those in alpine and arctic habitats, are frequently the focus of single-species studies due to their predicted vulnerability to changes in climate. In addition, two factors pertaining to arctic and alpine species allow us to more easily isolate climatic drivers as causal mechanisms for

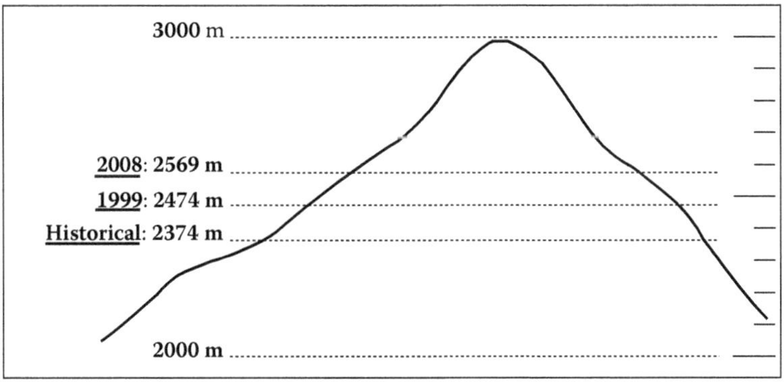

FIGURE 5.2 Upslope range retraction of the American pika (*Ochotona princeps*) on mountain ranges within the Great Basin ecoregion of the western United States (data collected by E. Beever and J. Wilkening). The average minimum elevation of 25 populations with historical records (1898–1956) was 2374 m. By 2008, nine of these 25 populations had gone extinct, and the average minimum elevation of individuals within persisting populations had risen nearly 200 m. Almost half of this rise occurred within the last decade (Beever et al. 2012).

distributional shifts. First, the natural history of many arctic and alpine mammal species has been well studied. Second, the habitats of these species are typically less disturbed by human land conversion practices compared to other areas.

One such sensitive alpine mammal is the American pika (*Ochotona princeps*), a species known to avoid even normal summer temperatures by retreating into crevices among the rocks of its talus habitat (Smith 1974). This species' specificity for alpine talus is perhaps driven by its low vagility and thermal sensitivity (Smith and Weston 1990). In their study of *O. princeps* in the Great Basin region of the United States, Beever et al. (2003) found elevation (serving as a proxy for climate) to be one of the best predictors of local population extirpations. Six of 25 populations documented before 1990 were found locally extinct in surveys from 1994 to 1999. The lower elevational limit of extant populations has risen dramatically over the last century, and especially during the last decade (Figure 5.2). The most important predictor of population extinction was the maximum elevation of talus near each site—potentially a local refuge or source of rescue through recolonization. Extinct populations tended to lack this elevational refuge, and also experienced 19.6% less annual precipitation and 7.7–10.2% higher mean daily maximum temperatures during summer (June 1–August 31), based on PRISM (Daly et al. 1994) values during 1961–1990. In addition, extinct populations were generally located on smaller habitat patches, within grazed allotments, and closer to access by road.

The correlational nature of recent climate change research is exemplified in this admittedly "exploratory" study (Beever et al. 2003). While elevation may be a reasonable proxy for climatic variables, it is less clear that grazing and distance to roads are appropriately orthogonal proxies for anthropogenic drivers of local extinction. Because roads and grazing are more common at lower elevations, the predictive ability of these variables may reflect effects of elevation rather than—or in addition to—effects of human activity.

Subsequent studies of this system reflect a more mechanistic approach. To explain patterns of loss, Beever et al. (2010) posited alternative mechanisms of thermal stress that might be experienced by pikas in this region, including chronic heat-stress (e.g., high average summer temperature), acute heat-stress (e.g., a daily maximum above 28°C), and acute cold-stress (e.g., a daily minimum below –5°C). To estimate the thermal stress experienced by pikas in the 25 locations discussed previously, Beever et al. used thermal sensors to record temperatures within the crevices of the talus. Sensors recorded temperatures every two to four hours during 2005–2006. These data were used not only to characterize recent temperatures, but also to model temperatures within each site over the past six decades. These hindcasts were developed by regressing data from the in-situ sensors on data from nearby weather stations within the Historical Climate Network. Hindcasts for each site were divided into two periods, 0–30ybp and 30–60ybp, in order to characterize both the prevailing climate (e.g., summer temperature averaged over both periods) and climate change (e.g., the difference between periods in average summer temperature). In all, Beever et al. derived 12 potential predictors of extinction: four metrics of thermal stress, each calculated from data on recent climate, prevailing climate, and climate change. Of these 12 variables, the best predictor was chronic heat-stress (the average of daily maximum and minimum temperatures during the summer months). But the potential for effects of cold stress were also apparent in this analysis: the number of daily minima below –5°C was a better predictor of extinction than the number of daily minima below 0°C, suggesting increasingly detrimental effects of extreme cold. In a post-hoc analysis to investigate this trend, Beever et al. found that the number of daily minima below –10°C was by far the best predictor of local extinction (Figure 5.3).

Ironically, this cold-adapted species may be stressed most by cold temperatures, at least within the Great Basin. The rationale for testing a cold-stress hypothesis was twofold: winter snowpack has been in decline throughout the western United States (Mote et al. 2005), and there is some evidence that pika mortality is higher where winters are colder and snowpacks are thinner (Smith 1978; for similar evidence related to other ochotonids, see Smith et al. 2004; Morrison and Hik 2007). Beever et al. hypothesized that pikas might die of exposure to extremely low temperatures without the thermal insulation provided by sufficient snow cover, especially if snowpacks are in decline due to global warming. Although they did not find strong evidence for effects of climate change, their evidence for effects of both chronic heat-stress and acute cold-stress suggest that future climate change will impact pikas if summer temperatures continue to rise and snowpacks continue to thin. It is also important to keep in mind that any apparent effects of temperature may be indirect (e.g., allowing the emergence of a disease) rather than directly causing thermal stress (Beever et al. 2010). A better understanding of the relationship between climate and the distribution of the American pika awaits a synthesis of related research across the species' range (Millar and Westfall 2010).

As with broadly "alpine" species, tracking distributional change for sensitive "arctic" species is also complex, involving unique species traits and habitat features in that region. In the Arctic, range limits are largely dictated by bioenergetics (Humphries et al. 2004). Relatively few species can survive year-round in extreme northern climates, and those that do so have adaptations allowing for energy storage

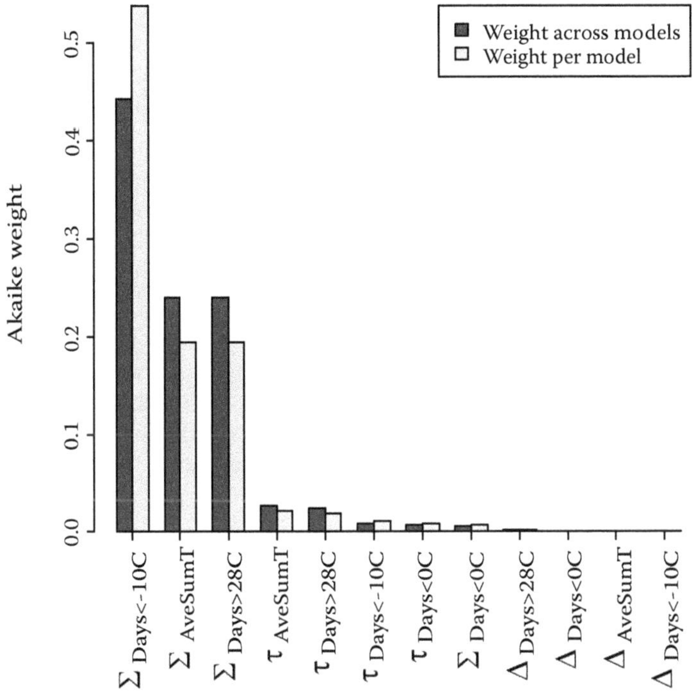

FIGURE 5.3 Relative support for each of 12 potential predictors of pika extinction within the system described in Figure 5.2 (Beever et al., in press). Relative support is shown in the form of rescaled Akaike weights, derived from an information-theoretic analysis (Burnham and Anderson 2002). Predictor descriptions: **Σ** represents cumulative or average metrics calculated from the prevailing climate, 0–60ybp; **Δ** represents changes in each metric over two periods, 30–60ybp versus 0–30ybp; **τ** represents recent climate metrics calculated from thermal sensors *in situ* during 2005–2006; AveSumT = average summer temperature (average of the daily minima and maxima for June 1–August 31), a potential metric of chronic heat-stress; Days < –10°C, 0°C = number of daily minima below –10°C and 0°C, potential metrics of acute cold-stress; Days > 28°C = number of daily maxima above 28°C, a potential metric of acute heat-stress. Because the number (n) of models fitted varied across predictors ($4 \le n \le 27$), raw weights (dark bars) are contrasted with weights per model fitted (light bars). Weights were rescaled to sum to 1.0 (Beever et al. 2010).

and survival over extended fasting periods (Humphries et al. 2004). These adaptations have allowed arctic species to exploit a largely uninhabitable region of the world, and yet these same traits may put species at risk as arctic climates change.

For polar bears (*Ursus maritimus*) and other arctic mammals, survival is tightly linked to annual temperature fluctuations and subsequent effects on sea-ice stability (Laidre et al. 2008; Stirling et al. 1999; Derocher et al. 2004). Sea ice serves a critical role in the annual energetic balance of polar bears, whose primary food source is seal pups hunted along the ice-sea edge (DeMaster and Stirling 1981). This hunting occurs in spring and continues until sea ice melts (DeMaster and Stirling 1981). The energy stores accumulated must sustain individuals throughout the summer fasting period (DeMaster and Stirling 1981).

Polar bear summer fasting periods are extending in length due to climatic changes. Data from monitoring programs in Canada show a trend toward earlier breakup of sea ice in the last several decades of the twentieth century (Stirling et al. 1999; Stirling et al. 2004), in correlation with increasing spring air temperature (Skinner et al. 1998). Earlier sea-ice breakup leads to a longer summer fasting period for polar bears and is also thought to cause increased bear movement as individuals track sea-ice habitat and optimal hunting grounds (Stirling et al. 1999; Derocher et al. 2004). These increases in energy use have led to documented decreases in individual fitness and reproductive success within Hudson Bay populations (Stirling et al. 1999; Stirling and Parkinson 2006). Despite indications of population decline, increased movement due to the need for habitat and hunting sites has led to some local reports of increases in polar bear range (Stirling and Parkinson 2006). Studies tracking the annual movement of individuals, however, indicate that increased movement does not lead to a net range shift, as bears demonstrate high summer site fidelity (Stirling et al. 2004). If the range limits of *U. maritimus* continue to remain in place, the species will be forced to adapt in situ to different habitat or hunting styles, or will go locally extinct.

Our case studies of pikas in the alpine and polar bears in the Arctic help highlight the complex nature of climatic changes and their cascading effects on species population biology. It is likely that warming average temperatures directly impact neither pikas nor polar bears. Instead, different climatic drivers, such as exposure to extreme cold events in the case of pikas, or retreating sea ice affecting hunting distances and the bioenergetics of polar bears, are likely causing population declines. Although pikas are likely losing habitat over a large portion of their range (Figure 5.2), so far polar bears are maintaining their original range while declining in abundance.

PART 3: MOVING TOWARD A MECHANISTIC UNDERSTANDING

O. princeps and *U. maritimus* were picked as case studies because they are among a handful of mammals for which we have information on how climate directly or indirectly affects distribution and diversity. For the vast majority of other mammal groups, such detailed information linking climate, population biology, thermal biology, spatial ecology, and how these factors interact is lacking. Because of this, our predictions about species responses to future accelerating climatic changes are likely to be coarse at best and wrong at worst. We now turn to methods that begin to close the gap between simple correlation studies that provide relatively weak inferences and more process-oriented approaches that provide stronger inferences. To do so requires replicated observations that include appropriate climatic measurements and population monitoring over a large set of mammal species.

We discuss four improvements that will provide the kinds of data that are necessary for stronger inferences relating climatic changes to latitudinal and elevational species responses: (1) a focus on long-term ecological monitoring projects associated with high-resolution climatic data; (2) better capture of all relevant biological data, including samples that can reveal a species' physiological response to climate; (3) use of new methods in statistical modeling to infer which factors are most important for determining species response; (4) use of better methods for sharing legacy and new data generated from such projects. We discuss each of these improvements in turn.

Long Term Ecological Monitoring Projects Capturing Appropriate Climatic Data

The first major step toward a mechanistic understanding of how climate change affects mammalian range shifts is to properly study these two processes simultaneously. Today's scientific community is well equipped to take such a step. In the past, the climatic portion of this research has been limited by the location and reliability of weather stations. Stations accepted as components of the Historical Climate Network (http://cdiac.ornl.gov/epubs/ndp/ushcn/ushcn.html) provide greater confidence in the reliability of weather data, but these weather stations do not cover the landscape to the degree needed to fully understand climatic links with mammalian range shifts. Modern technology and modeling techniques are closing these gaps.

Discontinuities in weather data can be managed via extrapolation or interpolation through modeling (e.g., the PRISM project, http://www.prism.oregonstate.edu/), the deployment of automated data loggers, or a combination of these techniques. Although strict extrapolation is a cost-effective and valid option, any modeling procedure without proper ground-truthing can be problematic. Many mammals are dependent on microclimates that occur at too fine a scale to be detected or properly modeled by these techniques. Deploying data loggers to measure local climate increases the reliability and sensitivity of climatic measurements; however, the increased cost and labor associated with this technique limits its applicability. The third of these options, combining information from data loggers and weather stations to model climatic patterns, may be the best option. This technique essentially provides the ground-truthing needed to improve extrapolation models, while not requiring full coverage of the landscape in data loggers. As demonstrated in Beever et al. 2010, these techniques can also be used to increase the temporal extent as well as the spatial resolution of historical climate data.

Better Capture of All Relevant Biological Field Data

In order to take full advantage of improved climatic data, we must monitor populations with equal thoroughness. As we argue below, Long Term Ecological Research (LTER) sites are essential but not sufficient to carry out monitoring of mammal responses to climatic changes. While LTER sites are increasing in number and many have compiled mammal population data for several decades, the temporal depth and spatial extent of these datasets varies among sites. Generally, range edges do not coincide with LTER sites. Thus, further monitoring at appropriate range boundaries should be initiated to properly document range shifts. Due to the cyclical nature of many (particularly small) mammal population dynamics, demographic monitoring must be consistent and long term. In addition, to truly understand range-shift dynamics and correctly diagnose range contractions and expansions, monitoring across the full geographic range of focal species is critical. Such range-wide monitoring will require careful selection of key populations and habitats. In the best-case scenario, the number and distribution of sites surveyed will be adequate to support statistical analyses intended to determine the relative importance of hypothesized site occupancy covariates.

While demography should be the focus of such monitoring efforts, mechanistic understanding of climatic impacts on mammals requires additional information. Although close monitoring of range and climatic shifts allows for greater confidence in perceived trends, such relationships would remain correlational. The second critical step in understanding the mechanisms driving range shifts is to monitor the factors listed in Table 5.1. These factors, or subsets thereof, can be considered as hypotheses for the mechanisms driving range shifts. While some of these factors, such as physiological response to temperature, may be direct, many of the hypothesized mechanisms for climate-driven range shifts are indirect. As the polar bear example demonstrates, indirect effects may be more important than simple physical stress, particularly for homeothermic species such as mammals. As a result, ecological and physiological indicators must be selected based on the natural history of the mammal in question. Indicators will vary with species natural history, but are likely to be factors such as individual fitness, changes in community structure, and landscape change. Ideally, individual and population health should be directly measured through the collection of appropriate samples for physiological and genetic analyses. Such direct physical data could vastly improve our ability to determine the causes of mammalian range shifts.

Using State-of-the-Art Approaches to Inference

Once the proper data are collected, the next step toward proper inference is data analysis. Inferring the cause(s) of range-shift is largely a post-hoc business, poorly suited to controlled experiments and hypothesis testing. Although we may be able to determine the physiological limits of a species through manipulation, it may be difficult to perform controlled experiments *in situ* and in a way that allows for natural behaviors that may mitigate effects of climate change. In most cases, we must infer causality from observational data—observations of species response and potential drivers. Under the traditional hypothesis-testing approach, we might hypothesize that the distribution of a species is affected by climate, and collect data on species occurrence and climate to test against the null hypothesis that there is no effect of climate on the species' distribution. Given a significant relationship between our metrics of climate and range-shift, we could reject the null and accept our alternative hypothesis. But this approach to ecological study has been outdated for at least a decade. At least since the publication of *The Ecological Detective* (Hilborn and Mangel 1997), ecologists have been privy to more direct and useful approaches to inference. These approaches allow us to ignore less rational (null) hypotheses and to actually calculate the relative support for any number of rational hypotheses.

Currently, the preferred approach to inferring causation from ecological observations involves developing a set of rational and a priori hypotheses, formulating each hypothesis as a mathematical model, and using data to estimate the relative support for each model (for recent reviews and tutorials, see Burnham and Anderson 2002; Hobbs and Hilborn 2006; Clark 2007; Bolker 2008; and Anderson 2008). By developing hypotheses prior to data analysis, we avoid the temptation to dredge for patterns in the data, an approach that is likely to yield spurious relationships. By formulating hypotheses as mathematical models we define each hypothesis precisely and generate quantitative predictions that can be compared with observations. We can estimate the relative support for different models in the way that we would estimate support for a null hypothesis, by calculating the probability of the data given each model. This is called the likelihood approach, because the probability of the data given a model is defined as the "likelihood" of the model given the data. The likelihood is proportional to the full probability of the model given the data, which is what we would really like to know. But by comparing likelihoods, we compare the relative probabilities of alternative models, which is often sufficient.

The Bayesian approach goes one step further by calculating the full probability of each model given the data—calculations that were prohibitively difficult for all but the simplest models prior to the development of modern computers. The Bayesian approach is also particularly useful for building complex mechanistic models, for including multiple types of evidence in support of alternative hypotheses, and for estimating the probability of various types of evidence encountered at various points within a process.

Given a set of alternative models, we can draw on either likelihood or Bayesian approaches to calculate an "information criterion" for each model, a statistic that balances model fit against the number of parameters in the model. For a given fit, models with more parameters offer less information per parameter, and thus receive a penalty for over-fitting. Calculating an information criterion for each model provides a basis for model selection (determining the "best" model) as well as model averaging (developing predictions based on a weighted average of models with relatively high support).

In their research on the American pika, Beever et al. (2010) used an information criterion to compare support for different metrics of physiological stress. This approach, driven by a priori hypotheses and supported by new data, allowed a formal ranking of individual predictors of range retraction. Each predictor was ranked according to a weighted average of the information in all models based on that predictor. While not a panacea, this "information-theoretic" approach (Burnham and Anderson 2002) provides a relatively robust alternative to common methods of inference in a field abounding with observational data.

Better Methods for Sharing Legacy and New Data

A theme that clearly emerges in this chapter is that we still know very little about mammalian response to climate change. A key deficit is the limited availability of high-quality data on past and present climates and species occurrence. Even for well-studied organisms such as pikas in well-studied areas like western North America, accumulating legacy occurrence data useful for studies in global change biology has proven, based on our experiences, difficult. The task of collating raw occurrences from natural history collections databases has improved with the advent of aggregating mechanisms such as the Mammal Information Network System (MaNIS, http://manisnet.org; Stein and Wieczorek 2004) and Global Biodiversity Information Facility (http://data.gbif.org), but this still leaves the onerous task of vetting data records to determine those with precise dates and geographic locations. This has proven to require intensive effort. In addition, the vast majority of results from biotic surveys remain locked away in nondigital format. Many of these studies represent systematically collected data that would be incredibly valuable if they were digitized and made available to the research community as baselines against which to measure change. Such surveys may provide especially valuable information about species absences.

There is also a strong need to develop mechanisms that increase the quality, and help determine fitness of use, of legacy biodiversity data. One such exemplary tool is Biogeomancer (http://biogeomancer.org), which provides a means to convert textual descriptions of localities (e.g., 5 miles southwest of Boulder, Colorado) into computer-readable geographic coordinates along with a measurement of geographic uncertainty (e.g., latitude 39.9637418, longitude –105.3371878, radius of geographic uncertainty of 7191 m for text description above). Geographic uncertainty measurements are fundamental, providing a standardized assessment of the precision of legacy locality records. Those records with high precision are much more useful for local and regional biodiversity studies, while low-precision records (e.g., those referenced to county or even state) may only be useful for continental or global-scale analyses.

Equally important will be quickly publishing not only research results from field-based studies, but also the raw data that have been captured. All observational data should be rapidly and freely made available to the broader community so others may use those data for further analyses not envisioned by the initial data collectors. We argue for the importance of global repositories for such observational data. The quality of these data, in terms of sampling methodology and precision, continues to increase. We must make available not only our observational data but all quantitative data coming in from climate data loggers, along with quantitative phenotypic or phenological data as such information becomes available. In order for this endeavor to be successful, there needs to be a shift in academic culture concerning the value of data curation as balanced against competing needs such as publishable research results.

SUMMARY

Biologists continue to document that mammal species distributions respond to climatic changes in directions predicted by thermal biology. However, the many correlational studies on this topic do not facilitate an assessment of the causal links between changes in climate and changes in mammal distributions. The goal of this chapter was to first summarize existing theory and data concerning how terrestrial mammal species respond to climatic change and then provide a roadmap for more process-oriented studies of species distributional responses. In particular, we stress the need to better document the actual factors to which species respond. Species are less likely to respond to "warming" than to shifts in the timing and magnitude of changes in temperature and precipitation. The chapter was organized into three main parts. First, we discussed four main factors that are thought to determine the rate and magnitude of distributional responses. Those factors are (1) the dispersal capability and habitat specializations of the mammal in question; (2) physiological tolerances to climatic conditions; (3) the rate and magnitude of climatic changes; and (4) the topographic heterogeneity within a species range. Focusing on these factors, we then developed some simple predictions regarding the rate and magnitude of response, with particular emphasis on response along both latitudinal and altitudinal gradients. In the second part of the chapter, we drew from studies documenting contemporary range shifts that show that mammal species distributions do respond to climatic changes as predicted. However, we still cannot infer mechanism(s) from most of the available, correlational studies. That is, we do not know which climatic parameters are driving the distributional shift for any of the species that have been studied. In the final part of the chapter, we outlined methods for capturing the necessary data and, utilizing the necessary new techniques, to begin moving toward more process-oriented and mechanistic studies in which the relative importance of key factors can be quantified.

REFERENCES

Anderson, D.R. 2008. *Model-Based Inference in the Life Sciences*. Springer, New York.

Barnosky, A.D., E.A. Hadly, and C.J. Bell. 2003. Mammalian response to global warming on varied temporal scales. *Journal of Mammalogy* 84: 354–368.

Beever, E.A., P.F. Brussard, and J. Berger. 2003. Patterns of apparent extirpation among isolated populations of pikas (Ochotona princeps) in the Great Basin. *Journal of Mammalogy* 84: 37–54.

Beever, E.A., C. Ray, P.W. Mote, and J.L. Wilkening. 2010. Testing alternative models of climaate-mediated exterpations. *Ecological Applications* 20: 164–178.

Blois, J.L. and E.A. Hadly. 2009. Mammalian response to Cenozoic climatic change. *Annual Review of Earth and Planetary Sciences*. 37: 8.1–8.28.

Bolker, B.M. 2008. *Ecological Models and Data in R*. Princeton University Press, Princeton, New Jersey.

Burnham, K.P., and D.R. Anderson. 2002. *Model Selection and Inference: A Practical Information-Theoretic Approach*. 2nd edition. Springer-Verlag, New York.

Clark, J.S. 2007. *Models for Ecological Data: An Introduction*. Princeton University Press, New Jersey.

Dalén, L., et al. 2007. Ancient DNA reveals lack of postglacial habitat tracking in the arctic fox. *Procedures of the National Academy of Science USA* 104: 6726–6729.

Daly, C., R.P. Neilson, and D.L. Philips. 1994. A statistical–topographic model for mapping climatological precipitation over mountainous terrain.

DeMaster, D.P., and I. Stirling. 1981. *Ursus maritimus. Mammalian Species* 145: 1–7.

Derocher, A.E., N.J. Lunn, and I. Stirling. 2004. Polar bears in a warming climate. *Integrative and Comparative Biology* 44: 163–176.

Ditto, A.M., and J.K. Frey. 2007. Effects of ecogeographic variables on genetic variation in montane mammals: Implications for conservation in a global warming scenario. *Journal of Biogeography* 34: 1136–1149.

Easterling, D.R., G.A. Meehl, C. Parmesan, S.A. Changnon, T.R. Karl, and L.O. Mearns. 2000. Climate extremes: Observations, modeling, and impacts. *Science* 289: 2068–2074.

Etterson, J.R., and R.G. Shaw. 2001. Constraint to adaptive evolution in response to global warming. *Science* 294: 151–154.

Guralnick, R.P. 2006a. Do flatland and mountain dwelling species show different structuring of their experienced environment over latitude? A western versus central-eastern North America rodent multispecies comparison. *Diversity and Distributions* 12: 731–741.

———. 2006b. The legacy of past climate and landscape change on species' current experienced climate and elevation ranges across latitude: A multispecies study utilizing mammals in western North America. *Global Ecology and Biogeography* 15: 505–518.

Guralnick, R.P., A.W. Hill, and M. Lane. 2007. Toward a collaborative, global infrastructure for biodiversity assessment. *Ecology Letters* 10: 663–672.

Hilborn, R., and M. Mangel. 1997. *The Ecological Detective: Confronting Models with Data.* Princeton University Press, New Jersey.

Hobbs, N.T., and R. Hilborn. 2006. Alternatives to statistical hypothesis testing in ecology: A guide to self-teaching. *Ecological Applications* 16: 5–19.

Houghton, J.T., Y. Ding, D.J. Griggs, M. Noguer, P.J. van der Linden, X. Dai, K. Maskell, and C.A. Johnson (Eds.). 2001. *Climate Change 2001: The Scientific Basis.* Cambridge University Press, United Kingdom.

Humphries, M.M., J. Umbanhowar, and K.S. McCann. 2004. Bioenergetic prediction of climate change impacts on northern mammals. *Integrative and Comparative Biology* 44: 152–162.

Jannett, F.J., M.R. Broschart, L.H. Grim, and J.P. Schaberl. 2007. Northerly range extensions of mammalian species in Minnesota. *American Midland Naturalist* 158: 168–176.

Jump, A.S., and J. Peñuelas. 2005. Running to stand still: Adaptation and the response of plants to rapid climate change. *Ecology Letters* 8: 1010–1020.

Laidre, K.L., I. Stirling, L.F. Lowry, O. Wiig, M.P. Heide-Jorgensen, and S.H. Ferguson. 2008. Quantifying the sensitivity of arctic mammals to climate-induced habitat change. *Ecological Applications* 18: S97–S125.

Millar, C.L. and R.D. Westfall. 2010. Distribution and climatic relationships of the American pica. (Ochotona princeps) in the Sierra Nevada and Great Basin, USA; Periglacial landforms as refugia in warming climates. *Arctic, Antarctic, and Alpine Research* 42: 76:88.

Moritz, C., J.L. Patton, C.J. Conroy, J.L. Parra, G.C. White, and S.R. Beissinger. 2008. Impact of a century of climate change on small-mammal communities in Yosemite National Park, USA. *Science* 322: 261–264.

Morrison, S., and D. Hik. 2007. Demographic analysis of a declining pika *Ochotona collaris* population: Linking survival to broad-scale climate patterns via spring snowmelt patterns. *Journal of Animal Ecology* 76: 899–907.

Mote, P.W., A.F. Hamlet, M.P. Clark, and D.P. Lettenmaier. 2005. Declining mountain snowpack in western North America. *Bulletin of the American Meteorological Society* 86: 39–49.

Myers, P., B.L. Lundrigan, S.M.G. Hoffman, A.P. Haraminac, and S.H. Seto. 2009. Climate-induced changes in small mammal communities of the Northern Great Lakes Region. *Global Change Biology* 15: 1434–1454.

Parmesan, C. 2006. Ecological and evolutionary responses to recent climate change. *Annual Review of Ecology, Evolution and Systematics 37*: 637–669.

Peterson, A.T. 2003. Projected climate change effects on Rocky Mountain and Great Plains birds: Generalities of biodiversity consequences. *Global Change Biology* 9: 647–655.

Ray, C. 2001. Maintaining genetic diversity despite local extinctions: Effects of population scale. *Biological Conservation* 100: 3–14.

Root, T.L., J.T. Price, K.R. Hall, S.H. Schneider, C. Rosenzweig, and J.A. Pounds. 2003. Fingerprints of global warming on wild animals and plants. *Nature* 421: 57–60.

Sekercioglu, C.H., S.H. Schneider, J.P. Fay, and S.R. Loarie. 2008. Climate change, elevational range shifts, and bird extinctions. *Conservation Biology* 22: 140–150.

Shoo, L.P., S.E. Williams, and J.-M. Hero. 2006. Detecting climate change induced range shifts: Where and how should we be looking? *Austral Ecology* 31: 22–29.

Skinner, W.R., R.L. Jefferies, T.J. Carleton, R.F. Rockwell, and K.F. Abraham. 1998. Prediction of reproductive success and failure in lesser snow geese based on early season climatic variables. *Global Change Biology* 4: 3–16.

Smith, A.T. 1974. The distribution and dispersal of pikas: Influences of behavior and climate. *Ecology* 55: 1368–1376.

———. 1978. Comparative demography of pikas (*Ochotona*): Effect of spatial and temporal age-specific mortality. *Ecology* 59: 133–139.

Smith, A.T., W. Li, and D.S. Hik. 2004. Pikas as harbingers of global warming. *Species* 41: 4–5.

Smith, A.T., and M.L. Weston. 1990. *Ochotona princeps. Mammalian Species* 352: 1–8.

Stein, B.R., and J. Wieczorek. 2004. Mammals of the world: MaNIS as an example of data integration in a distributed network environment. *Biodiversity Informatics* 1: 14–22.

Stirling, I., N.J. Lunn, and J. Iacozza. 1999. Long-term trends in the population ecology of polar bears in western Hudson Bay in relation to climatic change. *Arctic* 52: 294–306.

Stirling, I., N.J. Lunn, J. Iacozza, C. Elliott, and M. Obbard. 2004. Polar bear distribution and abundance on the southwestern Hudson Bay coast during open water season, in relation to population trends in annual ice patterns. *Arctic* 57: 15–26.

Stirling, I., and C.L. Parkinson. 2006. Possible effects of climate warming on selected populations of polar bears (*Ursus maritimus*) in the Canadian Arctic. *Arctic* 59: 261–275.

Walther, G.-R., et al. 2002. Ecological responses to recent climate change. *Nature* 416: 389–395.

Warren, M.S., et al. 2001. Rapid responses of British butterflies to opposing forces of climate and habitat change. *Nature* 414: 65–69.

Section III

Higher-Level Ecological Relationships to Climate Change

6 Effects of Climate Change on the Elevational Limits of Species Ranges

Robert J. Wilson and David Gutiérrez

CONTENTS

BACKGROUND

Elevational patterns in species distributions and diversity represent promising sources of evidence for the ecological effects of climate change. Elevational patterns in diversity are sometimes argued to be a consequence of climatic limits to individual species distributions, and considerable empirical data have been collected to inform this debate (Rahbek 1995; Lomolino 2001; McCain 2005). Apart from this wealth of theory and baseline evidence, there are also strong arguments why species should be expected to shift their distributions more quickly across elevational rather than latitudinal gradients as the climate warms. First, temperature and other climate variables change over much shorter geographical distances with elevation than with latitude, and indeed in the Tropics there are only weak latitudinal climate gradients (Colwell et al. 2008), so that species do not need to shift elevations as far as latitudes to remain in suitable climate space. Second, human impact is generally less pronounced in mountain environments than at lower elevations (Ellis and Ramankutty 2008; Nogués-Bravo et al. 2008). As a result, species should be able to track the distribution of suitable climates more readily by colonizing natural habitats in mountains rather than by moving across lowlands dominated by intensive human land use, where the distributions of many species are failing to keep up with climate change (Warren et al. 2001; Menéndez et al. 2006).

Indeed, some of the first and most widely cited examples of ecological responses to climate change refer to elevational changes in species distributions. Parmesan (1996, 2005) showed that the average elevation of populations of the butterfly *Euphydryas editha* in western North America shifted upward by 124 m during the twentieth century (the average latitude also shifted north by 92 km), as isotherms shifted uphill by 105 m. Pounds et al. (1999) recorded increasing domination of the bird and herptile fauna of Monteverde National Park in Costa Rica by species associated with lower elevations, consistent with increasingly dry local conditions between the 1980s and 1998. Grabherr et al. (1994) found that plant species richness had increased on mountaintops in the Alps, with estimated rates of elevational expansion of <1–4 m per decade, compared with 8–10 m decadal upward shifts in isotherms. In another example where plant distributions did not appear to be keeping pace with climate change, Wardle and Coleman (1992) recorded uphill shifts in the treeline of 5–30 m during the twentieth century in New Zealand. Meta-analysis revealed an average observed uphill shift in distributions of 6.1 m per decade (Parmesan and Yohe 2003), consistent with an effect of climate warming, but likely not keeping pace with rates of climate change (the twentieth century increase in terrestrial surface temperature of 0.6–0.7°C would represent an upward shift in isotherms of *c.* 10 m per decade, for a 6°C/km lapse rate). Lapse rates in the tropics, especially humid locations, are lower, which would translate to higher range shifts to compensate for changes in temperature (Colwell et al. 2008; Sekercioglu et al. 2008). In a review of ecological responses to recent climate change, Parmesan (2006) noted that uphill range shifts had been rather poorly documented compared with latitudinal shifts. The majority of documented range shifts also referred to poleward or uphill *expansions*, rather than to range *contractions* at trailing edges (Parmesan and Yohe 2003; Rosenzweig et al. 2007).

Why so few early examples reported either elevation range shifts or lower latitude/elevation contractions appears to reflect partly the fact that few researchers were looking for them, and partly that the data used were unable to detect these kinds of change (Parmesan 2006; Thomas et al. 2006). Two general approaches have been used to gather evidence for elevational shifts. Atlas distribution data have been compared between different time periods, and elevational gradients with prior information on species distributions have been resampled systematically. Both types of comparison present challenges in ensuring that changes to sampling effort or regime do not bias results (see the following; see also Rowe 2005 for museum specimens, Bergamini et al. 2009 for herbarium specimens, and Tingley and Beissinger 2009 for a review). One problem particularly with atlas data is that one presence in a new grid square or region, or on a new mountain summit, is sufficient to detect a range expansion, whereas contracting species may decline gradually to patchy refugia (Wilson et al. 2004), so that it is difficult to document unequivocally the extinction of a species from a region. As a result, few such declines could be documented using coarse-scale atlas data. In fact, elevation range shifts may have helped to obscure range contractions at species' lower latitudinal margins, if species shifted their distributions uphill rather than to higher latitudes, and such changes to distributions would again not be detected by coarse-scale data (Hill et al. 2002). Furthermore, attributing observed declines to climate change is difficult in the many landscapes where intensive human land use represents an additional threat to species, particularly at lower elevations (Thomas et al. 2006).

In the last five years, however, an increasing number of uphill range shifts have been identified, often driven by local extinctions at the downslope limits of species distributions, and at faster rates than suggested by earlier meta-analyses. Here, we describe the approaches that have been used to detect such range shifts, focusing on evidence from the apparent breeding ranges of species, rather than temporary migratory movements. Thus, we interpret the range shifts as changes to the distribution of populations of species, reflecting local extinctions in zones of range contraction and colonizations in zones of range expansion.

REPORTED ELEVATIONAL SHIFTS IN SPECIES DISTRIBUTIONS

Evidence now exists that plants, invertebrates, and ectothermic and endothermic vertebrates have shifted to higher elevations associated with recent climate warming in tropical, temperate, and polar latitudes (Tables 6.1 and 6.2). Four types of evidence have been presented for uphill range shifts.

1. *Average elevation of species distributions.* The average elevation of a species' range could shift upward because of extinctions at low elevations or colonizations at high elevations, either at the margins of or within the species' range. Different types of data lend themselves to different methods for estimating average elevation. Studies using atlas data have calculated the mean elevation of grid squares with distribution records in repeated surveys (Hill et al. 2002; Konvicka et al. 2003; Hickling et al. 2006). Repeat sampling along the same elevation gradients can be used to calculate the midpoint of a species' range (Raxworthy et al. 2008), although such an approach is not sensitive to changes in abundance or occupancy between upper and lower limits (Shoo et al. 2005a, 2005b). Alternatively, the mean elevation weighted by the abundance or cover of species may be an appropriate technique for discrete sampling elevations, even when relatively few (5–20) locations have been sampled (Kelly and Goulden 2008; Chen et al. 2009). When many sites are sampled, optimum elevations with the highest probability of occurrence for species can be modelled based on presence and absence records; such models can be applied to repeat surveys in the same region (Wilson et al. 2005) or to national-scale site-based data sliced into discrete time periods (Lenoir et al. 2008).

   Reported uphill shifts in the average or optimum elevations of species distributions are rather consistent among taxa and regions. European studies based on atlas data report absolute uphill shifts (averaged across species) of 20–60 m over the 25–30 years at the end of the twentieth century, using data for butterflies (Hill et al. 2002; Konvicka et al. 2003) and a wide range of ectothermic groups (arthropods and fish; Hickling et al. 2006).

TABLE 6.1
Community-Level Evidence for Upward Shifts in Elevational Distributions

		Reported Change[a]				
Taxon	**Location (Lat°)**	**Ave**	**Low**	**Upp**	**Comp**	**References**
Alpine plants	Italy, Switzerland, Austria (46–47N)			+	+	Grabherr et al. 1994; Keller et al. 2000; Pauli et al. 2001, 2007; Walther et al. 2005; Parolo and Rossi 2008; Vittoz et al. 2008
	Norway (61N)		+	+	+	Klanderud and Birks 2003
Forest plants	France (43–48N)	+				Lenoir et al. 2008
Dominant plants	Spain (41N)		+	+	+	Peñuelas and Boada 2003
	Vermont, USA (44N)		+	+	+	Beckage et al. 2008
	California, USA (33N)	+			+	Kelly and Goulden 2008
Island flora	Marion Island (46S)			+	+	Le Roux and McGeoch 2008
Bryophytes	Switzerland (47N)	+		+	+	Bergamini et al. 2009
Tree line	New Zealand (45S)			+		Wardle and Coleman 1992
	Canada (51N)			+		Luckman and Kavanagh 2000
	Sweden (62N)			+		Kullman 2001, 2002
	Siberia, Russia (64N)			+		Moiseev and Shiyatov 2003
Lepidoptera	Britain (52–56N)	+	+	+		Hill et al. 2002
	Czech Republic (49N)	+	+	+		Konvicka et al. 2003
	Spain (40N)	+	+		+	Wilson et al. 2005, 2007
	Borneo (6N)	+			+	Chen et al. 2009
Arthropods, fish	Britain (52N)	+				Hickling et al. 2006
Herptiles, birds	Costa Rica (10N)				+	Pounds et al. 1999
Herptiles	Ecuador (1S)			+		Bustamante et al. 2005
	Peru (14S)			+		Seimon et al. 2007
	Madagascar (14S)	+	(+)	(+)		Raxworthy et al. 2008
Birds	SE Asia (0–20N)		(+)	+		Peh 2007
Mammals	California (37N)		+	(+)	+	Moritz et al. 2008

[a] Reported change in elevational range: Ave, average; Low, lower limits; Upp, upper limits; Comp, species composition. Symbols shown in parentheses show nonsignificant upward shifts from multispecies analyses.

TABLE 6.2
Evidence from Single Species Studies for Upward Shifts in Elevational Distributions Linked to Climate Change

Species and Evidence for Range Shift	Location (Lat°)	References
Extinction of Low-Elevation Historical Populations		
Euphydryas editha (butterfly)	Western North America (30–53N)	Parmesan 1996, 2005
Ochotona princeps (pika)	Western USA (38–43N)	Beever et al. 2003
Ovis canadensis (sheep)	California, USA (33–38N)	Epps et al. 2004
Erebia epiphron (butterfly)	Britain (53–56N)	Franco et al. 2006
Upper-Elevation Expansion		
Thaumetopoea pityocampa (moth)	Italy (46N)	Battisti et al. 2005, 2006
Changes to Elevational Distribution/Abundance Patterns		
Salmo trutta (fish)	Switzerland (47N)	Hari et al. 2006
Chrysomela aeneicollis (beetle)	California, USA (36N)	Dahlhoff et al. 2008
Aporia crataegi (butterfly)	Spain (40N)	Merrill et al. 2008

Studies based on resampled sites or gradients show absolute uphill shifts of 62–68 m for temperate flora (Kelly and Goulden 2008; Lenoir et al. 2008), and tropical fauna (Raxworthy et al. 2008; Chen et al. 2009) over the past 10–40 years. There was an average increase in optimum elevation of 119 m for butterflies in central Spain over the same period of time (Wilson et al. 2005). Estimated uphill range shifts per decade (Figure 6.1a) are somewhat less for studies based on atlas data (7.7–20 m per decade) than for site or gradient studies (17–37 m per decade), because the average elevations of coarse-scale grid squares may underestimate elevations actually occupied by species, particularly in mountainous landscapes (Konvicka et al. 2003). Observed range shifts compare with uphill shifts in isotherms of 5–15 m per decade for the twentieth century, or of approximately 30–70 m per decade since the 1960s from regional studies.

Some studies have also begun to test for shifts in the elevational ranges of endotherms. A study in southwest France showed no significant change in breeding bird distributions between 1973–1980 and 2000–2002 (average shift = −2.4 m ±SE 8.0, n = 24 species; Archaux 2004). However, some European birds may not show marked elevational shifts in response to climate change if they are wide ranging, with relatively broad climatic tolerances. Furthermore, changes both to temperature and precipitation may mean that ranges do not shift in a predictable uphill direction. In California, four elevational transects that were sampled in 1911–1929 and again in 2003–2008 showed strong evidence that birds were shifting their ranges to track either temperature, precipitation, or both, but not necessarily uphill (Tingley et al. 2009).

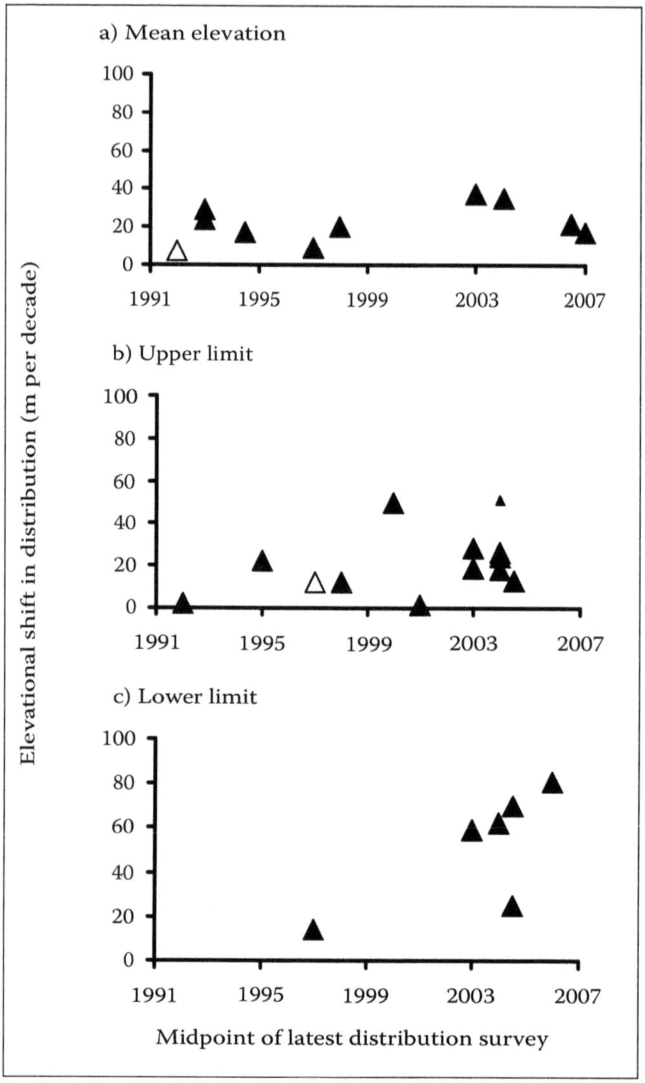

FIGURE 6.1 Rates of uphill range shifts reported in recent studies. Estimated mean uphill shifts (m per decade) are shown for (a) mean elevations, (b) upper elevations, and (c) lower elevations against the midpoint of the latest distribution survey in each study (to show how recently distribution shifts may have occurred). Large symbols show community-level studies (Table 6.1); small symbols show single species studies (Table 6.2). Open symbols show data from ~10-km resolution atlas surveys (Hill et al. 2002; Konvicka et al. 2003; Hickling et al. 2006). Closed symbols show site- or species-based repeat distribution surveys (data from Grabherr et al. 1994; Parmesan 1996, 2005; Kullman 2001, 2002; Klanderud and Birks 2003; Peñuelas and Boada 2003; Battisti et al. 2005, 2006; Walther et al. 2005; Wilson et al. 2005; Franco et al. 2006; Beckage et al. 2008; Kelly and Goulden 2008; Le Roux and McGeoch 2008; Lenoir et al. 2008; Merrill et al. 2008; Moritz et al. 2008; Parolo and Rossi 2008; Raxworthy et al. 2008; Bergamini et al. 2009; Chen et al. 2009).

2. *Upper limits of species distributions.* Atlas data suggest increases in the upper elevations occupied by butterfly species (Hill et al. 2002; Konvicka et al. 2003), although analyses using the average elevations of ≥10 km grid squares are likely to underestimate uphill expansions because of topographic variation within grid cells. Other multispecies evidence for upslope extensions by invertebrates is not forthcoming (Table 6.1). In contrast, there is abundant information that plant species are extending their distributions to higher elevations, predominantly from studies on the flora of European mountain summits (references in Table 6.1), from boundaries between dominant vegetation communities in Europe and North America (Peñuelas and Boada 2003; Beckage et al. 2008) and from the extension of tree lines in temperate and polar mountain ranges (Kullman 2001, 2002; Moiseev and Shiyatov 2003; for a comprehensive review see Harsch et al. 2009). Repeat surveys of herptile distributions along elevation gradients in the Tropics show increases in upper range limits for some species over the last 10–40 years (Bustamante et al. 2005; Raxworthy et al. 2008). In Peru, three anuran species colonized elevations of 5244–5400 m (the highest known amphibian populations in the world) following recent glacial retreat (Seimon et al. 2007). Repeat surveys of a 3000-m elevation gradient in California showed that four species of small mammal had extended their upper limits to higher elevations, but a further three species showed significant contractions at upper range margins (Moritz et al. 2008). Reported upper elevation limits for Southeast Asian birds based on historical (pre-1971) and recent (1971–1999) data suggest that 84 out of 306 species have shifted their distributions upward, by an average of 399 m (Peh 2007).

   Approximately twice as many published studies provide estimates for decadal rates of change in upper limits of species' ranges than for average or lower elevations (Figure 6.1). Reported decadal expansion rates range from 1 to 4 m (Grabherr et al. 1994; Peñuelas and Boada 2003) to *c.* 50 m (Kullman 2002; Battisti et al. 2005), with the majority in the range of 10–30 m per decade (e.g., Klanderud and Birks 2003; Walther et al. 2005; Beckage et al. 2008; Parolo and Rossi 2008; Raxworthy et al. 2008). There is a suggestion that analyses based on more recent resurveys of alpine flora indicate faster rates of change (Walther et al. 2005; Pauli et al. 2007), which might be expected based on increased rates of warming since the 1970s (Rosenzweig et al. 2007). Increasing rates of uphill expansion are not clear from published data on all taxonomic groups (Figure 6.1b), but are likely obscured by differences in data quality and the ecological characteristics of different systems. Nevertheless, the studies showing the fastest rates of uphill expansion have demonstrated the importance of very recent conditions for observed range shifts. For example, Kullman (2002) showed that in 2000, saplings above the former elevation range of seven tree and shrub species in the Swedish Scandes were aged between 7 and 12 years, corresponding to strong winter and summer warming since 1988.

3. *Lower limits of species distributions.* If climate change causes low-elevation range margins to shift upward, then the outcome is a reduction in species'

distribution sizes (Wilson et al. 2005) and an increase in extinction risk (Williams et al. 2003; Sekercioglu et al. 2008). Despite the urgency of identifying whether such changes are occurring, far fewer studies have reported uphill contractions at downslope margins than expansions at upslope limits. Part of the reason for this discrepancy is that the studies that have identified uphill advances (e.g., tree-line elevations, alpine summit flora) may be unlikely to detect species at their low-elevation limits, although vegetation studies have noted declines in abundance of high-altitude species at lower elevation sites (Klanderud and Birks 2003; Peñuelas and Boada 2003; Pauli et al. 2007; Kelly and Goulden 2008).

For plants, observed low-elevation limits may shift upward slowly where long-lived individuals survive but fail to reproduce (Peñuelas and Boada 2003; Pauli et al. 2007). For short-lived ectothermic animals, low elevation range limits may shift uphill rapidly, associated with increasing temperatures. Four British butterfly species and ten Czech butterfly species showed extinctions from low elevation atlas grid squares between the 1970s and 1990s (Hill et al. 2002; Konvicka et al. 2003). In central Spain, the low-elevation limits of 16 butterfly species shifted uphill on average by 212 m between 1967–1973 and 2004, as annual mean temperature increased by 1.3°C (equivalent to a *c.* 225-m rise in isotherms) (Wilson et al. 2005). Thirty montane herptile species in Madagascar showed an average increase in low-elevation limit of 76 m (59 m when correcting data for changes to sampling effort) between 1993 and 2003, although the majority of species showed no change or a slight decrease in sampled lower limit (Raxworthy et al. 2008).

Endotherms also show some evidence for uphill shifts in lower range limits. Of 28 small mammal species whose elevation ranges were sampled in Yosemite National Park in 1914–1920 and 2003–2006, ten showed significant increases in lower limits (mean +499 m), while two showed significant decreases (mean –744 m) (Moritz et al. 2008). However, only ten out of 306 generalist bird species in Southeast Asia showed apparent uphill shifts in low-elevation limits between two published field guides (Peh 2007). Reported range expansions by bird species at both high- and low-elevation limits in Europe appear to represent responses of overall population size to increased habitat favorability across the elevation range (Archaux 2004), possibly related to climate conditions in one case (Tryjanowski et al. 2005). Shifts in low-elevation limits are in the range of 25–81 m per decade where reported (apart from one study using atlas data: +14 m per decade; Hill et al. 2002), tallying well with isotherm shifts of approximately 30–70 m per decade since the 1960s (Figure 6.1c). However, data for decadal rates should be treated with some caution, as neither climatic change nor biotic responses are likely to be gradual in nature. For example, fieldwork for two recent European studies showing uphill shifts in butterfly distributions was conducted in 2004 and 2005 (Wilson et al. 2005; Franco et al. 2006), following the hot summer of 2003. Some species probably suffered increased rates of low elevation extinctions in 2003, rather than declining gradually from low elevations over the previous decades (C.D. Thomas, pers. comm.; R.J. Wilson and D. Gutiérrez, pers. obs.).

4. *Changes to species composition.* Uphill shifts in species' ranges imply elevational changes to species richness and composition, and many studies have identified systematic changes of this kind (Table 6.1). Measurements of the cover of different alpine plants show that species associated with low elevations are increasing, and in some cases species associated with high elevations are becoming less abundant (Keller et al. 2000; Klanderud and Birks 2003; Pauli et al. 2007; Vittoz et al. 2008). Faunal studies show declines in the distribution or abundance of species associated with high elevations, and increases or no change in species associated with low elevations (Pounds et al. 1999; Moritz et al. 2008; Rovito et al. 2009). Two studies have quantified these elevational changes in community composition. There was an uphill shift of 42 m in the mean elevational associations of moth species per sampling location on Mount Kinabalu (Borneo) between 1965 and 2007 (Chen et al. 2009). This upward shift was rather less than that reported for individual species, for two probable reasons: first, species that had colonized the sampled elevations from below by the second survey could not be included in the analysis; second, upper elevation expansions were truncated by the tops of the mountain, or by the upper elevation limits of moths' larval host plants. Butterfly community composition in central Spain, quantified using multivariate statistics, shifted uphill by 290 m between 1967–1973 and 2004–2005, as isotherms shifted upward by *c.* 225 m (Wilson et al. 2007).

IDENTIFICATION OF UPHILL RANGE SHIFTS

Studies of changes to species' ranges have employed historical baseline data that were collected for a variety of reasons, using a variety of approaches. As a result, means to account for changes to sample effort may be required (e.g., Rowe 2005; Tingley and Beissinger 2009). Where point counts or quadrats have been established, these can be repeated with the same sampling effort (e.g., Archaux 2004; Pauli et al. 2007). But in many other cases repeat surveys designed to detect current species distributions are likely to employ more rigorous sampling protocols and greater sampling effort, and sampling effort may increase most at relatively inaccessible high elevations. Recent atlas surveys may also include larger numbers of distribution records than historical data (Hill et al. 2002). Simply based on increases in sampling effort, range extensions should be expected at both high- and low-elevation limits (Shoo et al. 2006). Therefore, techniques must be employed to ensure that reported uphill range expansions are independent of sampling bias and that increased sampling effort does not obscure low-elevation contractions. Most commonly, data from more-intensive (usually recent) surveys have been subsampled so that the total number of samples from each site, region, or elevation interval are the same in both surveys (Hill et al. 2002; Wilson et al. 2005, 2007; Lenoir et al. 2008; Raxworthy et al. 2008; Chen et al. 2009). Alternatively, analyses have been restricted to abundant, widespread, or well-recorded taxa in both surveys

(Peh 2007), or to heavily recorded atlas grid squares (Hickling et al. 2006). Moritz et al. (2008) used daily trapping records for small mammals in 1914–1920 and 2003–2006 to model detectability of species (Mackenzie et al. 2006), and therefore to calculate unbiased estimates of species absence from different elevation intervals in each time period.

The mean elevation for a given species draws on information from throughout the species' range, whereas sampled upper and lower limits are based on relatively few datapoints, and may therefore be more subject to sampling error. As a result, average elevations may be more sensitive than high or low elevation boundaries to directional shifts in species distributions, and less subject to variation in sampling effort (Shoo et al. 2006). The effects on perceived range shifts of sampling effort and the measure of change (upper or lower limit, or average elevation) can be illustrated using data for butterflies in the Sierra de Guadarrama (central Spain) in 1967–1973 and 2004–2005 (Table 6.3). Here, data are analyzed as presence or absence in six 200-m elevation bands from <600 m to >1600 m, each corresponding to >7% of the landscape (Wilson et al. 2007). The total number of samples varied with elevation and was generally greater in the second survey. For those species with more than 20 individuals sampled in both time periods ($n = 67$), minimum 200-m elevation band calculated using raw data did not increase significantly between surveys (Wilcoxon test for paired samples, $Z = -1.77$, $P = 0.077$), but maximum ($Z = -3.52$, $P < 0.001$) and mean ($Z = -3.58$, $P < 0.001$) both showed significant increases. When data were resampled so that there were equal numbers of visits to each 200-m elevation band in each survey (see Wilson et al. 2007), there were significant increases in low elevation limit ($Z = -2.19$, $P = 0.029$) and mean elevation ($Z = -2.25$, $P = 0.025$), but not in upper 200-m band ($Z = -0.32$, $P = 0.75$) ($n = 66$, because one species went extinct from all 200-m bands in resampled data). Failure to resample these data would have led to the erroneous conclusion that the uphill shift in mean elevation was driven by upslope extension of species ranges, rather than by contractions at the downslope limits. Analysis at the scale of six separate 200-m intervals demonstrates the upward shift in species distributions, offering the possibility that even relatively coarse-scale data on elevational ranges can be used to detect uphill range shifts.

ATTRIBUTION OF RANGE SHIFTS TO EFFECTS OF CLIMATE CHANGE

Once range shifts in the direction predicted by climate change have been identified, there are further problems associated with their attribution to climate change rather than other environmental drivers (Thomas et al. 2006). Most of the studies in Tables 6.1 and 6.2 address alternative explanations, and find these unable or unlikely to fully explain observed changes to species distributions. Possible approaches at the stage of sampling or analysis are to exclude sites whose habitat is seen to be degraded (Parmesan 1996; Chen et al. 2009), or to limit analyses to a particular habitat type (e.g., woodland in Lenoir et al. 2008) or to protected areas (Moritz et al. 2008), although in both cases vegetation change may still influence species range shifts.

TABLE 6.3
Elevational Distributions of Butterfly Species in the Sierra de Guadarrama (Central Spain) in Two Time Periods (1967–1973 and 2004–2005)

	Elevation Band (m)[a]					
Species	**<800**	**800–1000**	**1000–1200**	**1200–1400**	**1400–1600**	**>1600**
Aglais urticae	+/–	+/(+)	+/(+)	+/+	+/+	+/+
Anthocharis belia	+/–	+/(+)	+/+	+/+	+/+	–/(+)
Anthocharis cardamines	+/–	+/+	+/+	+/+	+/+	+/(+)
Aporia crataegi	+/–	+/+	+/+	+/+	+/+	+/+
Argynnis adippe	–/–	+/–	+/+	+/+	+/+	+/+
Argynnis aglaja	–/–	+/–	+/+	+/+	+/+	+/+
Argynnis paphia	–/–	+/–	+/+	+/+	+/+	–/+
Boloria selene	–/–	+/–	+/–	+/(+)	+/(+)	+/(+)
Brenthis daphne	–/–	(+)/–	+/(+)	+/+	+/+	+/+
Celastrina argiolus	+/+	+/(+)	+/+	+/+	+/+	–/+
Coenonympha arcania	–/–	–/(+)	+/+	+/+	+/+	–/+
Cupido minimus	–/–	–/(+)	+/+	+/+	+/(+)	–/(+)
Cyaniris semiargus	–/–	+/–	+/+	–/+	+/(+)	–/(+)
Cynthia cardui	–/+	+/+	+/+	+/+	+/+	+/+
Erebia meolans	–/–	–/–	–/–	–/(+)	+/+	+/+
Erebia triaria	–/–	+/–	+/(+)	+/+	+/+	+/+
Euphydryas aurinia	–/+	+/+	+/+	+/+	+/+	+/(+)
Glaucopsyche alexis	+/+	+/+	+/+	+/+	+/+	+/(+)
Glaucopsyche melanops	+/+	+/+	–/(+)	–/(+)	+/(+)	–/–
Hesperia comma	–/–	+/(+)	+/+	+/+	+/+	+/+
Hipparchia alcyone	–/–	(+)/–	+/+	+/+	+/+	–/+
Hipparchia statilinus	–/+	+/+	+/+	+/+	+/+	+/+
Hyponephele lycaon	–/–	(+)/+	+/+	+/+	–/+	–/+
Inachis io	–/–	+/–	+/+	+/+	+/+	+/+
Iphiclides podalirius	+/+	+/+	+/+	+/+	–/+	–/+
Kanetisa circe	–/+	+/+	+/+	+/+	+/+	–/+
Laesopis roboris	+/+	+/+	+/+	+/(+)	+/–	–/–
Lasiommata maera	+/–	+/–	+/+	+/(+)	–/+	+/+
Lasiommata megera	+/+	+/+	+/+	+/+	+/+	–/+
Limenitis reducta	–/+	+/(+)	+/–	+/+	+/+	–/–
Lycaena alciphron	+/–	+/(+)	+/+	+/+	+/+	+/+
Lycaena tityrus	+/+	+/+	+/+	+/+	+/+	–/–
Lycaena virgaureae	–/–	(+)/–	–/–	(+)/+	+/+	+/+
Lysandra albicans	–/+	–/+	–/+	+/+	+/+	+/–
Maniola jurtina	+/+	+/+	+/+	+/+	+/+	–/+
Melanargia ines	+/–	+/(+)	+/(+)	–/(+)	–/–	–/+
Melanargia russiae	–/–	–/–	–/–	+/(+)	+/+	+/+
Melitaea cinxia	–/–	+/+	+/+	+/+	+/+	+/–
Melitaea phoebe	+/+	+/+	+/+	+/+	+/+	–/–

(*continued*)

TABLE 6.3 (continued)
Elevational Distributions of Butterfly Species in the Sierra de Guadarrama (Central Spain) in Two Time Periods (1967–1973 and 2004–2005)

	Elevation Band (m)[a]					
Species	<800	800–1000	1000–1200	1200–1400	1400–1600	>1600
Melitaea trivia	–/–	+/(+)	+/+	+/+	+/+	+/+
Nymphalis polychloros	+/+	+/(+)	+/+	+/+	+/+	+/(+)
Papilio machaon	+/+	+/(+)	+/+	+/+	+/(+)	–/+
Pararge aegeria	+/–	+/+	+/+	+/+	+/+	+/+
Parnassius apollo	–/–	+/–	+/–	+/(+)	+/+	+/+
Pieris napi	+/–	+/+	+/+	+/+	+/+	+/+
Polygonia c-album	+/+	+/(+)	+/–	+/+	+/+	+/+
Polyommatus icarus	+/+	+/+	–/+	+/+	+/+	+/+
Pseudophilotes panoptes	–/+	–/+	–/+	+/+	–/(+)	–/–
Pyronia bathseba	+/+	+/+	+/+	+/+	+/+	–/(+)
Pyronia cecilia	+/+	+/+	+/+	+/(+)	–/(+)	–/–
Pyronia tithonus	–/+	+/+	+/+	+/+	+/+	–/+
Satyrus actaea	–/–	(+)/–	+/–	+/–	+/+	+/+
Thecla quercus	–/+	+/+	+/+	+/+	+/+	–/(+)
Vanessa atalanta	+/–	+/–	+/(+)	+/+	+/+	+/+
Zerynthia rumina	–/–	+/+	+/+	+/+	+/+	–/(+)

[a] Columns show presence (+) or absence (–) of species in 200-m elevation intervals in 1967–1973 (left symbol) and 2004–2005 (right symbol). Distribution status: +/+ Survived; +/– Extinct; –/+ Colonized; –/– Absent. Symbols in parentheses show species recorded in raw data but absent from data resampled to equalize sampling effort among elevation intervals and between time periods. Twelve species (not shown in table) were observed in all six elevation intervals in both surveys, for both raw and resampled data: *Argynnis niobe, Argynnis pandora, Coenonympha pamphilus, Colias croceus, Gonepteryx rhamni, Hipparchia semele, Issoria lathonia, Lampides boeticus, Lycaena phlaeas, Melanargia lachesis, Pieris rapae, Pontia daplidice.*

Direct habitat change is unlikely to be the key driver where species with a wide range of habitat associations all show uphill shifts in distributions; for example, the mean elevations of desert, chaparral, and montane plants all increased at similar rates in southern California between 1977 and 2006–2007 (Kelly and Goulden 2008).

Human land use intensity is generally greater at low elevations (Nogués-Bravo et al. 2008); hence it may be difficult to disentangle local extinctions caused by climate change from those caused by direct habitat degradation. One approach is to revisit sites with historical records for focal species, and to record whether apparently suitable habitats and populations of the species remain. This approach can be effective for habitat specialist species, where limiting resources can be clearly identified (see examples in Table 6.2). Studies revisiting historical populations of butterfly species have documented population extinctions from low-elevation sites where host plant species survive, suggesting direct effects of climate change (Parmesan 1996, 2005;

Franco et al. 2006; Merrill et al. 2008) (Table 6.2). In other studies, models for population survival versus extinction that show significant effects of altitude once habitat effects have been accounted for also imply an influence of climate change on extinction risk (Beever et al. 2003; Epps et al. 2004). In contrast, for three high-latitude/-elevation butterfly species studied by Franco et al. (2006), extinctions were largely recorded from locations where suitable host plants were no longer present. A study of the fen plant *Swertia perennis* in northern Switzerland found that the low elevation limit had increased, but that low-elevation sites were now managed unsuitably for the species (Lienert et al. 2002). In these two cases, climate conditions may now be less suitable for species at low elevations, but direct effects of habitat degradation may be the major driver of local extinctions.

Uphill range expansions can also be identified by sampling suitable habitats or host plants for species above their recorded elevation range: when suitable host species or habitats were also present beforehand, climate change is likely to have promoted the range extension (Battisti et al. 2005, 2006; Dahlhoff et al. 2008).

MECHANISMS BEHIND UPHILL RANGE SHIFTS

Further support for the role of climate change in driving range shifts is provided by research identifying the mechanisms responsible for distribution change.

Population extinctions at low elevations in the butterfly *E. editha* have been linked to drier and less predictable rainfall conditions, which affect rates of larval host-plant senescence and larval mortality (McLaughlin et al. 2002; Parmesan 2005). Translocation of egg batches for the butterfly *Aporia crataegi* in central Spain showed that summer egg-hatch and larval survival increased with elevation, suggesting that populations may no longer be able to survive at elevations below 900 m, even though host-plant species remain (Merrill et al. 2008). Herptile extinctions in Costa Rica have been linked to increasingly dry conditions, and to uphill range expansions of the fungus *Batrachochytrium dendrobatidis*, which causes the disease chytridiomycosis in amphibians (Pounds et al. 1999, 2006). Temperatures at high elevations may be shifting toward the growth optimum for *B. dendrobatidis* (Pounds et al. 2006; but see Rohr et al. 2008 for a challenge to this hypothesis), and infections have been recorded up to 5348 m in anuran populations on recently deglaciated terrain in the Andes (Seimon et al. 2007). For the Brown Trout *Salmo trutta*, a quantitative physiological model for the effects of water temperature on population size was used to show that the thermal habitat range for the species shifted uphill by 130 m because of 0.1–1.1°C increases in water temperature between 1978–1987 and 1988–2002 in Switzerland (Hari et al. 2006). The proportion of *S. trutta* infected by temperature-dependent proliferative kidney disease also decreased with elevation. For this species, all populations below 600 m were declining before 1987, yet all populations below 1000 m have declined following rapid warming since 1987.

Fecundity and recruitment can also decline at low elevations because of increased temperatures, as shown for the beech *Fagus sylvatica* at its low-elevation limits in northeast Spain (Peñuelas and Boada 2003). However, temperature increases do not always lead to reductions in fecundity at low-latitude or -elevation range margins; if there is a bell-shaped physiological response to temperature, the response to warming depends on whether current conditions lie below, at, or above the thermal optimum for a species. Warmer temperatures in August have led to increased mass and greater reproductive output by adult female Common Lizard *Lacerta vivipara* at its southern range margin in France (Chamaillé-Jammes et al. 2006). An understanding of the genetic and physiological mechanisms underpinning the effects of climate on fecundity can provide valuable information on the capacity of species to adapt to change. For example, the willow beetle *Chrysomela aeneicollis* experienced low-elevation extinctions and high-elevation colonizations at its southern range margin in California between 1998 and 2007. Genetic polymorphism at the phosphoglucose isomerase (PGI) locus is associated with traits that confer temperature adaptation in fecundity in this species, opening up the possibility to research the population-level consequences of genetic variation in a changing climate (Dahlhoff et al. 2008).

The rates at which species extend their distributions upward depend on growth, fecundity, and survival at and beyond the high-elevation range limit, and rates of dispersal to higher elevations. In plants, seed-regenerating species may face relatively little dispersal limitation over elevation gradients (Molau and Larsson 2000), and may therefore track suitable conditions faster than vegetatively propagating species (Kullman 2002). In the Alps, species with lighter diaspores have expanded their distributions more quickly uphill (Parolo and Rossi 2008), but dispersal and recruitment limitation may significantly slow tree-line expansions (Dullinger et al. 2004). Observed uphill shifts in tree lines may represent growth in height by individuals that were already present at high elevations (Kullman 2001), although warmer winters since 1987 in Sweden have also led to increased high-elevation germination and seedling establishment (Kullman 2002). Nevertheless, recruitment at tree lines may depend more on natural disturbance and the invasibility of existing vegetation than thermal conditions (Cullen et al. 2001; Daniels and Veblen 2003; Dullinger et al. 2004). In some cases, the tree line may reflect human patterns of disturbance and may expand upward because of land use abandonment rather than climate warming (Dullinger et al. 2004; Batllori and Gutiérrez 2008). Microclimate and soil conditions also influence local or regional rates of uphill expansion (Holtmeier and Broll 2005), and tree lines have been shown to shift upward faster on sheltered or south-facing slopes than on exposed or north-facing sites, because of more favorable conditions for growth (Luckman and Kavanagh 2000; Kullman 2001).

The mechanisms behind upward range extension by the pine processionary moth *Thaumetopoea pityocampa* in Italy have been studied using experiments on the effects of temperature on survival and dispersal. The amount of time in winter above threshold temperatures for larval feeding has increased, leading to increased survival and uphill extensions by 110 m on a north-facing aspect and 230 m on a south-facing aspect, related to differences in insolation (Battisti et al. 2005). In addition, the extreme hot summer of 2003 had five times as many nights above the threshold temperature for female flight at the upper-elevation range limit, resulting in dispersal and colonization of extreme high-elevation sites (Battisti et al. 2006). Warm winter conditions enabled these new high-elevation colonies to survive into 2004. For *T. pityocampa*, a widely available larval host plant (*Pinus sylvestris*) permitted the uphill range extension; for many specialist insect species, the upper limits of ranges may be constrained by upper range limits of host plants (e.g., Merrill et al. 2008; Chen et al. 2009).

MODELING EFFECTS OF UPHILL RANGE SHIFTS ON BIODIVERSITY

Climate envelope models have been widely used to predict changes to species distributions based on different climate change and dispersal scenarios, and to model the consequences for extinction rates (e.g., Williams et al. 2003; Thomas et al. 2004). These models have usually used coarse-scale data on climate and species distributions to produce coarse-scale predictions. Fine-scale topographic variation in mountainous regions means that many high-elevation species are restricted to much narrower elevation ranges (or ranges of climate conditions) than is evident from coarse-scale grid cells; as a result, many bioclimate models may have underestimated the threat posed by climate change to the persistence of montane species (Trivedi et al. 2008). Conversely, coarse-scale climate surfaces might overlook local thermal refugia produced by processes such as cold-air drainage or shading from local topography (see Luoto and Heikkinen 2008; Randin et al. 2009).

The equivalent of climate envelope models for mountain regions have used the current elevation ranges of species as measures of their thermal associations, and have predicted future regional range sizes assuming that species maintain these thermal associations as the climate warms (e.g., Wilson et al. 2005; Raxworthy et al. 2008). Inevitably, because of the roughly conical shape of mountains, these models predict reductions in regional distribution sizes as species are forced to higher elevations. There may be few opportunities for such species to colonize other climatically suitable mountain ranges, because such areas are often separated by lowland expanses that are unsuitable both in terms of habitat and climate (Ohlemüller et al. 2006, 2008). A model incorporating the elevation ranges of landbird species (87% of all avian species) predicted that 400–550 landbird species would face extinction from 2.8°C warming, based on reductions in species distribution areas (Sekercioglu et al. 2008). However, it remains necessary to identify the causal influence of climate over species' ranges to be sure whether estimates of extinction risk from bioclimate models are likely to be accurate; the models could either over- or underestimate threat because of local adaptation (Harte et al. 2004), biotic interactions (Davis et al. 1998), or the effects of spatial scale (Trivedi et al. 2008; Luoto and Heikkinen 2008; Randin et al. 2009), among other factors (see Pearson and Dawson 2003 for a general critique).

Climate change is likely to have a major impact on tropical biodiversity through its effects on species' elevational ranges. In the Tropics, weak latitudinal thermal gradients mean that species are much less likely to shift to higher latitudes than higher elevations. As a result, three key processes might threaten biodiversity in the Tropics (Colwell et al. 2008): *lowland biotic attrition*, where there is a net loss in species at low elevations because no species associated with hotter conditions are available to colonize when others contract to higher elevations; *range-shift gaps*, where future elevational ranges of species do not overlap with their current ranges; and *mountaintop extinctions*, because no climatically suitable elevations remain after climate change. Assuming that species' thermal associations remain the same,

Colwell et al. (2008) used the elevational ranges of 1902 plant and insect species over a 30- to 2900-m gradient in Costa Rica to show that mountaintop extinctions were unlikely over 0–5°C warming (equivalent to 0- to 1000-m upslope shifts in isotherms). However, for 2–3°C warming (400- to 600-m isotherm shifts), 50–80% of epiphytes, Rubiaceae, and ants and 10–20% of moths would need to shift their distributions upslope to elevations not overlapping with their current range. Such range-shift gaps may be particularly likely in the Tropics, where lower variation in ambient temperature than at temperate latitudes may select for narrower thermal tolerance, and hence narrower elevational ranges (Janzen 1967; Ghalambor et al. 2006). Lowland biotic attrition was also predicted to affect 30–80% of taxa analyzed, given a 500-m isotherm shift. As a result, Colwell et al. (2008) predict that "tropical lowland biotas may face a level of net lowland biotic attrition without parallel at higher latitudes (where range shifts may be compensated for by species from lower latitudes)." Although there is an opportunity at temperate latitudes (which does not exist in the tropical lowlands) for warm-adapted species to colonize communities from which cool-adapted species are lost, the existing evidence for butterfly communities in central Spain suggests that the barriers to colonization caused by lack of habitat availability may either delay or prevent this replacement from occurring (Wilson et al. 2007). Hence, lowland biotic attrition may be a consequence of climate change at a wide range of latitudes. Natural habitats need to be protected across elevational ranges to allow species to shift their distributions uphill as the climate warms, but protection of surrounding lowland habitats may be critical to prevent species richness declines in mountain foothills (Chen et al. 2009).

It is by no means certain that species will uniformly maintain their thermal associations as the climate warms, which is one assumption underlying climate envelope models of future ranges. However, if species' elevational ranges in different parts of their geographic distributions show similar thermal associations, this could support the validity of climate envelope predictions by implying limited local thermal adaptation. Thermal associations of butterflies in the Sierra de Guadarrama (central Spain) significantly predicted the elevational ranges of the same species elsewhere in central and northern Spain, but were much more reliable for abundant species with clearly defined elevational ranges than for more localized or less abundant species, for which local differences in habitat or interacting species are likely to play a key role (D. Andrés, D. Gutiérrez, and R. J. Wilson, unpublished data). Comparisons of this kind among the elevational ranges of species in different regions could help to identify ecological characteristics of species affecting the reliability of climate envelope approaches to modeling future distributions.

Species are likely to show individualistic responses to climate warming, depending on the roles of climate, habitat, and interacting species in limiting their distributions. Recent documented shifts in elevational ranges vary widely among species in the same communities (e.g., Wilson et al. 2005 [Table 3]; Le Roux and McGeoch 2008; Moritz et al. 2008; Raxworthy et al. 2008; Bergamini et al. 2009). Given the strong evidence that species are generally shifting upslope, the challenge now is to understand the heterogeneity of species responses and the implications of the changes in community composition that follow. Evidence of the mechanisms that determine the range limits of species can also increase the reliability of modeled estimates of change. The elevational distributions of 48 out of 53 bird species sampled in California tracked changes to temperature or precipitation between 1911–1929 and 2003–2008, and 77% of species ranges tracked the variable predicted from a priori models of species range sensitivity to climatic variables (Tingley et al. 2009). Interestingly, the elevational associations of species influenced the climatic factors to which they responded; those species that tracked only changes to precipitation occurred in the lowlands (mean range center = 916 m), while those tracking only temperature were associated with the highlands (mean range center = 1944 m) (Tingley et al. 2009). The observed variation in range-limiting factors may reflect the roles of precipitation in limiting Net Primary Productivity (NPP) at low elevations, and of energy in limiting NPP at high elevations (see also Hawkins et al. 2003).

Ecophysiological models may be valuable in producing testable predictions about how the elevational ranges of species will respond to climate change (Hodkinson 1999). For example, for low-elevation margins of tropical lowland species to contract as the climate warms, current species distributions must be situated near their thermal optima, and this is suggested by evidence for tropical ectotherms (Deutsch et al. 2008) and plants (Clark et al. 2003; Feeley et al. 2007). Experimentally derived indices of physiological stress caused by climate warming could provide quantitative assessments of extinction risk at species' low-elevation range margins (e.g., Bernardo and Spotila 2006), but the capacity of species to expand to high elevations may depend more on the prior availability of suitable resources or conditions for colonization (e.g., Dullinger et al. 2004; Merrill et al. 2008). Future species ranges will depend both on the fine-scale effects of microclimate on growth, survival and fecundity, and the landscape-scale effects of land use on the distributions of suitable habitats. In addition, differences in the environmental cues that influence the phenology of interacting species may determine whether species are able to track otherwise tolerable climatic conditions (for examples concerning herbivorous insects and their host plants, see Hodkinson 1999). In practice, models based on topoclimatic factors may provide reasonable best-guess predictions of species' future ranges, especially where climate varies over a fine scale and relatively large areas of natural habitat remain. However, available evidence suggests that even over steep elevation gradients there may be a greater time lag in upslope colonizations than in downslope extinctions (Wilson et al. 2007).

SYNTHESIS, FUTURE DIRECTIONS, AND CONSERVATION

There is now strong evidence that species distributions have shifted uphill associated with recent climate change. There are indications that rates of range contraction from low-elevation margins may be keeping closer pace with climate warming than high-elevation range extensions, which may be constrained by habitat availability. A number of studies have begun to identify the climatic mechanisms driving these range shifts, and to predict future effects of climate change on the biodiversity of elevational gradients. In particular, recent discussion of the likely impact of elevational range shifts on tropical biodiversity represents an important contribution to the debate (Colwell et al. 2008; Sekercioglu et al. 2008), and further such research may be vital to assess the implications of climate change for tropical biodiversity (Chen et al. 2009).

The research systems described in this chapter will allow continued monitoring of shifts in species' elevational ranges as the climate warms. Indeed, repeat surveys of alpine plants already suggest that rates of upslope shifts have increased in the last two decades (Walther et al. 2005; Pauli et al. 2007). The many published studies of elevational patterns in diversity (e.g., Rahbek 1995; Lomolino 2001; McCain 2005) provide further potential sources of baseline data with which to test the generality of the range shifts we describe, and to address a number of unresolved questions. Many recent studies show how systems with data on the past and/or current elevational ranges of species present the opportunity to test for change (see Tingley and Beissinger 2009). Where possible, such systems should continue to be monitored (or new elevational transects should be established) to provide better insight into processes such as metapopulation dynamics at range boundaries (e.g., Epps et al. 2004), changes to community structure (see examples in Table 6.1), or the biotic interactions that influence species range shifts (e.g., Hodkinson 1999). Sufficient data do not yet exist to rigorously analyze differences among taxonomic groups; intraregional studies for a range of taxa might help to confirm whether uphill shifts are faster, more widespread, or more predictably uphill (related to temperature) in ectotherms than endotherms, or whether differential shifts in the distributions of interacting species are leading to host release from natural enemies or are constraining species ranges relative to shifts in suitable climate space. Studies of the elevational ranges of the same species in different regions may help to determine the coherence of thermal associations across species distributions, and hence the reliability of climate envelope models or the feasibility of assisted colonization programs to adapt conservation to climate change (which requires consideration of the local adaptation of species to climatic conditions; Hoegh-Guldberg et al. 2008). Uphill shifts in species distributions caused by climate change appear unwelcome for biodiversity, yet they present an opportunity to test the factors governing the limits to species ranges, and thereby to plan adaptive measures for conservation. We therefore urge further ecological research into the effects of climate change on the elevational limits and extent of species distributions.

Observed elevational range shifts also allow some general recommendations to be drawn about approaches to conservation in the face of climate change. Species should be more able to track suitable climates over elevational rather than latitudinal gradients, so conservationists must ensure that sufficient favorable habitat is conserved at a range of elevations to allow them to do so. In this context, land use is critical. Species such as plants and butterflies associated with alpine meadows are threatened by land-use changes such as afforestation and the abandonment of traditional farming (e.g., Tasser and Tappeiner 2002; Van Swaay et al. 2006; Becker et al. 2007). Dense forestry can reduce rates of movement between populations of mountain butterflies (Roland et al. 2000) and might therefore reduce colonization rates in metapopulations or act as a barrier to uphill range shifts. In contrast, whereas grazing is important for maintaining open habitats, excessive grazing pressure in upland Britain restricts some upland plants to locations inaccessible to sheep (Thompson and Brown 1992). Some species may only be restricted to mountains because human land-use intensity is now so great at lower elevations (Nogués-Bravo et al. 2008). It is now essential to minimize threats such as these to the high-elevation habitats to which species are being progressively restricted by climate change.

Whether most species will shift predictably toward mountain summits, as well as the area requirements for long-term persistence by mountain species, are also uncertain (e.g., see Randin et al. 2009; Tingley et al. 2009). To counter this uncertainty, conservation managers need to use the heterogeneity in topography and vegetation at their disposal to maintain as wide a variety of microclimates as possible. This might involve ensuring that regional protected areas include representative samples at different altitudes of habitats ranging from open to wooded, as well as a wide range of aspects, slopes, and exposures (hilltops and valleys). Such an approach should increase the likelihood that suitable microclimates are available to species, even if their altitudinal or vegetation associations shift as the climate changes. For example, the butterfly *Parnassius apollo* in central Spain is associated with more shrubby habitats at lower elevations and more open habitats at high elevations, probably related to prevailing climatic conditions (Ashton et al. 2009). Another instructive example comes from the mountains of Utah, where small mammal species associated with cool and moist habitats have increased in abundance and distribution relative to xeric species, despite warming temperatures and increased drought in recent decades (Rowe 2007). The driver of this change appears to be reduced livestock grazing pressure since the early twentieth century, showing how land management could mitigate against the effects of climate warming.

Climate change could alter the pressures or priorities of conservation in mountains for reasons ranging from socioeconomic to ecological. For example, in some regions pressure from ski developments may be alleviated as snowfall becomes less reliable, but reduced snowfall could itself have negative consequences for many alpine plants (Giménez-Benavides et al. 2007). High-altitude plant communities could also be threatened by increased competition from new high-elevation colonists such as shrubs and trees (e.g., Sanz-Elorza et al. 2003). Indeed, managers of mountain habitats may need to develop strategies for conserving new arrivals or emergent communities, following the kinds of uphill range shifts described in this chapter.

Overall, the composition of ecological communities may be increasingly dynamic as a result of climate change, particularly in mountains where species can track climatic conditions over relatively small distances. Such dynamism presents both new opportunities and new challenges for conservation managers, who may need to set increasingly flexible targets and prioritize the conservation of heterogeneity as a means to conserve mountain biodiversity.

ACKNOWLEDGMENTS

Víctor J. Monserrat, Javier Gutiérrez, Sonia B. Díez, David Andrés, David Martínez, Rosa Agudo, Miriam Jiménez, and Irene Martínez assisted with research on Spanish butterfly distributions, which was funded by the Spanish Ministry for Education and Science (grant reference CGL2005-06820/BOS) and the British Ecological Society (ECPG0410). Erik Beever and three anonymous referees gave helpful comments on the text. John Hopkins, Ilya Maclean, and the 2009 class of module BIOM4012 at the University of Exeter provided useful thoughts on the conservation implications of elevational range shifts.

SUMMARY

Climate warming is causing species distributions to shift to higher latitudes and elevations. Until recently, poleward shifts were more widely documented than upslope movements, but there is now convincing evidence that species are also shifting their ranges to higher elevations. Depending on the type of data and analysis, it has been possible to show uphill shifts in the lower limits, upper limits, or average elevations of species' ranges, which are robust to changes in sampling effort over time. Likewise, there is evidence from many studies that climate change, rather than direct habitat modification, is likely to be the main driver of the observed changes. Comparison of observed rates of uphill range shifts with rates of climate warming suggest negligible time lags between climate warming and extinction at warm margins (trailing edges) of species distributions. Expansions at the upslope limits (leading edges) can also occur rapidly, but may be constrained by the availability of suitable habitats, or by maximum regional elevations. A number of empirical studies have begun to pinpoint the mechanisms that determine rates and limitations to elevation range shifts. These mechanistic studies, combined with observed changes to species distributions, provide empirical support for models that suggest pronounced consequences for biodiversity. Here, we review recent evidence for uphill species range shifts, before considering the information that will be needed and the approaches that can be taken to predict the future effects of climate change and the consequences for biodiversity and conservation.

REFERENCES

Archaux, F. 2004. Breeding upward when climate is becoming warmer: no bird response in the French Alps. *Ibis* 146:138–144.

Ashton, S., Gutiérrez, D., and R. J. Wilson. 2009. Effects of temperature and elevation on habitat use by a rare mountain butterfly: implications for species responses to climate change. *Ecol. Entomol.* 34:437–446.

Batllori, E., and E. Gutiérrez. 2008. Regional tree line dynamics in response to global change in the Pyrenees. *J. Ecol.* 96:1275–1288.

Battisti, A., Stastny, M., Buffo, E., and S. Larsson. 2006. A rapid altitudinal range expansion in the pine processionary moth produced by the 2003 climatic anomaly. *Global Change Biol.* 12:662–671.

Battisti, A., Stastny, M., Netherer, S., et al. 2005. Expansion of geographic range in the pine processionary moth caused by increased winter temperatures. *Ecol. Appl.* 15: 2084–2096.

Beckage, B., Osborne, B., Gavin, D. G., Pucko, C., Siccama, T., and T. Perkins. 2008. A rapid upward shift of a forest ecotone during 40 years of warming in the Green Mountains of Vermont. *Proc. Nat. Acad. Sci. U.S.A.* 105:4197–4202.

Becker, A., Körner, C., Brun, J.-J., Guisan, A., and U. Tappeiner. 2007. Ecological and land use studies along elevational gradients. *Mt. Res. Dev.* 27:58–65.

Beever, E. A., Ray, C., Wilkening, J. L., Brussard, P. F., and P. W. Mote. 2011. Contemporary climate changes alters the pace and drivers of extinction. *Global Change Biol.* 17(6):2054–2070.

Beever, E. A., Brussard, P. F., and J. Berger. 2003. Patterns of apparent extirpation among isolated populations of pikas (*Ochotona princeps*) in the Great Basin. *J. Mammal.* 84:37–54.

Bergamini, A., Ungricht, S., and H. Hofman. 2009. An elevational shift of cryophilous bryophytes in the last century—an effect of climate warming? *Diversity Distrib.* 15:871–879.

Bernardo, J., and J. R. Spotila. 2006. Physiological constraints on organismal response to global warming: mechanistic insights from clinally varying populations and implications for assessing endangerment. *Biol. Lett.* 2:135–139.

Bustamante, M. R., Ron, S. R., and L. A. Coloma. 2005. Cambios en la diversidad en siete comunidades de Anuros en los Andes de Ecuador. *Biotropica* 37:180–189.

Chamaillé-Jammes, S., Massot, M., Aragón, P., and J. Clobert. 2006. Global warming and positive fitness response in mountain populations of common lizards *Lacerta vivipara*. *Global Change Biol.* 12:392–402.

Chen, I.-C., Shiu, H.-J., Benedick, S., et al. 2009. Elevation increases in moth assemblages over 42 years on a tropical mountain. *Proc. Nat. Acad. Sci. U.S.A.* 106:1479–1483.

Clark, D. A., Piper, S. C., Keeling, C. D., and D. B. Clark. 2003. Tropical rain forest tree growth and atmospheric carbon dynamics linked to interannual temperature variation during 1984–2000. *Proc. Nat. Acad. Sci. U.S.A.* 100:5852–5857.

Colwell, R. K., Brehm, G., Cardelús, C. L., Gilman, A. C., and J. T. Longino. 2008. Global warming, elevational range shifts, and lowland biotic attrition in the wet tropics. *Science* 322:258–261.

Cullen, L. E., Stewart, G. H., Duncan, R. P., and J. G. Palmer. 2001. Disturbance and climate warming influences on New Zealand *Nothofagus* tree-line population dynamics. *J. Ecol.* 89:1061–1071.

Dahlhoff, E. P., Fearnley, S. L., Bruce, D. A., et al. 2008. Effects of temperature on physiology and reproductive success of a montane leaf beetle: implications for persistence of native populations enduring climate change. *Physiol. Biochem. Zool.* 81:718–732.

Daniels, L. D., and T. T. Veblen. 2003. Regional and local effects of disturbance and climate on altitudinal treelines in Northern Patagonia. *J. Veg. Sci.* 14:733–742.

Davis, A. J. Jenkinson, L. S., Lawton, J. H., Shorrocks, B., and S. Wood. 1998. Making mistakes when predicting shifts in species range in response to global warming. *Nature* 391:783–786.

Deutsch, C. A., Tewksbury, J. J., Huey, R. B., et al. 2008. Impacts of climate warming on terrestrial ectotherms across latitude. *Proc. Nat. Acad. Sci. U.S.A.* 105:6668–6672.

Dullinger, S., Dirnböck, T., and G. Grabherr. 2004. Modelling climate change-driven treeline shifts: relative effects of temperature increase, dispersal and invasibility. *J. Ecol.* 92:241–252.

Ellis, E. C., and N. Ramankutty. 2008. Putting people in the map: anthropogenic biomes of the world. *Front. Ecol. Environ.* 6:439–447.

Epps, C. W., McCullough, D. R., Wehausen, J. D., Bleich, V. C., and J. R. Rechel. 2004. Effects of climate change on population persistence of desert-dwelling mountain sheep in California. *Conserv. Biol.* 18:102–113.

Feeley, K. J., Wright, S. J., Supardi, M. N. N., Kassim, A. R., and S. J. Davies. 2007. Decelerating growth in tropical forest trees. *Ecol. Lett.* 10:461–469.

Franco, A. M. A., Hill, J. K., Kitschke, C., et al. 2006. Impacts of climate warming and habitat loss on extinctions at species' low-latitude range boundaries. *Global Change Biol.* 12:1545–1553.

Ghalambor, C. K., Huey, R. B., Martin, P. R., Tewksbury, J. J., and G. Wang. 2006. Are mountain passes higher in the tropics? Janzen's hypothesis revisited *Integr. Comp. Biol.* 46:5–17.

Giménez-Benavides, L., Escudero, A., and J. M. Iriondo. 2007. Reproductive limits of a late-flowering high-mountain Mediterranean plant along an elevational climate gradient. *New Phytologist* 173:367–382.

Grabherr, G., Gottfried, M., and H. Pauli. 1994. Climate effects on mountain plants. *Nature* 369:448.

Hari, R. E., Livingstone, D. M., Siber, R., Burkhardt-Holm, P., and H. Güttinger. 2006. Consequences of climatic change for water temperature and brown trout populations in Alpine rivers and streams. *Global Change Biol.* 12:10–26.

Harsch, M. A., Hulme, P. E., McGlone, M. S., and R. P. Duncan. 2009. Are treelines advancing? A global meta-analysis of treeline response to climate warming. *Ecol. Lett.* 12:1040–1049.

Harte, J., Ostling, A., Green, J. L., and A. Kinzig. 2004. Climate change and extinction risk. *Nature* 430:U3.

Hawkins, B. A., Field, R., Cornell, H. V., et al. 2003. Energy, water, and broad-scale geographic patterns of species richness. *Ecology* 84:3105–3117.

Hickling, R., Roy, D. B., Hill, J. K., Fox, R., and C. D. Thomas. 2006. The distributions of a wide range of taxonomic groups are expanding polewards. *Global Change Biol.* 12:450–455.

Hill, J. K., Thomas, C. D., Fox, R., et al. 2002. Responses of butterflies to twentieth century climate warming: implications for future ranges. *Proc. R. Soc. Lond. B* 269:2163–2171.

Hodkinson, I. D. 1999. Species response to global environmental change or why ecophysiological models are important: a reply to Davis et al. *J. Anim. Ecol.* 68:1259–1262.

Hoegh-Guldberg, O., Hughes, L., Mcintyre, S., et al. 2008. Assisted colonization and rapid climate change. *Science* 321:345–346.

Holtmeier, F.-K., and G. Broll. 2005. Sensitivity and response of northern hemisphere altitudinal and polar treelines to environmental change at landscape and local scales. *Global. Ecol. Biogeogr.* 14:395–410.

Janzen, D. H. 1967. Why mountain passes are higher in the tropics. *Am. Nat.* 101:233–249.

Keller, F., Kienast, F., and M. Beniston. 2000. Evidence of response of vegetation to environmental change on high-elevation sites in the Swiss Alps. *Regional Env. Change* 1:70–77.

Kelly, A. E., and M. L. Goulden. 2008. Rapid shifts in plant distribution with recent climate change. *Proc. Nat. Acad. Sci. U.S.A.* 33:11823–11826.

Klanderud, K., and H. J. B. Birks. 2003. Recent increases in species richness and shifts in altitudinal distributions of Norwegian mountain plants. *Holocene* 13:1–6.

Konvicka, M., Maradova, M., Benes, J., Fric, Z., and P. Kepka. 2003. Uphill shifts in distribution of butterflies in the Czech Republic: effects of changing climate detected on a regional scale. *Global Ecol. Biogeogr.* 12:403–411.

Kullman, L. 2001. 20th century climate warming and tree-limit rise in the southern Scandes of Sweden. *Ambio* 30:72–80.

———. 2002. Rapid recent range-margin rise of tree and shrub species in the Swedish Scandes. *J. Ecol.* 90:68–77.

Le Roux, P. C., and M. A. McGeoch. 2008. Rapid range expansion and community reorganization in response to warming. *Global Change Biol.* 14:2950–2962.

Lenoir, J., Gégout, J. C., Marquet, P. A., De Ruffray, P., and H. Brisse. 2008. A significant upward shift in plant species optimum elevation during the 20th century. *Science* 320:1768–1771.

Lienert, J., Fischer, M., and M. Diemer. 2002. Local extinctions of the wetland specialist *Swertia perennis* L. (Gentianaceae) in Switzerland: a revisitation study based on herbarium records. *Biol. Conserv.* 103:65–76.

Lomolino, M. V. 2001. Elevation gradients of species-density: historical and prospective views. *Global Ecol. Biogeogr.* 10:3–13.

Luckman, B. H., and T. Kavanagh. 2000. Impact of climate fluctuations on mountain environments in the Canadian Rockies. *Ambio* 29:371–380.

Luoto, M., and R. K. Heikkinen. 2008. Disregarding topographical heterogeneity biases species turnover assessments based on bioclimatic models. *Global Change Biol.* 14:483–494.

Mackenzie, D. I., Nichols, J. D., Royle, J. A., Pollock, K. H., Bailey, L. L., and J. E. Hines. 2006. *Inferring Patterns and Dynamics of Species Occurrence*. San Diego: Elsevier.
McCain, C. M. 2005. Elevational gradients in diversity of small mammals. *Ecology* 86:366–372.
McLaughlin, J. F., Hellmann, J. J., Boggs, C. L., and P. R. Ehrlich. 2002. Climate change hastens population extinctions. *Proc. Nat. Acad. Sci. U.S.A.* 99:6070–6974.
Menéndez, R., González Megías, A., Hill J. K., et al. 2006. Species richness changes lag behind climate change. *Proc. R. Soc. Lond. B* 273:1465–1470.
Merrill, R. M., Gutiérrez, D., Lewis, O. T., Gutiérrez, J., Díez, S. B., and R. J. Wilson. 2008. Combined effects of climate and biotic interactions on the elevational range of a phytophagous insect. *J. Anim. Ecol.* 77:145–155.
Moiseev, P. A., and S. G. Shiyatov. 2003. Vegetation dynamics at the treeline ecotone in the Ural highlands, Russia. In *Alpine Biodiversity in Europe—A Europe-Wide Assessment of Biological Richness and Change*, Eds. L. Nagy, G. Grabherr, C. Körner, and D. B. A. Thompson, 423–435. Ecological Studies Vol. 167. Berlin: Springer.
Molau, U., and E.-L. Larsson. 2000. Seed rain and seed bank along an alpine altitudinal gradient in Swedish Lapland. *Can. J. Bot.* 78:728–747.
Moritz, C., Patton, J. L., Conroy, C. J., Parra, J. L., White, G. C., and S. R. Beissinger. 2008. Impact of a century of climate change on small-mammal communities in Yosemite National Park, USA. *Science* 322:261–264.
Nogués-Bravo, D., Araújo, M. B., Romdal, T., and C. Rahbek. 2008. Scale effects and human impact on the elevational species richness gradients. *Nature* 453:216–220.
Ohlemüller, R., Anderson, B. J., Araújo, M. B., et al. 2008. The coincidence of climatic and species rarity: high risk to small-range species from climate change. *Biol. Lett.* 4:568–572.
Ohlemüller, R., Gritti, E. S., Sykes, M. T., and C. D. Thomas. 2006. Toward European climate risk surfaces: the extent and distribution of analogous and non-analogous climates 1931–2100. *Global Ecol. Biogeogr.* 15:395–405.
Parmesan, C. 1996. Climate and species' range. *Nature* 382:765–766.
———. 2005. Detection at multiple levels: *Euphydryas editha* and climate change. In *Climate Change and Biodiversity*, Eds. T. E. Lovejoy and L. Hannah, 55–60. New Haven, CT: Yale University Press.
———. 2006. Ecological and evolutionary responses to recent climate change. *Ann. Rev. Ecol. Evol. Syst.* 37:637–669.
Parmesan, C., and G. Yohe. 2003. A globally coherent fingerprint of climate change impacts across natural systems. *Nature* 421:37–42.
Parolo, G., and G. Rossi. 2008. Upward migration of vascular plants following a climate-warming trend in the Alps. *Basic Appl. Ecol.* 9:100–107.
Pauli, H., Gottfried, M., and G. Grabherr. 2001. High summits of the Alps in a changing climate: the oldest observation series on high mountain plant diversity in Europe. In *"Fingerprints" of Climate Change*, Eds. G. R. Walther, C. A. Burga, and P. J. Edwards, 139–149. New York: Kluwer Academic/Plenum Publishers.
Pauli, H., Gottfried, M., Reier, K., Klettner, C., and G. Grabherr. 2007. Signals of range expansions and contractions of vascular plants in the high Alps: observations (1994–2004) at the GLORIA*master site Schrankogel, Tyrol, Austria. *Global Change Biol.* 13:147–156.
Pearson, R.G., and T. P. Dawson. 2003. Predicting the impacts of climate change on the distribution of species: are bioclimate envelope models useful? *Global Ecol. Biogeogr.* 12:361–371.
Peh, K. S. H. 2007. Potential effects of climate change on elevational distributions of tropical birds in Southeast Asia. *Condor* 109:437–441.
Peñuelas, J., and M. Boada. 2003. A global change-induced biome shift in the Montseny mountains (NE Spain). *Global Change Biol.* 9:131–140.
Pounds, J. A., Bustamante, M. R., Coloma, M. A., et al. 2006. Widespread amphibian extinctions from epidemic disease driven by global warming. *Nature* 439:161–167.

Pounds, J. A., Fogden, M. P. L., and J. H. Campbell. 1999. Biological response to climate change on a tropical mountain. *Nature* 398:611–615.

Rahbek, C. 1995. The elevational gradient of species richness: a uniform pattern? *Ecography* 18:200–205.

Randin, C. F., Engler, R., Normand, S., et al. 2009. Climate change and plant distribution: local models predict high-elevation persistence. *Global Change Biol.* 15:1557–1569.

Raxworthy, C. J., Pearson, R. G., Rabibisoa, N., et al. 2008. Extinction vulnerability of tropical montane endemism from warming and upslope displacement: a preliminary appraisal for the highest massif in Madagascar. *Global Change Biol.* 14:1703–1720.

Rohr, J. R., Raffel, T. R., Romansic, J. M., McCallum, H., and P. J. Hudson. 2008. Evaluating the links between climate, disease spread, and amphibian declines. *Proc. Nat. Acad. Sci. U.S.A.* 105:17436–17441.

Roland, J., Keyghobadi, N., and S. Fownes. 2000. Alpine *Parnassius* butterfly dispersal: effects of landscape and population size. *Ecology* 81:1642–1653.

Rosenzweig, C., Casassa, G., Karoly, D. J., et al. 2007. Assessment of observed changes and responses in natural and managed systems. In *Climate Change 2007: Impacts, Adaptation and Vulnerability. Contribution of Working Group II to the Fourth Assessment Report of the Intergovernmental Panel on Climate Change*, Eds. M. L. Parry, O. F. Canziani, J. P. Palutikof, P. J. van der Linden, and C. E. Hanson, 79–131. United Kingdom: Cambridge University Press.

Rovito, S. M., Parra-Olea, G., Vásquez-Almazán, C. R., Papenfuss, T. J., and D. B. Wake. 2009. Dramatic declines in neotropical salamander populations are an important part of the global amphibian crisis. *Proc. Nat. Acad. Sci. U.S.A.* 106:3231–3236.

Rowe, R. J. 2005. Elevational gradient analyses and the use of historical museum specimens: a cautionary tale. *J. Biogeog.* 32:1883–1897.

———. 2007. Legacies of land use and recent climatic change: the small mammal fauna in the mountains of Utah. *Amer. Nat.* 170:242–257.

Sanz-Elorza, M., Dana, E. D., González, A., and E. Sobrino. 2003. Changes in the high-mountain vegetation of the Central Iberian Peninsula as a probable sign of climate warming. *Ann. Botany* 92:273–280.

Seimon, T. A., Seimon, A., Daszak, P., et al. 2007. Upward range extension of Andean anurans and chytridiomycosis to extreme elevations in response to tropical deglaciation. *Global Change Biol.* 13:288–299.

Sekercioglu, C. H., Schneider, S. H., Fay, J. P., and S. R. Loarie. 2008. Climate change, elevational range shifts, and bird extinctions. *Conserv. Biol.* 22:140–150.

Shoo, L. P., Williams, S. E., and J.-M. Hero. 2005a. Climate warming and the rainforest birds of the Australian Wet Tropics: using abundance data as a sensitive predictor of change in total population size. *Biol. Conserv.* 125:335–343.

———. 2005b. Potential decoupling of trends in distribution area and population size of species with climate change. *Global Change Biol.* 11:1469–1476.

———. 2006. Detecting climate change induced range shifts: where and how should we be looking? *Austral Ecol.* 31:22–29.

Tasser, E., and U. Tappeiner. 2002. Impact of land use changes on mountain vegetation. *Appl. Veg. Sci.* 5:173–184.

Thomas, C. D., Cameron, A., Green, R. E., et al. 2004. Extinction risk from climate change. *Nature* 427:145–148.

Thomas, C. D., Franco, A., and J. K. Hill. 2006. Range retractions and extinction in the face of climate warming. *Trends Ecol. Evol.* 21:415–416.

Thompson, D. B. A., and A. Brown. 1992. Biodiversity in montane Britain: habitat variation, vegetation diversity and some objectives for conservation. *Biodivers. Conserv.* 1:179–208.

Tingley, M. W., and S. R. Beissinger. 2009. Detecting range shifts from historical species occurrences: new perspectives on old data. *Trends Ecol. Evol.* 24:625–633.

Tingley, M. W., Monahan, W. B., Beissinger, S. R., and C. Moritz. 2009. Birds track their Grinnellian niche through a century of climate change. *Proc. Nat. Acad. Sci. U.S.A.* 106:19637–19643.

Trivedi, M. R., Morecroft, M. D., Berry, P. M., and T. P. Dawson. 2008. Spatial scale affects bioclimate model projections of climate change impacts on mountain plants. *Global Change Biol.* 14:1089–1103.

Tryjanowski, P., Sparks, T. H., and P. Profus. 2005. Uphill shifts in the distribution of the white stork *Ciconia ciconia* in southern Poland: the importance of nest quality. *Divers. Distrib.* 11:219–223.

Van Swaay, C., Warren, M. S., and G. Loïs. 2006. Biotope use and trends of European butterflies. *J. Insect Conserv.* 10:189–209.

Vittoz, P., Bodin, J., Ungricht, S., Burga, C., and G. R. Walther. 2008. One century of vegetation change on Isla Persa, a nunatak in the Bernina massif in the Swiss Alps. *J. Veg. Sci.* 19:671–680.

Walther, G. R., Beissner, S., and C. A. Burga. 2005. Trends in the upward shift of alpine plants. *J. Veg. Sci.* 16:541–548.

Wardle, P., and M. C. Coleman. 1992. Evidence for rising upper limits of four native New Zealand forest trees. *New Zeal. J. Bot.* 30:303–314.

Warren, M. S., Hill, J. K., Thomas, J. A., et al. 2001. Rapid responses of British butterflies to opposing forces of climate and habitat change. *Nature* 414:65–69.

Williams, S. E., Bolitho, E. E., and S. Fox. 2003. Climate change in Australian tropical rainforests: an impending environmental catastrophe. *Proc. R. Soc. Lond. B* 270:1887–1892.

Wilson, R. J., Gutiérrez, D., Gutiérrez, J., Martínez, D., Agudo, R., and V. J. Monserrat. 2005. Changes to the elevational limits and extent of species ranges associated with climate change. *Ecol. Lett.* 8:1138–1146.

Wilson, R. J., Gutiérrez, D., Gutiérrez, J., and V. J. Monserrat. 2007. An elevational shift in butterfly species richness and composition accompanying recent climate change. *Global Change Biol.* 13:1873–1887.

Wilson, R. J., Thomas, C. D., Fox, R., Roy, D. B., and W. E. Kunin. 2004. Spatial patterns in species distributions reveal biodiversity change. *Nature* 432:394–396.

7 Climate Change and Sandy Beach Ecosystems

Alan R. Jones

CONTENTS

> Sea-level rise will have impacts on soft-sediment shorelines and intertidal ecosystems, which will be especially vulnerable to change with additional impacts from extreme events. (CSIRO 2002)

INTRODUCTION

In extent, sandy beaches are major ecosystems comprising about 50–60% of Earth's coastlines (Bird 1996). That these beaches have iconic status and large socioeconomic value is well understood (Batley and Cocks 1992; Klein et al. 2004; Blackwell 2007). Less appreciated is the fact that they are diverse, living ecosystems (McLachlan and Brown 2006) that have intrinsic value and provide many ecological goods and services to humans (Schlacher et al. 2008; Defeo et al. 2009). This failing is largely due to the scant attention ecologists have afforded beaches in contrast with other coastal systems such as rocky shores, sea grasses, mangroves, and coral reefs (Fairweather 1990).

This knowledge deficit is unfortunate because beaches are highly vulnerable to a "coastal squeeze" between burgeoning human populations on the terrestrial side and growing marine environmental changes engendered by climate change. The former imposes numerous developmental, recreational, and pollution pressures on beach ecosystems (Brown and McLachlan 2002; Schlacher et al. 2008; Defeo et al. 2009); the latter threatens environmental challenges including habitat changes (e.g., sand erosion) and the overlying water (e.g., temperature, pH, and circulation) (Jones et al. 2007). In some circumstances, the combination of climate-induced sea-level rise, erosion, and engineering solutions (e.g., seawalls) has the potential to remove sandy beaches entirely. This would be most unfortunate from both ecocentric and anthropocentric points of view. Consequently, the need for informed, scientifically based management is clear if the worst consequences of climate change are to be avoided.

This chapter briefly introduces sandy-beach habitats, their biota and ecology, and linkages to adjacent marine, terrestrial, and estuarine ecosystems (see McLachlan and Brown 2006 for a recent comprehensive ecological text). Attention is then given to the vulnerability of beaches to pressures imposed by climate change, the likely effects of these pressures, the concept of resilience (Walker and Salt 2006), and ecosystem-based approaches (Curtin and Prellezo 2010) to the management and conservation of sandy shores. Herein, the term "beach" is used synonymously with "beach face," defined as "the intertidal slope of the beach between the high- and low-tide marks" by McLachlan and Brown (2006, p. 325).

SANDY BEACHES AS HABITATS

Sandy beaches are unstable, three-dimensional habitats dominated by sand-sized sediments that are strongly influenced by tides, waves, currents, biogenesis, and geological inheritance (Short 1999). Because sedimentary instability excludes attached plants, beach animals are partly dependent on adjacent ecosystems for primary production (see text under Linkages) and lack the secondary habitat options provided by vegetation in other ecosystems. However, as beach sediments are relatively coarse, interstitial spaces occur between the sand particles, and these are inhabited by numerous species of meiofauna and photosynthesizing microflora.

Another habitat feature of beaches is that spatial and temporal abiotic variability can be large. In particular, the effects of season and tide on patterns of desiccation and temperature are important and many species display seasonally and tidally related behavioral adaptations (Brown and McLachlan 1990). Temporal variation is enhanced by the erosive and abrasive power of periodic storms, and spatial variability occurs both within and among beaches. For example, individual beaches exhibit both horizontal (across- and along-shore) and vertical physicochemical gradients in factors such as temperature, moisture content, dissolved oxygen, and sediment grade.

Variation among beaches allows them to be classified morphodynamically into reflective and dissipative beaches (Short and Wright 1984), although, in reality, there is a continuum between the two extremes with most beaches being intermediate (Short 1999). Reflective beaches are narrow and steep with waves breaking at the shoreline. Consequently, they lack a surf zone (the subtidal area where breaking waves approach the beach face), and the swash zone (where broken waves advance up the beach face and retreat as backwash) is narrow and energetic. The sand is coarse with little organic matter and is well drained and oxygenated. By comparison, dissipative beaches are wide and flat with an extensive surf zone, finer sand, abundant organic matter, and a wide, low-energy swash zone that gives filter-feeders greater access to planktonic food. Their vertical chemical gradients are steeper (McLachlan and Turner 1994) and most organisms live in the oxygenated layer near the surface. Reflective and dissipative beaches are more common in tropical and temperate regions, respectively (McLachlan and Brown 2006).

Not only do these habitat features provide the ecological setting, some also promote the vulnerability of beaches to climate change. For example, the instability of the sand facilitates the predicted erosion, desiccation may be worsened by temperature increases, and lower water pH may affect both calcareous sediments and calcium-dependent species.

BIOTA AND ECOLOGY

Contrary to popular belief, sandy beaches are not ecological deserts but support hundreds of species (McLachlan and Brown 2006). A few of these are large and mobile and use the beach or dunes for nesting and feeding (e.g., seabirds and turtles) or are opportunistic visitors (e.g., bears and kangaroos). Most species, however, are small and inconspicuous. Some, such as ocypodid ghost crabs, talitrid sandhoppers, and wrack-dependent isopods, amphipods, and insects, inhabit the upper beach. Others, the great majority, live intertidally beneath the sand surface. They include primary-producing, microscopic algae (mostly episammic and epipelic diatoms), decomposers (bacteria and fungi), and invertebrates (mostly nematodes, crustaceans, polychaetes, and mollusks). The invertebrates are often grouped on the basis of size, the meiofauna being less than half a millimeter long, the macrofauna larger. The former occupy interstitial spaces between sandgrains and comprise far more species than the macrofauna (Brown 2001). Feeding types include scavengers, predators, and deposit- and filter-feeders.

Sandy-beach species exhibit physiological, morphological, and behavioral adaptations to the habitat (Brown and McLachlan 1990). These include orientation mechanisms (Scapini 2006), mobility and burrowing, and rhythmic tidal migrations (Enright 1961; Forward 1986; Hacking 1996; Jones et al. 1998). The life histories of many macrofaunal species are metamorphic with planktonic, dispersive larvae. Exceptions include peracarid crustaceans, a factor that may influence their ability to migrate as climate-change pressures take effect. In addition, many species have calcium-based exoskeletons and their ability to adapt or acclimate to reduced pH imposed by climate change is unknown.

To date, many sandy-beach ecologists have addressed structural questions via descriptive-correlative studies at various spatial scales. At small, within-beach scales, an important distributional paradigm is that of across-beach zonation (see McLachlan and Jaramillo 1995 for a review). Zonation schemes are based either on physical factors (e.g., degree of saturation of the sediments) or biological distributions. However, the term "zonation" may be inappropriate because the across-shore distributional boundaries of sets of species may not coincide (Hacking 1996). Furthermore, many species migrate tidally across the shore (Brown and McLachlan 1990) causing any biologically defined "zones" to vary with tide height. There is, however, a separation of high-shore air breathers from lower-shore water breathers with the low shore being richer in species than the high shore (Hacking 1996; James and Fairweather 1996).

At larger, among-beach scales, there are ecological differences associated with beach geomorphology. Dissipative beaches have more species, greater abundance and biomass, and more soft-bodied species (e.g., polychaete worms) than reflective beaches (McLachlan et al. 1993). In extreme cases, reflective beaches may support no macrofaunal species at all (McLachlan et al. 1993). Patterns of species richness are correlated with composite indices such as the Beach Index of McLachlan and Dorvlo (2005), which is based on tide range, beach slope, and particle size.

Because most patterns are correlated with physical factors rather than biological factors, it is likely that physical factors are most important in structuring the ecological assemblages, i.e., they are "physically controlled" rather than "biologically accommodated" (Brown and McLachlan 1990; Jaramillo and McLachlan 1993). Indeed, Brown et al. (2000) state that wave action is the super-parameter controlling South African sandy-beach ecology, suggesting the applicability of the "autecological hypothesis" (Noy-Meir 1979; McLachlan 1990) in which species in physically controlled habitats respond individualistically to physicochemical factors.

Similarly, the ecological differences between reflective and dissipative beaches were attributed to the hydrodynamic force in the swash zone, i.e., the "swash exclusion hypothesis" (McLachlan et al. 1993). Recently, this has been extended into the "multicausal severity hypothesis" (Brazeiro 2001) and the "habitat harshness hypothesis" (Defeo et al. 2003). The former incorporates the additional physical factors of grain size, organic matter, and erosion-accretion dynamics, while the latter combined the autecological hypothesis with the swash exclusion hypothesis. It predicts a "decrease in the performance of sandy beach macrofauna descriptors at the community (lower species richness, diversity, abundance) and population (lower abundance, growth in weight and length, fecundity, reproductive output, and higher mortality rates) levels from dissipative to reflective conditions" (Defeo et al. 2003, p. 353). A consequence of physical control is that the loss of any particular species would have little effect on others, unlike biologically accommodated assemblages where cascading effects may follow the loss of key species (e.g., Schiel et al. 2004). If true, this may moderate the ecological effects of climate change on beaches.

Although the claim of physical control is plausible, it is also questionable because the explanatory models have not been tested via controlled experimentation, largely because of feasibility problems in an energetic and unstable environment. Moreover, recent studies have indicated the influence of biological factors, especially on dissipative beaches (Defeo and McLachlan 2005). The autecological predictions of the habitat harshness hypothesis were not supported empirically for two Uruguayan species by Defeo et al. (2003) who suggested (p. 352) that "biotic mechanisms could be of utmost importance in dissipative beaches." This conclusion is supported by studies that have suggested or demonstrated the effects of competition in the yellow clam *Mesodesma mactroides* (Braziero and Defeo 1999) and the density-related effects on burrowing in *Donax* spp. (McLachlan 1998; Dugan et al. 2004). In addition, predation (by shore birds, onuphid polychaetes, and ghost crabs) and trophic cascades may be influential. For example, on a southern California beach, the quantity of macrophyte wrack was significantly correlated with the species richness, abundance, and biomass of wrack-associated fauna and the abundance of predators such as shore birds (Dugan et al. 2003). Furthermore, upward cascading reductions can occur if poorly designed beach nourishment interventions suppress macrofaunal recovery (Peterson et al. 2006).

Consequently, both physical and biological factors play ecological roles on sandy beaches, although the relative importance of various factors and mechanisms in explaining distributional patterns may vary with beach type and is uncertain. This knowledge deficit hinders attempts to predict the effects of events such as climate change on beach ecosystems. This is further complicated by the existence of strong linkages between beaches and adjacent habitats.

HABITAT LINKAGES

Sandy beaches have functional, ecological linkages to adjacent ecosystems, especially sand dunes and the surf zone (Brown and McLachlan 1990). Indeed, they are sometimes considered a single ecological unit (James 2000a) or a single geomorphic unit termed the "littoral active zone" in which the exchange of sand and organic matter is particularly important (Short and Hesp 1982; McLachlan and Brown 2006). Beaches are also linked to estuaries in terms of sediments, nutrients, and shared species (Schoeman and Richardson 2002; Sherman et al. 2002; Gladstone et al. 2006). In addition, nutrient-rich groundwater flows through beaches from terrestrial aquifers to the sea.

Trophic links are particularly important because beaches, being unstable, have no large attached plants to provide food for animals. Instead, much primary production is imported via surf-zone phytoplankton, stranded macrophytes, carrion, and terrestrial vegetation (Koop et al. 1982; Dugan et al. 2003). Phytoplankton are consumed by beach-dwelling filter-feeders, and stranded vegetation/detritus is fragmented by small invertebrates. Plant fragments and the nutrient-rich feces of these invertebrate consumers are decomposed by beach bacteria, and nutrients are transported back to the sea to fuel primary production (McLachlan et al. 1981). For example, the feces of the filter-feeding bivalve *Donax serra* generate up to 24% of the nutrients used by phytoplankton blooms (McLachlan et al. 1981). Consequently, the trophic links among the surf zone, kelp beds, rocky shores, and beaches can be strong (Soares et al. 1997). Links also exist with estuaries, which provide a trophic subsidy to beach invertebrates (Schlacher and Connolly 2009). For example, estuarine discharge can influence the abundance of bivalve mollusks on beaches (Donn 1987; Schoeman and

Richardson 2002). In addition, some fish species use surf zones as nurseries (Lasiak 1981; Lenanton and Caputi 1989; J. M. Leis pers. com.) and feed on beach invertebrates (Du Preez et al. 1990).

Sandy beaches are also linked to other ecosystems through the movements of animals that occupy the beach for only part of their life cycle. For example, the beach is an essential transitory element in the long-distance migrations of birds and turtles (Bouchard and Bjorndal 2000; Colombini and Chelazzi 2003). Also, many invertebrates occupy the beach as adults but their dispersive larvae have an oceanic existence as plankton.

Because of these links, climate-change factors may affect beaches indirectly through their effects on adjacent systems. For example, if near-shore planktonic productivity changes because of climate-induced hydrological changes (Barth et al. 2007; Rost et al. 2008), the consequences for intertidal invertebrates will be large. Also, the supply of sediment to beaches will fall if there is increased damming of rivers to compensate for less rain (Sherman et al. 2002; Finkl and Walker 2004) and if increased acidity reduces the offshore biological production of carbonate sediments.

CLIMATE CHANGE

Climate change encompasses several different environmental changes. These have been categorized hierarchically by Cocks and Crossland (1991) into primary changes (higher atmospheric CO_2 concentrations and air temperatures), consequent second-order effects (increased storminess, warmer water with reduced pH, changed rainfall affecting water and sediment flows to the coast), third-order (sea-level rise and changed hydrology), and lower-order effects (saltwater intrusion into aquifers, changes to streamflow, beach erosion, and coastal retreat). Each of these factors has the potential to affect beaches directly and indirectly. Moreover, the disturbances generated by climate change fall into the categories of disturbance that are of particular concern (Dovers et al. 1996; Salafsky et al. 2002; Jones 2003b). These include

- Disturbances of very large scale and/or momentum. These change the global ecological theater with potentially severe consequences for ecosystems and humans alike (Vitousek et al. 1997; IPCC 2002; Stenseth et al. 2002). Erosion and acidification are climate-related examples relevant to beaches.
- Disturbances that persist, i.e., press disturbances (Bender et al. 1984). Increased acidity constitutes a press disturbance. It is likely that the ecological effects will be great and persistent (Orr et al. 2005; Turley et al. 2005; Vezina and Hoegh-Guldberg 2008) unless the steady increase in acidity is sufficiently slow for physiological acclimation or evolutionary adaptation to occur. Beach biota are particularly vulnerable to increased acidity because so many species, especially mollusks and crustaceans, are calcium dependent.
- Disturbances that do not persist, i.e., pulse disturbances (Bender et al. 1984). Storms and pollution are examples. Once these abate, ecological recovery is likely to occur (Ansell 1983). Consequently, pulse disturbances are less serious than press disturbances, although the immediate impacts may be large.

- Ramp disturbances (Lake 2000) are long-term events whose intensity changes over time. Potentially, these are of great importance, especially if they incorporate positive feedback loops that cause accelerating ecological effects. For example, the combination of steadily rising sea levels and storminess have the potential to erode dunes and destroy stabilizing vegetation, thus enabling further erosion to occur.
- Interactions among disturbances may combine with each other or with contaminants (Schiedek et al. 2007) to produce greater effects than each disturbance alone. For example, storms, raised sea levels, and reduced sediment supply will act in concert to exacerbate beach erosion. Apart from the loss of beach habitat, these factors may combine to negatively affect the nesting of turtles (Poloczanska et al. 2009). In addition, calcifying species with acid-weakened shells may be especially vulnerable to the physical effects of increased storminess and vehicular traffic. Also, coastal aquifer depletion may cause land subsidence that interacts with climate change to magnify the effects of sea-level rise (Galbraith et al. 2002). These interactive disturbances are more difficult to assess scientifically than single factors, and are therefore poorly understood and probably underrated as risks.

Managing these disturbances requires the assessment of risk and vulnerability. Risk is a function of the probability of an adverse event and its consequences; vulnerability is a function of exposure to pressure(s), sensitivity to change, and the capacity to adapt (IPCC 1992). The vulnerability of the beach habitat is largely dependent on the rates of sea-level rise, coastal erosion, and frequency of extreme events (Voice et al. 2006). Reduced sediment supply from impounded rivers is also a factor. On this basis, together with the biological effects of climate stresses and the dependencies on vulnerable adjacent systems, beaches are clearly "especially vulnerable" (CSIRO 2002). Indeed, dunes and beachfronts were included among the "vital areas" that would be lost or experience dysfunction via multiorder effects (Cocks and Crossland 1991).

The greatest threats will apply to urban sandy beaches because, unlike nonurban beaches, it is unlikely that they will be allowed to retreat at the expense of societal assets. If these assets are to be protected by seawalls, the very existence of beaches will be jeopardized as seas rise (Cowell et al. 2006). Thus, the coastal squeeze will tighten as stresses due to climate and human population grow further.

Although the exposure of all ecosystems to climate change is clear, knowledge of the ecological effects is an undeveloped but rapidly evolving field.

ECOLOGICAL EFFECTS OF CLIMATE CHANGE

Predictions about overall ecological effects are necessarily imprecise because several forcing factors (i.e., sea level, storminess, wave climate, water temperature, water chemistry, and physical beach processes) are involved, they may interact in complex ways, and some effects may be indirect. Moreover, there is uncertainty about the magnitude of change in these factors, prompting various physicochemical scenarios (IPCC 2007; Brown and McLachlan 2002; CSIRO 2002; International Scientific Steering Committee 2005; Cowell et al. 2006). Nonetheless, there is little doubt that climate change is already affecting both terrestrial and marine ecosystems in many ways (Kennedy et al. 2002; Hughes 2003; Schiel et al. 2004; Lovejoy and Hannah 2005; Harley et al. 2006; Vezina and Hoegh-Guldberg 2008; Brierley and Kingsford 2009; Belant et al. 2010; Najjar et al. 2010). The actual biological processes involved are difficult to identify but include growth, distribution, reproduction, activity rates, recruitment, and mortality (Drinkwater et al. 2010). Moreover, some forecasts and speculations have been made about the consequences of climate-driven biodiversity changes on the functioning of rocky-shore ecosystems (Hawkins et al. 2009), on the structure of calcifying assemblages (Kurihara 2008), on the development of marine invertebrate larvae (Dupont et al. 2008), and on the biomass of intertidal, estuarine invertebrates (Fujii and Raffaelli 2008).

An important question involves the ability of species to acclimate or adapt to climate change, given that climate-related extinctions have already occurred in some habitats (Walther et al. 2002; Hughes 2003). Species able to migrate poleward may evade temperature stress but not other stresses (e.g., increased storminess and acidity). Furthermore, surviving species are likely to live in assemblages whose composition has been altered by climate change—another evolutionary pressure. Taken together, these evolutionary pressures may well have profound ecological effects. Although there is little climate-related research on sandy-beach species directly (Jones et al. 2007), some broad ecological effects of climate change on beaches can be suggested from general principles and relevant research from other habitats.

Temperature

By 2070, annual average air temperatures are projected to rise by 1.0–6.0°C relative to 1990, and changes in extreme temperatures are also expected (IPCC 2007). Presumably, rises in water temperature would be less than those for the air, and the extremes would be less severe. Nevertheless, because temperature is an important ecological factor (Krebs 1978) and some species are stenothermal (Kennedy et al. 2002), even small temperature changes may have severe ecological consequences. For example, Leemans and van Vliet (2005) judged that, to be ecologically acceptable, temperature rises should not exceed 1.5°C, and the rate of change should be less than 0.5°C per decade.

In terms of biogeography, warming is likely to cause a poleward shift in intertidal species' distributional boundaries with the replacement of coldwater species by those from warmer waters (Kennett and Stott 1991). In addition, along-shore distributions may be altered by a "squeeze" effect (Harley et al. 2006). Here, abiotic stress may alter across-shore zonation patterns such that a given species may be displaced (squeezed) by another species moving into its zone. These processes would change the assemblage composition on any shore, although this is not necessarily a cause for concern because variation in the composition of biotic assemblages is normal and assemblage diversity and function may change little.

The species most affected would be those now living close to their upper thermal limit and unable to acclimate or adapt. They would become locally extinct, and their persistence would depend on migration to cooler areas. This may be difficult for intertidal species (IPCC 2001), especially those lacking dispersive larval stages, e.g., peracarid crustaceans, which are an important component of sandy beach macrofauna (McLachlan and Brown 2006). Narrow-range endemic species would be at particular risk. For example, in southeast Australia, 3.7% of coastal marine invertebrates are endemic to the region and some are limited to cool-temperate waters (O'Hara 2002). At even greater risk would be the species unable to migrate to cooler waters such as those endemic to southeastern Tasmania (Edgar et al. 1997). Such species may become extinct as predicted for terrestrial species in equivalent circumstances (Hughes 2003). Consequently, the distributional effects of higher temperatures may vary with latitude and dispersive ability.

While some ecosystems have already experienced substantial temperature-related changes (Walther et al. 2002; Hughes 2003), the effects of temperature rise on sandy beaches may be "subtle rather than dramatic" (Brown and McLachlan 2002, p. 71) because water temperatures will rise less than the air, many species can burrow to evade extreme heat, and the change may be sufficiently gradual to allow acclimation. However, ecological processes such as photosynthesis, decomposition, and nutrient recycling will probably be accelerated by temperature rises as a product of the Q_{10} law. This would enhance productivity, which, in turn, may permit larger populations and greater species richness. Alternatively, the faster decomposition of beach wrack may enhance the deoxygenation of the underlying sand with deleterious effects on some species. In addition, reproductive processes may be affected. For example, the sex ratio in the painted turtle (*Chrysemys picta*) is temperature dependent and a rise of 2–4°C may compromise the production of male offspring (Janzen 1994).

To complicate matters, the ecological effects of temperature change may be both direct and indirect with modulation from species' interactions. For example, temperature can have not only direct effects on the survival of intertidal barnacles, but also indirectly change competitive and predatory relationships (Poloczanska et al. 2008). Other indirect effects of temperature change on the beach biota involve the plankton and oxygen. For example, a rise of 0.6°C was associated with major changes in planktonic ecosystems in the North Sea (Richardson and Schoeman 2004). Given that plankton is a source of food for some adult beach species and/or their larvae, it is likely that the beach biota will be affected by planktonic changes. Also, increased water temperatures will depress dissolved oxygen tensions—a factor likely to be influential in some demersal, stratified situations (Kennedy et al. 2002). However, this is less likely in the energetic, wave-driven intertidal habitat.

Finally, temperature can affect species interactions in food webs (Kennedy et al. 2002). For example, if warming and storms generate more stranded macrophyte wrack, cascading effects on the wrack-dependent fauna are likely (Dugan et al. 2003). Such effects apparently occurred on Californian rocky coasts where community changes were driven largely by changes in abundance of several key species, especially habitat-forming algae (Schiel et al. 2004).

Sea Level

Global sea levels were estimated to rise between 18 and 59 centimeters by 2100 (IPCC 2007), but more recent predictions claim rises of at least one meter per century (quoted in Brahic 2008). These rises, together with increased storminess, will cause geomorphic adjustments to coasts. In particular, sand will be eroded from the upper beach and deposited on the near-shore bottom (Cowell and Thom 1994). This, according to the "Bruun Rule" (Bruun 1988), will cause shorelines to retreat horizontally at 50–100 times the vertical sea-level rise (but note criticism of the Bruun Rule by Cooper and Pilkey 2004). By 2100, recession would range from 4.5 to 88 meters in Australia (CSIRO 2002), but recent stochastic modeling suggests this is an underestimate (Cowell et al. 2006). Consequently, climate change will accelerate the current global trend of beach retreat (70% of beaches are eroding), stable beaches (20–30%) will begin to retreat, and the number of accreting beaches (<10%) will decrease (Burkett et al. 2001). This may affect beach ecosystems little if a slow landward retreat is the only change. However, if there are accompanying habitat changes (e.g., to dunes, wave climate, grain size, beach area, and slope) or habitat losses (e.g., narrower beaches), great ecological effects would occur on a given beach. For example, major intertidal habitat loss was projected to have severe consequences for California shorebirds (Galbraith et al. 2002), and model simulations for the Humber estuary suggested large losses of intertidal area and invertebrate biomass (Fujii and Raffaelli 2008). Alternatively, erosion may be retarded if the sediment supply is increased as the combination of sea-level rise and increased storminess erodes cliff faces.

If dunes suffer rapid erosion, vegetation would be lost. This loss would further destabilize dunes, creating a destructive feedback loop. Such losses would directly affect vegetation-dependent dune fauna and indirectly affect beach fauna via reduced supply of organic matter to the upper beach. In addition, turtles and seabirds that nest in dunes would be severely affected. If particle size and beach slope (i.e., morphodynamic state) change, the richness, abundance, and kinds of macrofauna will change because steep, coarse-grained, reflective beaches are impoverished compared with flat, fine-grained, dissipative beaches. Similar ecological consequences may arise if climate change enhances along-shore sediment transfer via changes in the directional weightings of the wave climate, i.e., "Gordon Effects" (Cowell and Thom 1994). Shoreline stability is even more sensitive to Gordon Effects than Bruun Effects (Cowell and Thom 1994).

The ecological consequences of sea-level rise may be exacerbated by engineering responses such as seawalls and groynes. The latter accumulate sand in their vicinity but also lead to downdrift erosion (Peterson et al. 2000). This downdrift erosion is likely to increase under conditions of higher sea levels, and more frequent storms will lead to loss of significant areas of beach habitat including entire beaches (Speybroeck et al. 2006). Where seawalls are used to defend societal assets in urban areas, beaches may be lost entirely as in the "jersification" of the coast (Finkl and Walker 2004). However, the meager literature on the ecological effects of seawalls shows conflicting results. For example, while the ocypodid ghost crab *Ocypode cordimanus* appears to be badly affected in New South Wales (Barros 2001), the macrofauna of a Chilean beach showed no significant effects (Jaramillo et al. 2002).

Storminess/Wave Energy

Climate change will intensify storms and amplify wave energy (IPCC 2001) with several consequences for beach species. First, there may be a greater metabolic cost of maintaining position in the swash zone, with flow on effects to survival and reproduction. However, the burrowing anomuran crab *Emerita analoga* showed no difference in oxygen uptake (a measure of physiological stress) across a range of wave exposures on the same beach (Lastra et al. 2004). Either the premise was false or acclimation to changes in wave energy occurred.

Second, increased wave energy may change the morphodynamic state (grain size and slope) of the beach. If beaches were to become more reflective, their macrobenthic assemblages would become impoverished as explained previously. In addition, changes in grain size and beach slope may affect settlement and recruitment. For example, the South African wedge clam *D. serra* settles as spat in the surf zone, preferring areas with fine sediment. Juveniles subsequently recruit into the adult population on the intertidal beach, which is influenced by the beach slope (Schoeman and Richardson 2002). Furthermore, if a changed beach slope caused the abundance of *Donax* to fall, there may be cascading trophic consequences for shorebirds. For example, Peterson et al. (2006) found that fewer shorebirds (e.g., sanderlings) foraged on beaches after a reduction in the density of their prey, *Donax* spp. In addition, a change in sediment composition can affect the foraging efficiency of shorebirds by reducing their ability to probe with their bills and manipulate prey (Quammen 1982; Peterson et al. 2006). Other climate-related population changes involve the sandhopper *Talitrus saltator* on the Baltic coast (Weslawski et al. 2000), which now occurs in fewer localities and at lower densities compared with previous records—a pattern correlated with increases in storm frequency, winter severity, and sea level.

Third, storms dislodge kelp, which is deposited on beaches as wrack (Griffiths and Stenton-Dozey 1981). If bigger storms produce more beach wrack, there may be contrasting faunal effects (McGwynne et al. 1988). There would be more localized hypoxia via decomposition. If the redox discontinuity layer rises, or if "methane geysers" develop underneath dense wrack accumulations, infaunal invertebrate populations would decrease (McLachlan 1985; Hacking, unpublished data). In addition, stranded kelp affects both the feeding and burrowing efficiency of the wedge clam, *D. serra* (Soares et al. 1996). However, because this is a pulse disturbance, recovery should occur after dispersal of the wrack. In contrast, species that depend on wrack for food or habitat are likely to benefit from denser wrack. These include insect larvae, amphipods, and their predators (Dugan et al. 2003). Furthermore, wrack piles insulate the underlying sand from temperature extremes and maintain high humidity. Beach fauna that have no active mechanism for water retention such as the amphipod *Orchestia gammarellus* are favored by this.

Fourth, increased hydrodynamic energy may cause increased near-shore turbidity and decreased surf-zone photosynthesis. If so, this would reduce the food supply for beach fauna, especially on dissipative beaches. Increased hydrodynamic energy may also enhance abrasion, which may be deleterious for some species, especially those with shells weakened by acidity. Finally, a likely consequence of larger storms is the increased risk of shipwrecks and oil pollution with consequent impacts (Jones

2003a; de la Huz et al. 2005). However, this may be a declining threat with the imminent advent of peak oil.

Oceanography

Climate change may alter both horizontal and vertical current systems via raised water temperatures (Pearce 2005). If so, numerous species would probably be affected because many migrate either as larvae (most invertebrates) or as adults (e.g., turtles). These changes may arise via changes at large scales (e.g., changes in major current systems) and small scales. The latter includes possible changes in near-shore currents responsible for the deposition of wrack and carrion and local larval transport and retention. Furthermore, more energy may boost coastal upwelling (IPCC 2001). This would transport more nutrients to the photic zone and enhance photosynthesis, thus increasing the food supply for beach filter-feeders.

Rainfall

Rainfall is likely to increase in some regions and decrease in others (CSIRO 2002). These changes may affect beaches because of their linkages with dunes and lagoons. For example, rainfall may influence the growth of dune vegetation with consequences for dune stability and the exchanges of sand and organic matter with the beach. In addition, rainfall changes will alter the discharges from estuaries and the frequency of natural openings of coastal lagoons. This may change the supply of sediment and nutrients from terrestrial catchments to the sea and beach. Also, rainfall changes would alter the erodability of sand and the level of the water table of beaches. In turn, this is likely to affect the distribution or survival of burrowing species of the upper beach via desiccation.

Acidification, Calcification, Acidosis

While CO_2 in the atmosphere is relatively inert, it becomes highly reactive when dissolved in seawater, this causing the pH of ocean waters to fall as atmospheric CO_2 levels rise. In fact, the pH of surface waters is already 0.1 units less than preindustrial levels and is projected to decline between 0.14 and 0.35 units over the twenty-first century (IPCC 2007)—changes not seen for about 20 million years (Haugan and Drange 1996; Brewer 1997). This acidification can potentially affect many biogeochemical conditions and processes (Orr et al. 2005; Turley et al. 2005), and its importance has prompted a theme section in a leading marine journal (Vezina and Hoegh-Guldberg 2008).

Because lowered pH will reduce the concentration of carbonate, there may be both physical and biological consequences. The former involves the reduction of sediment supply, especially on carbonate beaches. The latter involves biological calcification rates, which will decrease in oceanic organisms ranging from test-forming phytoplankton to scleractinian corals (Feely et al. 2004). Beach taxa such as mollusks and crustaceans are also dependent on calcification because they have robust protective shells. The crustacean exoskeleton contains chitin, lipids, and proteins strengthened by either amorphous or crystalline calcium carbonate. Mollusk shells have crystals of calcium carbonate dispersed in an organic matrix (Lowenstam 1981; Harper 2000). These calcium carbonate crystals take the form of either aragonite or calcite, there being considerable variation within molluskan families in crystal type (Taylor and Reid 1990). Because aragonite is at least 50% more soluble in acidic seawater than calcite, mollusks with aragonite shells may be particularly susceptible to acidification (Mucci 1983; Feely et al. 2004). Because the shell thickness affects the susceptibility to breakage (Vermeij 1979), mollusks will presumably become more vulnerable to predators (e.g., fishes, crabs, and shorebirds), abrasion from storms, and also to crushing by off-road vehicles. It is thus reasonable to assume that many beach species will be at risk if their calcium metabolism is impaired.

An elevated level of CO_2 can have other marine biological effects. First, it may promote photosynthesis (Melillo et al. 1993) and thus the nutritional supply to beach fauna. Second, elevated CO_2 leads to the acidosis of tissues and body fluids with negative consequences for growth and reproduction, especially in lower marine invertebrates because they have a weak capacity for physiological compensation (Pörtner et al. 2004; Pörtner 2008). Third, acidification can reduce the development, growth, and survival of planktonic, invertebrate larvae (Dupont et al. 2008; Kurihara 2008), a situation with probable consequences for beach-dwelling adults. Finally, the possibility of long-term negative effects is suggested by the role of oscillations in aquatic CO_2 in the Permian Triassic mass-extinction events (Bambach et al. 2002; Berner 2002).

Alternatively, elevated CO_2 and acidity may pose few problems for species that can acclimate or adapt, that have a specialized tolerant physiology, or that are preadapted to environments with wide natural variations in CO_2 (Pörtner et al. 2004). Sandy beaches may be such environments because interstitial CO_2 concentrations are naturally variable, i.e., they increase during high tide and decrease during low tide (Pearse et al. 1942). Species living interstitially, especially nematodes and harpacticoids, can be tolerant of both varying pH and temperature (Wieser et al. 1974).

Interacting Stressors

More complicated are the effects of interactions among climate-change factors and with other anthropogenic pressures. For example, if the shells of calcifying species are weakened by lowered pH, they may be more susceptible to vehicles and heavier storms. In addition, nonclimate factors such as pollution may act to lower the resilience of assemblages to the pressures of climate change. Also, the interaction of increased wave energy with sea-level rise and spring tides may combine to exacerbate the erosion of beaches and dunes with probable ecological impacts. While the intertidal beach invertebrates may recover rapidly from storms (Ansell 1983), their effects on dunes would persist longer if attached dune plants and their dependent fauna were lost (Brown and McLachlan 2002; DLWC 2001). Birds such as the Little Tern, Pied Oystercatcher, Beach Stone-Curlew, Hooded Plover, and Red-Capped Plover nest in the supralittoral (Annette Harrison, pers. comm.) as do the Green Turtle and Leathery Turtle. Consequently, these species would be at risk and most are already listed as threatened.

In summary, sandy-beach ecosystems are particularly vulnerable to climate change, which is already changing their physicochemical and biological components in complex ways across various scales in space and time. Of the several climate factors involved, those that promote sea-level rise and erosion seem to be the most important because they threaten the very existence of the intertidal beach habitat, especially in urban areas where societal assets are protected by hard engineering. Elsewhere, this threat may be minimal if setbacks are available and beaches are able to retreat slowly. Another factor of great potential importance is acidification because many beach species have calcium-based shells. Given the vulnerability of beaches to a range of climate pressures and the limited ecological knowledge available, the advent of climate change will intensify the challenges already faced by coastal scientists, policy makers, and especially managers.

MANAGEMENT

> The challenge for the 21st century, then, is to understand the vulnerabilities and resilience of ecosystems so that we can find ways to reconcile the demands of human development with the tolerances of nature. (World Resources Institute 2000)

General Considerations

Because environmental management is meaningless in the absence of clear goals, it is now common practice to develop a vision statement, objectives consistent with the vision, strategies for reaching the objectives, and monitoring performance indicators to measure success. A suggested vision statement for beaches is

> Beaches maintained in a near-pristine state supporting fully diverse, functioning ecosystems and sustainable low-impact human uses.

This vision is not only consistent with the principles of ecologically sustainable development (ESD), but also with both ecocentric and anthropocentric ethics. It amounts to a win-win solution because the value of most beaches to humans derives from their natural state. The ESD objectives for ecosystems include the protection of biodiversity and ecological processes. More specific objectives should apply to particular beaches or regions, but such clarity is rarely present for sandy beaches (James 2000b). Of course, the emphasis will vary depending on the context. In pristine areas, primary objectives may involve the maintenance of the habitat for recreational and conservation purposes. In urban areas, the protection of societal assets is becoming increasingly important with the rise of sea levels.

In attempting to achieve desired visions and objectives, two broad management strategies concerning climate change exist: mitigation and adaptation. Mitigation slows or even reverses climate change by reducing emissions; adaptation accepts the reality of climate change and copes with unavoidable impacts with various technological options available for the coastal zone (Klein et al. 2001). Major investment in both kinds of strategy is needed now (International Scientific Steering Committee 2005).

Mitigation is a long-term approach because the greenhouse effect already has great momentum. Suggested approaches include reducing demand for energy by human behavioral modification, the rapid development and deployment of cleaner-energy technologies, and the geosequestration of CO_2. Less accepted but arguably more important would be addressing the underlying drivers of climate change, i.e., growth in human populations and economies—factors usually ignored in ecological management. Unfortunately, the task is daunting with the human population increasing by over 70 million per year (UNFPA 2005) and energy-demanding economic growth is espoused by most politicians and economists. For example, Australia's last Prime Minister, John Howard, said, "The idea that we can address climate-change matters successfully at the expense of economic growth is not only unrealistic but also unacceptable" (reported in Breusch 2006). Consequently, mitigation will depend largely on human behavioral changes and technical fixes. Any slowing of climate change will be ecologically beneficial because it will allow more time for evolutionary adaptation to occur. Coastal adaptive approaches include various engineered defenses, legislating for setbacks, and the minimization of nonclimate pressures (e.g., pollution, off-road vehicles) in an attempt to enhance the resilience of ecosystems.

Ecological resilience has been defined as "returning to the reference state (or dynamic) after a temporary disturbance" (Grimm and Wissel 1997, p. 323). It is one of the components of stability, the others being constancy (staying essentially unchanged) and persistence (persistence through time of an ecological system). Unfortunately, these terms are plagued by difficulties and ambiguities (Grimm and Wissel 1997). For example, constancy is similar to resistance (the ability to resist disturbance without substantial change) and resilience is similar to recovery (return to a reference state following ecological impacts) and elasticity (speed of return to a reference state). Resilience is used here to incorporate the concepts of recovery and resistance because these are issues of management interest. Note that the recovery component applies to pulse disturbances such as storms in which the disturbance is temporary and ecological recovery is likely. Importantly, climate change also imposes press disturbances (e.g., higher temperatures and acidity), which persist. Recovery to the original regime may not occur unless acclimation or adaptation takes place. If not, resistance becomes an important management question, e.g., what fall in pH can be tolerated before an alternative state arises?

To further complicate the concept of resilience, it has evolved to incorporate socioeconomic components (IPCC 2001) and has become a management buzzword (see, e.g., Walker and Salt 2006; Gotts 2007). It has arisen because of two false assumptions in natural resource policy. The first assumption is that ecosystem responses to human use are linear, predictable, and controllable. The second is that human and natural systems can be treated independently. In this context, resilience science acknowledges humans as components of ecosystems and focuses on social-ecological systems (SESs) in an integrated manner (e.g., Folke 2006; Walker et al. 2006). It addresses questions of sustainability in a complex and changing world. A resilient system is one whose ecological and socioeconomic subsystems can either resist stresses/disturbances or else recover in an acceptable time frame.

Several general principles apply to enhancing the resilience of beach SESs. First, all the important pressures must be identified and addressed. Second, preventing multiple stresses that have synergistic effects enhances resilience. In fact, the only adaptive management strategy for press stresses such as lowered pH may be the removal of additional stresses such as pollution or off-road vehicles (ORVs). Third, management strategies must be ecosystem based (EBM). Such management recognizes the holistic, inter-relatedness of ecosystems at various scales in space and time. It also recognizes the interdependence among ecological, social, economic, and institutional perspectives; considers cumulative effects; and applies the precautionary principle (Dovers and Handmer 1995). A key goal of EBM is to sustain the ability of ecosystems to deliver all ecosystem services rather than the current preoccupation with the short-term provision of a single service (McLeod et al. 2005).

Fourth, maintaining biodiversity (to promote adaptation) and multiple examples of each habitat type (to provide insurance and colonists following impacts) enhances resilience. Fifth, effective management needs substantial scientific knowledge of the ecosystem in question and its coupling with socioeconomic systems. This need is magnified by the probability that climate change will have novel, unpredictable ecological effects (Schneider and Root 1996) or cause abrupt, nonlinear changes (Burkett et al. 2005). Existing knowledge will thus be inadequate. In these circumstances, active adaptive management is highly recommended (Walters 1986; Folke et al. 2002). Briefly, this is a learn-by-doing approach that views policies as experiments elucidating the processes affecting resilience. It requires flexible social institutions that promote learning and adaptation.

Although the EBM/resilience approach is in its infancy, some suggestions regarding beach resilience can be made.

Beach Resilience

A great threat from climate change involves beach erosion and retreat because this will threaten both social systems (through loss of economy, amenity, and assets) and ecological systems (via loss of beach habitat). Consequently, the integrity of the sediment budget is of prime concern to beach ecosystem resilience. Adaptive measures to manage the sediment budget include the protection, vegetation, and stabilization of dunes, and maintaining the supply of sediment from coastal rivers. Dune management is crucial because dunes provide a protective buffer and reservoir of sand that is important at times of intense storms and erosion.

Unfortunately some dunes have already been impaired by human activities (Bird 1996; Tomlinson 2002; Alonso and Cabrera 2002) and numerous dams now reduce the supply of sediment to the coast (Sherman et al. 2002). Moreover, dune management would only retard coastal retreat. Other adaptive strategies involving managed retreat (e.g., rolling easements or setbacks that allow the landward migration of the coastline) will be necessary. Unfortunately, setbacks involving buyouts would be extremely expensive and may be socially unacceptable in urban areas. Here, society may protect coastal assets via engineering solutions such as large-scale beach nourishment and/or the construction of seawalls. The latter, however, would cause beaches to become narrower, reducing biotic diversity and abundance (Dugan et al. 2008), or they may even be lost entirely (Pilkey and Wright 1989). Given accepted conservation objectives that entail the protection of biodiversity and ecological processes, seawalls would fail as an ecological management strategy. Moreover, seawall protection may only be a short-term solution as sea levels continue to rise. In the long term, "our coastal communities may have to rethink their location and may be forced to consider a retreat from the beach" (Tomlinson 2002).

An alternative to seawalls is the soft-engineering approach of beach nourishment—a process whereby beaches are raised by the importation of sand, often from offshore. Nourishment has become popular because it can maintain beaches in a seminatural state (Finkl and Walker 2004). Thus it is more attractive than seawalls for societal purposes and, moreover, well-designed projects enable ecological recovery to be fast, i.e., weeks to months rather than years (Speybroeck et al. 2006; Jones et al. 2008). Nevertheless, drawbacks exist, e.g., engineering is expensive, ecological impacts occur at both borrow and deposition sites, repetition is needed, and the supply of sand may run out. In addition, replacement sediments should match the original beach sediments; otherwise, ecological recovery is retarded or prevented (Nelson 1993; Bilodeau and Bourgeois 2004; Peterson and Bishop 2005; Peterson et al. 2006).

Management recommendations for nourishment made by Speybroeck et al. (2006) include

- Importing sediments and creating beach profiles that match the original beach conditions as closely as possible
- The avoidance of sediment compaction
- Careful timing of operations to minimize biotic impacts and enhance recovery
- The selection of locally appropriate techniques
- The implementation of several small projects rather than a single large project
- Interspersion of nourished beach sections with unaffected areas
- Repeated application of sediment in shallow layers (<30 cm) rather than single pulses that kill the fauna by deep burial

In addition, the positioning of dredged sand is a management issue. For example, the deposition of sand on ebb-tide deltas may be ecologically preferable to direct deposition on beaches (Bishop et al. 2006). Similarly, the across-shore positioning of spoil (from high tide to shallow subtidal) is likely to be an ecological factor. Furthermore, periodic renourishment (and consequent economic and ecological costs) will be minimized if the beach profile is engineered to be stable (Blott and Pye 2004). Certainly, the choice of option will depend on local economic considerations (Knogge et al. 2004) and different approaches may be integrated (Finkl and Walker 2004).

Concerning other consequences of climate change (e.g., higher temperatures, lower pH, and changed hydrodynamics), management options are limited to mitigation and minimizing additional stresses such as pollution, beach grooming, and ORVs. These not only impose ecological impacts in their own right (Godfrey and Godfrey 1980; Defeo et al. 2009; Schlacher et al. 2007), but also may interact with climate factors (see previous text). Perhaps the best hope is that these climate changes will occur sufficiently slowly for natural acclimation and/or evolutionary adaptation to occur.

In general, the resilience of an impacted beach will be enhanced if protected, unaffected beaches exist sufficiently close to provide dispersive, colonizing larvae. Protected beaches should be selected according to the comprehensive/adequate/representative criteria used elsewhere. In particular, different morphodynamic beach types should be included in different biogeographic zones. Dissipative beaches are particularly rich in biota and may also provide planktonic, larval colonists to depauperate reflective beaches, i.e., the source-sink hypothesis (Defeo and McLachlan 2005). Moreover, adjacent terrestrial and marine areas should also be protected in order to accommodate the ecosystem linkages described previously.

SUMMARY

Sandy beaches are immensely important economically and culturally. Contrary to popular opinion, they provide numerous ecological services and support diverse ecological assemblages whose species are mostly small and buried, and evade attention. Beaches vary morphodynamically and this affects their biotic diversity (dissipative beaches are biologically richer than reflective beaches) as does tidal height (more species live low on the beach). Biotic assemblages appear to be largely physically controlled, especially on steep, reflective beaches, but biological relationships can also be present. In addition, beach ecosystems have strong linkages with adjacent ecosystems, especially dunes and the surf zone, concerning their sand budget, primary production, life histories, and migratory species. These linkages have strong implications for the scales at which management should operate.

Concern about beaches is increasing because of their vulnerability to burgeoning human terrestrial pressures and climate-related marine pressures. The latter include rises in temperature, sea level, storminess, and oceanic acidity. Most are large-scale press disturbances from which recovery will depend on acclimation or evolutionary adaptation. These disturbances may affect the biota singly or in combination and may be synergistic with existing stresses such as pollution and ORVs. Effects may be direct (e.g., reduced calcification) or indirect (e.g., changes to planktonic food supply).

Of particular concern is acidification because it is a press disturbance affecting all beaches, and many beach species are calcium dependent. Also important is the combination of storm surges and sea-level rise, which will exacerbate erosion rates. The most extreme effect can be the total loss of the beach habitat, especially if seawalls are deployed to protect urban coastal assets. Alternatively, in other areas, there may be a slow retreat of the coastline with few effects on beach ecosystems.

Concerning management, a perspective that seeks to promote resilience is appropriate. Resilience is a concept with both ecological and social components and treats these in an integrated manner. In terms of broad strategies, both mitigation and adaptation are needed. The former would seek to constrain the increase in atmospheric greenhouse gases by addressing the underlying causes (i.e., population and economic growth) and by applying appropriate low-carbon technologies. Adaptation strategies would recognize the linkages between beaches, dunes, and surf zones; maintain sand movement and storage; and allow for the landward migration of beaches. Failing this, the active maintenance of beaches in their present location by nourishment is far preferable to hard engineering (e.g., seawalls) from the nature conservation and socioeconomic points of view. In addition, resilience would be promoted by minimizing nonclimate stresses, which may act synergistically with climate factors. Because beach ecosystems are poorly understood, active adaptive management is recommended.

ACKNOWLEDGMENTS

I have learned much about sandy beach biology from the "Serans group" (Anton McLachlan, Thomas Schlacher, Felicita Scapini, David Schoeman, Mariano Lastra, and Jenifer Dugan) and from other prominent biologists such as Omar Defeo and Eduardo Jaramillo. Thanks are also due to William Gladstone and Nicole Hacking, who contributed to an earlier version of this paper; to Peter Cowell, Andy Short, Peter Roy, and Doug Lord for discussions on geomorphology; to Ross Sadlier, Helen Stoddart, Roger Springthorpe, and Anna Murray for taxonomic assistance; and especially to Robin Marsh for organizing the literature. An anonymous referee made many constructive comments.

REFERENCES

Alonso, J.A.-C., and J. Cabrera. 2002. Tourist resorts and their impact on beach erosion at Sotavento beaches, Fuerteventura, Spain. *Journal of Coastal Research* SI 36: 1–7.

Ansell, A.D. 1983. The biology of the genus *Donax*. In *Sandy Beaches as Ecosystems*, Eds. A. McLachlan and T. Erasmus, 607–635. The Hague, Netherlands: Junk.

Awosika, L.F., A.C. Ibe, and C.E. Ibe. 1993. Anthropogenic activities affecting sediment load balance along the West African coastline. In *Coastlines of Western Africa,* Eds. L.F. Awosika, A.C. Ibe, and P. Shroader, 26–39. New York: American Society of Civil Engineers.

Bambach, R.K., A.H. Knoll, and J.J. Sepkowski Jr. 2002. Anatomical and ecological constraints on Phanerozoic animal diversity in the marine realm. *Proceedings of the National Academy of Sciences of the United States* 99: 6845–6859.

Barros, F. 2001. Ghost crabs as a tool for rapid assessment of human impacts on exposed sandy beaches. *Biological Conservation* 97: 399–404.

Barth, J.A., B.A. Menge, J. Lubchenko, F. Chan, J.M. Bane, A.R. Kirinciich, M.A. McManus, K.J. Nielsen, S.D. Pierce, and L. Washburn. 2007. Delayed upwelling alters nearshore coastal ocean ecosystems in the northern California current. *Proceedings of the National Academy Sciences of the United States* 104: 3719–3724.

Batley, G.E., and K.D. Cocks. 1992. Defining and Quantifying National Coastal Resources. Working document 92/11. CSIRO, Canberra.

Belant, J.L., E.A. Beever, J.E. Gross, and J.L. Lawler. 2010. Special Section. Ecological responses to contemporary climate change within species, communities, and dcosystems. *Conservation Biology* 24: 7–9.

Bender, E.A., T.J. Case, and M.E. Gilpin. 1984. Perturbation experiments in community ecology: theory and practice. *Ecology* 65: 1–13.

Berner, R.A. 2002. Examination of hypotheses for the Permo-Triassic boundary extinction by carbon cycle modeling. *Proceedings of the National Academy of Sciences of the United States* 99: 4172–4177.

Bilodeau, A.L., and R.P. Bourgeois. 2004. Impact of beach restoration on the deep-burrowing Ghost Shrimp, *Callichirus islagrande. Journal of Coastal Research* 20: 931–936.

Bird, E.C.F. 1996. *Beach Management*. Chichester, UK: Wiley.

Bishop M.J., C.H. Peterson, H.C. Summerson, H.S. Lenihan, and J.H. Grabowski. 2006. Deposition and long-shore transport of dredge spoils to nourish beaches: impacts on benthic infauna of an Ebb-Tidal Delta. *Journal of Coastal Research* 22: 530–546.

Blackwell, B. 2007. The value of a recreational beach visit: an application to Mooloolaba beach and comparisons with other outdoor recreation sites. *Economic Analysis and Policy* 37: 77–98.

Blott, S.J., and K. Pye. 2004. Morphological and sedimentological changes on an artificially nourished beach, Lincolnshire, UK. *Journal of Coastal Research* 20: 214–233.

Bondesan, M., G.B. Castiglioni, C. Elmi, G. Gabbianelli, R. Marocco, P.A. Pirazzoli, and A. Tomasin. 1995. Coastal areas at risk from storm surges and sea level rise in northeasterrn Italy. *Journal of Coastal Rese*arch 11: 1354–1379.

Bouchard, S.S., and K.A. Bjorndal. 2000. Sea turtles as biological transporters of nutrients and energy from marine to terrestrial ecosystems. *Ecology* 81: 2305–2313.

Brahic, C. 2008. Sea level rises could far exceed IPCC estimates. *New Scientist* September 1.

Brazeiro, A. 2001. Relationship between species richness and morphodynamics in sandy beaches: what are the underlying factors? *Marine Ecology Progress Series* 224: 35–44.

Brazeiro, A., and O. Defeo. 1999. Effects of harvesting and density-dependence on the demography of sandy beach populations: the yellow clam *Mesodesma mactroides* of Uruguay. *Marine Ecology Progress Series* 182: 127–135.

Breusch, J. 2006. Greenhouse summit rejects fossil fuel cuts. *Financial Review* January 13: 1.

Brewer, P.G. 1997. Ocean chemistry of the fossil fuel CO_2 signal: the haline signal of "business as usual." *Geophysical Research Letters* 24: 1367–1369.

Brierley, A.S., and M.J. Kingsford. 2009. Impacts of climate change on marine organisms and ecosystems. *Current Biology* 19: R602–R614.

Brown, A.C. 2001. Biology of sandy beaches. In *Encyclopedia of Ocean Sciences*, Eds. J.H. Steele, S.A. Thorpe, and K.K. Turekian, Volume 5, 2496–2504. London: Academic Press.

Brown, A.C., and A. McLachlan. 1990. *Ecology of Sandy Shores*. Amsterdam: Elsevier.

———. 2002. Sandy shore ecosystems and the threats facing them: some predictions for the year 2025. *Environmental Conservation* 29: 62–77.

Brown, A.C., A. McLachlan, G.I.H. Kerley, and R.A. Lubke. 2000. Functional ecosystems: sandy beaches and dunes. In *Marine Biodiversity Status Report,* www.nrf.aczal/publications/marinerap. Access date 24 June 2006.

Bruun, P. 1988. The Bruun Rule of erosion by sea level rise: a discussion of large-scale two- and three-dimensional usages. *Journal of Coastal Research* 4: 627–648.

Burkett, V., J.O. Codignotto, D.L. Forbes, N. Mimura, R.J. Beamish, and V. Ittekkot. 2001. Coastal zones and marine ecosystems. In *Climate Change 2001: Impacts, Adaptation, and Vulnerability*. Contribution of Working Group II to the Third Assessment report of the Intergovernmental Panel on Climate Change, Eds. J.J. McCarthy, O.F. Canziani, N.A. Leary, D.J. Dokken, and K.S. White, 343–380, United Kingdom: Cambridge University Press.

Burkett, V.R., D.A. Wilcox, R. Stottlemyer, W. Barrow, D. Fagre, J. Baron, J. Price, J.L. Nielsen, C.D. Allen, D.L. Peterson, G. Ruggerone, and T. Doyle. 2005. Nonlinear dynamics in ecosystem response to climate change: case studies and policy implications. *Ecological Complexity* 2: 357–394.

Cocks, K.D., and C. Crossland. 1991. The Australian coastal zone: a discussion paper. Working Document 91/4. CSIRO, Canberra.

Colombini, I., and L. Chelazzi. 2003. Influence of marine allochthonous input on sandy beach communities. *Oceanography and Marine Biology* 41: 115–159.

Cooper, J.A.G., and O.H. Pilkey. 2004. Sea-level rise and shore-line retreat: time to abandon the Bruun Rule. *Global and Planetary Change* 43: 157–171.

Cowell, P.J., and B.G. Thom. 1994. Coastal impacts of climate change—modelling procedures for use in local government. *Proceedings of the First National Coastal Management Conference, Coast to Coast '94*. Hobart, 43–50.

Cowell, P.J., B.J. Thom, R.A. Jones, C.H. Everts, and D. Simanovic. 2006. Management of uncertainty in predicting climate-change impacts on beaches. *Journal of Coastal Research* 22: 232–245.

CSIRO. 2002. Climate change and Australia's coastal communities. www.dar.CSIRO.au/publications/coastalbroch2002.pdf Access date 24 June 2006.

Curtin, R., and R. Prellezo. In press. Understanding marine ecosystem based management: a literature review. *Marine Policy*. 35: 821–830, 2010.

de la Huz, R., M. Lastra, J. Junoy, C. Castellanos, and J.M. Vieitez. 2005. Biological impacts of oil pollution and cleaning in the intertidal zone of exposed sandy beaches: preliminary study of the "Prestige" oil spill. *Estuarine Coastal and Shelf Science* 65: 19–29.

Defeo, O. 2003. Marine invertebrate fisheries in sandy beaches: an overview. *Journal of Coastal Research* SI 35, 56–65.

Defeo, O., and A. McLachlan. 2005. Patterns, processes and regulatory mechanisms in sandy beach macrofauna: a multi-scale analysis. *Marine Ecology Progress Series* 295: 1–20.

Defeo, O, A. McLachlan, D. Schoeman, T. Schlacher, J. Dugan, A. Jones, M. Lastra, and F. Scapini. 2009. Threats to sandy beach ecosystems: a review. *Estuarine, Coast and Shelf Science* 81: 1–12.

DLWC. 2001. *Coastal Dune Management: A Manual of Coastal Dune Management and Rehabilitation Techniques*. Newcastle, Australia: NSW Department of Land and Water Conservation, Coastal Unit.

Donn, T.E. 1987. Longshore distribution of *Donax serra* in two log-spiral bays in the eastern Cape, South Africa. *Marine Ecology Progress Series* 35: 217–222.

Dovers, S.R., and J.W. Handmer. 1995. Ignorance, the precautionary principle and sustainability. *Ambio* 24: 92–97.

Dovers, S.R., T.W. Norton, and J.W. Handmer. 1996. Uncertainty, ecology, sustainability and policy. *Biodiversity and Conservation* 5: 1143–1167.

Drinkwater, K.F., G. Baugrand, M. Kaeriyama, S. Kim, G. Ottersen, R.I. Perry, H.-O. Portner, J.H. Povlovina, and A. Takasuka. 2010. On the processes linking climate to ecosystem changes. *Journal of Marine Systems* 79: 374–388.

Du Preez, H.H., A. McLachlan, J.F.K. Marais, and A.C. Cockroft. 1990. Bioenergetics of fishes in a high-energy surf-zone. *Marine Biology* 106: 1–12.

Dugan, J.E., D.M. Hubbard, M.D. McCrary, and M.O. Pierson. 2003. The response of macrofauna communities and shorebirds to macrophyte wrack subsidies on exposed sandy beaches of southern California. *Estuarine, Coastal and Shelf Science* 58S: 25–40.

Dugan, J.E., D.M. Hubbard, I. Rodil, D.L. Revell, and S. Schroeter. 2008. Ecological effects of coastal armoring on sandy beaches. *Marine Ecology* 29: 160–170.

Dugan, J.E., E. Jaramillo, D.M. Hubbard, H. Contreras, and C. Duarte. 2004. Competitive interactions in macroinfaunal animals of exposed sandy beaches. *Oecologia* 139: 630–640.

Dugan, J.E., O. Defeo, E. Jaramilb, A.R. Jones, M. Lastra, R. Nel, C.H. Peterson, F. Scapini, T. Schlager, and D.S. Schoeman. 2010. Give beach ecosystems their day in the sun. *Science* 329, (3 September 2010): 1146.

Dupont, S., J. Havenhand, W. Thorndyke, L. Peck, and M. Thorndyke. 2008. Near-future level of CO_2-driven ocean acidification radically affects larval survival and development in the brittlestar *Ophiothrix fragilis*. *Marine Ecology Progress Series* 373: 285–294.

Edgar, G.J., J. Moverly, N.S. Barrett, D. Peters, and C. Reed. 1997. The conservation-related benefits of a systematic sampling programme: the Tasmanian reef bioregionalisation as a case study. *Biological Conservation* 79: 227–240.

Enright, J.T. 1961. Pressure sensitivity of an amphipod. *Science* 133: 758–760.
Fairweather, P.G. 1990. Ecological changes due to our use of the coast: research needs versus effort. *Proceedings of the Ecological Society Australia* 16: 71–77.
Feely, R.A., C.L. Sabine, K. Lee, W. Berelson, J. Kleypas, V.J. Fabry, and F.J. Millero. 2004. Impact of anthropogenic CO_2 on the $CaCO_3$ system in the oceans. *Science* 305: 362–366.
Finkl, C.W., and H.J. Walker. 2004. Beach nourishment. In *The Encyclopedia of Coastal Science*, Ed. M. Schwartz, 37–54. Dordrecht, Netherlands: Kluwer Academic Press.
Folke, C. 2006. Resilience: the emergence of a perspective for social-ecological systems.
Folke, C., S. Carpenter, T. Elmqvist, L. Gunderson, C.S. Holling, and B. Walker. 2002. Resilience and sustainable development: building adaptive capacity in a world of transformations. *Ambio* 31: 437–440.
Forward, R.B. 1986. Behavioral responses of a sand-beach amphipod to light and pressure. *Journal of Experimental Marine Biology and Ecology* 102: 55–74.
Fujii, T., and D. Raffaelli. 2008. Sea-level rise, expected environmental changes, and responses of intertidal benthic macrofauna in the Humber estuary, UK. *Marine Ecology Progress Series* 371: 23–35.
Galbraith, H., R. Jones, R. Park, S. Herrod-Julius, B. Harrington, and G. Page. 2002. Global climate change and sea level rise: potential losses of intertidal habitat for shorebirds. *Waterbirds* 25: 173–183.
Gladstone W., N. Hacking, and V. Owen. 2006. Effects of artificial openings of intermittently opening estuaries on assemblages of macroinvertebrates in the entrance barrier. *Estuarine Coastal and Shelf Science* 67: 708–720.
Godfrey, P.J., and M. Godfrey. 1980. Ecological effects of off-road vehicles on Cape Cod. *Oceanus* 23: 56–67.
Gotts, N.M. 2007. Resilience, panarchy, and world-systems analysis. *Ecology and Society* 12: 1–24.
Griffiths, C.L., and J. Stenton-Dozey. 1981. The fauna and rate of degradation of stranded kelp. *Estuarine Coastal and Shelf Science* 12: 645–653.
Grimm, V., and C. Wissel. 1997. Babel, or the ecological stability discussions: an inventory and analysis of terminology and a guide for avoiding confusion. *Oecologia* 109: 323–334.
Hacking, N. 1996. Tidal movement of sandy beach macrofauna. *Wetlands (Australia)* 15: 55–71.
Harley, C.D.G., A.R. Hughes, K.M. Hultgren, B.G. Miner, C.J.B. Sorte, C.S. Thornber, L.F. Rodriguez, L. Tomanek, and S. L. Williams. 2006. The impacts of climate change in global marine systems. *Ecology Letters* 9: 228–241.
Harper, E.M. 2000. Are calcitic layers an effective adaptation against shell dissolution in the Bivalvia? *Journal of Zoology London* 251: 179–186.
Haugan, P.M., and H. Drange. 1996. Effects of CO_2 on the ocean environment. *Energy Conversion Management* 37: 1019–1022.
Hawkins, S.J., H.E. Sugden, N. Mieszkowska, P.J. Moore, E. Poloczanska, R. Leaper, R.J.H. Herbert, M.J. Genner, P.S. Moschella, R.C. Thompson, S.R. Jenkins, A.J. Southward, and M.T. Burrows. 2009. Consequences of climate-driven biodiversity changes for ecosystem functioning of north European rocky shores. *Marine Ecology Progress Series* 396: 245–259.
Hoegh-Guldberg, O., and J.F. Bruno. 2010. The impact of climate change on the world's marine ecosystems. *Science* June 18, 2010.
Hughes, L. 2003. Climate change and Australia: trends, projections and impacts. *Austral Ecology* 28: 423–443.
International Scientific Steering Committee. 2005. *Avoiding Dangerous Climate Change.* Report presented at the International Symposium on the Stabilisation of Greenhouse Gas Concentrations. Exeter, UK. February 1–3.

IPCC. 1992. A common methodology for assessing vulnerability to sea level rise, 2nd revision. In *Global Climate Change and the Rising Challenge of the Sea. Report of the Coastal Zone Management Subgroup*. The Hague, Netherlands: IPCC.

———. 2001. Intergovernmental Panel on Climate Change web site. www.grida.no/climate/ipcc_tar/wg2/index.htm (accessed March 2004).

———. 2002. Climate Change 2001: Synthesis report. 397. Cambridge University Press. Cambridge, New York.

———. 2007. Intergovernmental Panel on Climate Change web site. http://www.ipcc.ch/ipccreports/ar4-syr.htm (accessed December 2007).

James, R.J. 2000a. From beaches to beach environments: linking the ecology, human-use and management of beaches in Australia. *Ocean & Coastal Management* 43: 495–514.

———. 2000b. The first step for the environmental management of Australian beaches: establishing an effective policy framework. *Coastal Management* 28: 149–160.

James, R.J., and P.G. Fairweather. 1996. Spatial variation of intertidal macrofauna on a sandy ocean beach in Australia. *Estuarine Coastal and Shelf Science* 43: 81–107.

Janzen, F.J. 1994. Climate-change and temperature-dependent sex determination in reptiles. *Proceedings of the National Academy of Sciences, USA*. 91: 7484–7490.

Jaramillo, E., H. Contreras, and A. Bollinger. 2002. Beach and faunal response to the construction of a seawall in a sandy beach of south central Chile. *Journal of Coastal Research* 18: 523–529.

Jaramillo, E., and A. McLachlan. 1993. Community and population responses of the macroinfauna to physical factors over a range of exposed sandy beaches in south-central Chile. *Estuarine Coastal and Shelf Science* 37: 615–624.

Jones, A.R. 2003a. Assessing ecological recovery of amphipods on sandy beaches following oil pollution. *Journal of Coastal Research* 35: 66–73.

———. 2003b. Impacts on ecosystem health—what matters? In *Proceedings of the Airs Waters Places Transdisciplinary Conference on Ecosystem Health in Australia*, ed. G. Albrecht, 208–223. ISBN: 0-646-43100-5.

Jones, A.R., W. Gladstone, and N.J. Hacking. 2007. Australian sandy-beach ecosystems and climate change: ecology and management. *Australian Zoologist* 34: 190–202.

Jones, A.R., A. Murray, T. Lasiak, and R.E. Marsh. 2008. The effects of beach nourishment on the sandy-beach amphipod *Exoediceros fossor*: impact and recovery in Botany Bay, New South Wales, Australia. *Marine Ecology* 29 (Suppl. 1): 28–36.

Jones, A.R., A. Murray, and R.E. Marsh. 1998. A method for sampling sandy beach amphipods that tidally migrate. *Marine and Freshwater Research* 49: 863–865.

Kennedy, V.S., R.R. Twilley, J.A. Kleypas, J.H. Cowan Jr., and S.R. Hare. 2002. *Coastal and Marine Ecosystems & Global Climate Change: Potential Effects on U.S. Resources*. Prepared for the Pew Center on Global Climate Change, Arlington, VA.

Kennett, J.P., and L.D. Stott. 1991. Abrupt deep-sea warming, palaeoceanographic changes and benthic extinctions at the end of the Palaeocene. *Nature* 353: 225–229.

Klein, R.J.T., R.J. Nicholls, S. Ragoonaden, M. Capobianco, J. Aston, and E.N. Buckley. 2001. Technological options for adaptation to climate change in coastal zones. *Journal of Coastal Research* 17: 531–543.

Klein, Y.L., J.P. Osleeb, and M.R. Viola. 2004. Tourism-generated earnings in the coastal zone: a regional analysis. *Journal of Coastal Research* 20: 1080–1088.

Knogge, T., M. Schirmer, and B. Schuchardt. 2004. Landscape-scale socio-economics of sea-level rise. *Ibis* 146 (Suppl.): 11–17.

Koop, K., R.C. Newell, and M.I. Lucas. 1982. Biodegradation and carbon flow based on kelp (*Ecklonia maxima*) debris in a sandy beach microcosm. *Marine Ecology Progress Series* 7: 315–326.

Krebs, C.J. 1978. *Ecology. The Experimental Analysis of Distribution and Abundance,* 2nd edition. New York: Harper and Row.

Kurihara, H. 2008. Effects of CO_2-driven acidification on the early developmental stages of invertebrates. *Marine Ecology Progress Series* 373: 275–284.

Lake, P.S. 2000. Disturbance, patchiness and diversity in streams. *Journal of the North American Benthological Society* 19: 573–592.

Lasiak, T.A. 1981. Nursery grounds of juvenile teleosts: evidence from the surf zone of Kings Beach, Port Elizabeth. *South African Journal of Zoology* 77: 388–390.

Lastra, M., E. Jaramillo, J. López, H. Contreras, C. Duarte, and J.G. Rodŕiguez. 2004. Population abundances, tidal movement, burrowing ability and oxygen uptake of *Emerita analoga* (Stimpson) (Crustacea, Anomura) on a sandy beach of south-central Chile. *Marine Ecology* 25: 71–89.

Leemans, R., and A. van Vliet. 2005. *Responses of species to changes in climate determine climate protection targets.* Report presented at the International Symposium on the Stabilisation of Greenhouse Gas Concentrations. Exeter, UK. February 1–3.

Lenanton, R.C.J., and N. Caputi. 1989. The role of food supply and shelter in the relationship between fishes, in particular cnidoglanis macro capnalus valenciennes, and detached microphytes in the surf zone of sandy beaches. *Journal of Experimental Marine Biology and Ecology* 128: 165–76.

Lowenstam, H.A. 1981. Minerals formed by organisms. *Science* 211: 1126–1131.

Lovejoy, T.E., and L. Hannah, Eds. 2005. *Climate Change and Biodiversity.* New Haven, CT, and London: Yale University Press.

Macleod, K., and H. Leslie, Eds. 2009. Ecosystem-Based Management for the oceans. Washington: Island Press.

McGwynne, L.E., A. McLachlan, and J.P. Furstenburg. 1988. Wrack break-down on sandy beaches. Its impact on interstitial meiofauna. *Marine Environmental Research* 25: 213–232.

McLachlan, A. 1985. The biomass of macro- and interstitial fauna on clean and wrack-covered beaches in Western Australia. *Estuarine Coastal and Shelf Science* 21: 587–599.

———. 1990a. Dissipative beaches and macrofauna communities on exposed intertidal sands. *Journal of Coastal Research* 6: 57–71.

———. 1998. Interactions between two species of Donax on a high energy beach: an experimental approach. *Journal of Molluscan Studies* 64: 492–495.

McLachlan A., and A.C. Brown. 2006. *The Ecology of Sandy Shores.* Burlington, MA, Academic Press.

McLachlan, A., and A. Dorvlo. 2005. Global patterns in sandy beach macrofauna communities. *Journal of Coastal Research* 21: 674–687.

McLachlan, A., and E. Jaramillo. 1995. Zonation on sandy beaches. *Oceanography and Marine Biology: an Annual Review* 33: 305–335.

McLachlan, A., E. Jaramillo, T. E. Donn, and F. Wessels. 1993. Sandy beach macrofauna communities and their control by the physical environment: a geographical comparison. *Journal of Coastal Research* 15: 27–38.

McLachlan, A., and I. Turner. 1994. The interstitial environment of sandy beaches. *PZNI Marine Ecology* 15: 177–211.

McLachlan, A., T. Wooldridge, and A.H. Dye. 1981. The ecology of sandy beaches in southern Africa. *South African Journal of Zoology* 16: 219–231.

McLeod, K.L., J. Lubchenko, S.R. Palumbi, and A.A. Rosenberg. 2005. Scientific consensus statement on marine ecosystem-based management. Communication Partnership for Science and the Sea. http://www.compassonline.org/marinescience/solutions_ecosystem.asp

Melillo, J.M., A.D. McGuire, D.W. Kicklighter, B. Moor, C.J. Vorosmarty, and A.L. Schloss. 1993. Global climate change and terrestrial net primary production. *Nature* 363: 234–240.

Mucci, A. 1983. The solubility of calcite and aragonite in seawater at various salinities, temperature, and one atmosphere total pressure. *American Journal of Science* 283: 780–799.

Najjar, R.G, C.R. Pyke, M.B. Adams, D. Breitburg, C. Hershner, M. Kemp, R. Howarth, M.R. Mulholland, M. Paolisso, D. Secor, K. Sellner, D. Wardrop, and R. Wood. 2010. Potential climate-change impacts on the Chesapeake Bay. *Estuarine, Coastal and Shelf Science*, 86: 1–20.

———. 1993. Beach restoration in the southeastern US: environmental effects and biological monitoring. *Ocean and Coastal Management* 19: 157–182.

Noy-Meir, I. 1979. Structure and function of desert ecosystems. *Israel Journal of Botany* 28: 1–19.

O'Hara, T.D. 2002. Endemism, rarity and vulnerability of marine species along a temperate coastline. *Invertebrate Systematics* 16: 671–684.

Orr, J.C., V.J. Fabry, O. Aumont, et al. 2005. Anthropogenic ocean acidification over the twenty-first century and its impact on calcifying organisms. *Nature* 437: 681–686.

Pearce, F. 2005. Faltering currents trigger freeze fear. *New Scientist* 188: 6–7 (3 December).

Pearse, A.S., H.J. Humm, and G.W. Wharton. 1942. Ecology of sandy beaches at Beaufort, North Carolina. *Ecological Monographs* 12: 135–190.

Peterson, C.H., and M. Bishop. 2005. Assessing the environmental impacts of beach nourishment. *Bioscience* 55: 887–896.

Peterson, C.H., M.J. Bishop, G.A. Johnson, L.M. D'Anna, and L.M. Manning. 2006. Exploiting beach filling as an unaffordable experiment: benthic intertidal impacts propagating upwards to shore birds. *Journal of Experimental Marine Biology and Ecology* 338: 205–221.

Peterson, C.H., D.H.M. Hickerson, and G.G. Johnson. 2000. Short-term consequences of nourishment and bulldozing on the dominant large invertebrates of a sandy beach. *Journal of Coastal Research* 16: 368–378.

Pilkey, O.H., and H.L. Wright. 1989. Seawalls versus beaches. *Journal of Coastal Research*, Special Issue 4: 41–67.

Poloczanska, E.S., S.J. Hawkins, A.J. Southward, and M.T. Burrows. 2008. Modeling the response of populations of competing species to climate change. *Ecology* 89: 3138–3149.

Poloczanska, E.S., C.J. Limpus, and G.C. Hays. 2009. Vulnerability of marine turtles to climate change. *Advances in Marine Biology* 56: 151–211.

Pörtner, H.-O. 2008. Ecosystem effects of ocean acidification in times of ocean warming: a physiologist's view. *Marine Ecology Progress Series* 373: 203–217.

Pörtner, H.-O., M. Langenbuch, and A. Reipschläger. 2004. Biological impact of elevated ocean CO_2 concentrations: lessons from animal physiology and earth history. *Journal of Oceanography* 60: 705–718.

Quammen, M.L. 1982. Influence of subtle substrate differences on feeding by shorebirds on intertidal mudflats. *Marine Biology* 71: 339–343.

Richardson, A.J., and D.S. Schoeman. 2004. Climate impact on plankton ecosystems in the Northeast Atlantic. *Science* 305: 1609–1612.

Rost, B., I. Zondervan, and D. Wolf-Gladrow. 2008. Sensitivity to phytoplankton to future changes in ocean carbonate chemistry: current knowledge, contradictions and research directions. *Marine Ecology Progress Series* 373: 227–237.

Salafsky, N., R. Margolis, K.H. Redford, and J.G. Robinson. 2002. Improving the practice of conservation: a conceptual framework and research agenda for conservation science. *Conservation Biology* 16: 1469–1479.

Scapini, F. 2006. Keynote papers on sandhoppers orientation and navigation. *Marine and Fresh Water Behaviour and Physiology* 39: 73–85.

Schiedek, D., B. Sundelin, J.W. Readman, and R.W. Macdonald. 2007. Interactions between climate change and contaminants. *Marine Pollution Bulletin* 54: 1845–1856.

Schiel, D.R., J.R. Steinbeck, and M.S. Foster. 2004. Ten years of induced ocean warming causes comprehensive changes in marine benthic communities. *Ecology* 85: 1833–1839.

Schlacher, T.A., and R.M. Connolly 2009. Land-ocean coupling of carbon and nitrogen fluxes on sandy beaches. *Ecosystems* 12: 311–321.

Schlacher, T.A., D.S. Schoeman, J. Dugan, M. Lastra, A. Jones, F. Scapini, and A. McLachlan. 2008b. Sandy beach ecosystems: key features, management challenges, climate change impacts, and sampling issues. *Marine Ecology* 29: 70–90.

Schlacher, T.A., L.M.C. Thompson, and S. Price. 2007. Vehicles versus conservation of invertebrates on sandy beaches: quantifying direct mortalities inflicted by off-road vehicles (ORVs) on ghost crabs. *Marine Ecology—Evolutionary Perspective* 28: 354–367.

Schneider, S.H., and T.L. Root. 1996. Ecological implications of climate change will include surprises. *Biodiversity and Conservation* 5: 1109–1119.

Schoeman, D.S., and A.J. Richardson. 2002. Investigating biotic and abiotic factors affecting the recruitment of an intertidal clam on an exposed sandy beach using a generalized additive model. *Journal of Experimental Marine Biology and Ecology* 276: 67–81.

Sherman, D.J., K.M. Barron, and J.T. Ellis. 2002. Retention of beach sands by dams and debris basins in southern California. *Journal of Coastal Research* SI 36: 662–674.

Sherman, K., and A.M. Duda. 1999. An ecosystem approach to global assessment and management of coastal waters. *Marine Ecology Progress Series* 190: 271–287.

Short, A.D. 1999. Wave-dominated beaches. In *Handbook of Beach and Shoreface Morphodynamics*, Ed. A.D. Short, 173–203. Chichester: John Wiley and Sons.

Short, A.D., and P.A. Hesp. 1982. Wave, beach and dune interactions in southeastern Australia. *Marine Geology* 48: 259–284.

Short A.D., and L.D. Wright. 1984. Morphodynamics of high energy beaches: an Australian perspective. In *Coastal Geomorphology in Australia*, Ed. B.G. Thom. Academic Press, Sydney, Australia.

Soares, A.G., A. McLachlan, and T.A. Schlacher. 1996. Disturbance effects of stranded kelp on populations of the sandy beach bivalve *Donax serra* (Röding). *Journal of Experimental Marine Biology and Ecology* 205: 165–186.

Soares, A.G., T.A. Schlacher, and A. McLachlan. 1997. Carbon and nitrogen exchange between sandy beach clams (*Donax serra*) and kelp beds in the Benguela coastal upwelling region. *Marine Biology* 127: 657–664.

Speybroeck, J., D. Bonte, W. Courtens, T. Gheskiere, P. Grootaert, J.-P. Maelfait, M. Mathys, S. Provoost, K. Sabbe, W.M. Stienen, V. Van Lancker, M. Vincx, and S. Degraer. 2006. Beach nourishment: an ecologically sound coastal defence alternative? A review. *Aquatic Conservation: Marine and Freshwater Ecosystems* 16: 419–435.

Stenseth, N.C., A. Misterud, G. Ottersen, J.W. Hurrell, K.-S. Chan, and M. Lima 2002. Ecological effects of climate fluctuations. *Science* 297: 1292–1296.

Taylor, J.D., and D.G. Reid. 1990. Shell microstructure and mineralogy of the Littorinidae: ecological and evolutionary significance. *Hydrobiologia* 193: 199–215.

Tomlinson, R. 2002. Beaches our asset. Planning and management for natural variability on open coastlines. In *Proceedings of the Public Workshop Beach Protection and Management, Yeppoon 2002*, Ed. J. Piorewisz, 7–33. CRC for Coastal Zone, Estuary and Water Management and Faculty of Engineering and Physical Systems. CQU Press, Rockhampton, Australia.

Turley, C., J. Blackford, S. Widdicombe, D. Lowe, and P. Nightingale. 2005. Report presented at the International Symposium on the Stabilisation of Greenhouse Gas Concentrations. Exeter, UK. February 1–3.

UNFPA. 2005. *State of the World Population 2005*. United Nations Population Fund. ISBN 0-89714-750-2. 119 pp.

Vermeij, G.J. 1979. Shell architecture and causes of death of Micronesian reef snails. *Evolution* 33: 686–696.

Vezina, A.F., and O. Hoegh-Guldberg. 2008. Introduction to Theme Section: Effects of ocean acidification on marine ecosystems. *Marine Ecology Progress Series* 373: 199–201.

Vitousek, P.M., H.A. Mooney, J. Lubchenko, and J.M. Melillo. 1997. Human domination of Earth's ecosystems. *Science* 277: 494–499.

Voice, M., N. Harvey, and K. Walsh, Eds. 2006. *Vulnerability to Climate Change of Australia's Coastal Zone: Analysis of Gaps in Methods, Data and System Threshold.* Report to the Australian Greenhouse Office, Canberra, Australia. June.

Walker, B., L. Gunderson, A. Kinzig, C. Folke, S. Carpenter, and L. Scultz. 2006. A handful of heuristics and some propositions for understanding resilience in social-ecological systems *Ecology and Society* 11: 13.

Walker, B., and D. Salt. 2006. *Resilience Thinking*. Washington, DC: Island Press.

Walters, C.J. 1986. *Adaptive Management of Renewable Resources*. New York: McGraw-Hill.

Walther, G.-R., E. Post, P. Convey, A. Menzels, C. Parmesan, T.J.C. Beebee, J.-M. Fromentin, O. Hoegh-Guldberg, and F. Bairlein. 2002. Ecological responses to recent climate change. *Nature* 416: 389–395.

Weslawski, J.M., A. Stanek, A. Siewart, and N. Beer 2000. The sandhopper (*Talitrus saltator*, Montagu 1808) on the Polish Baltic coast. Is it a victim of increased tourism? *Oceanological Studies* 24: 77–87.

Wieser, W., J. Ott, F. Schiemer, and E. Gnaiger. 1974. An ecophysiological study of some meiofauna species inhabiting a sandy beach at Bermuda. *Marine Biology* 26: 235–248.

World Resources Institute. 2000. *A Guide to World Resources 2000–2001: People and Ecosystems: The Fraying Web of Life*. World Resources Institute, United Nations Development Program, United Nations Environment Program and World Bank. Washington. www.wri.org/wr2000. Access date June 24, 2006.

Section IV

Monitoring Ecological Consequences of Climate Change

8 Response of Western Mountain Ecosystems to Climatic Variability and Change: *A Collaborative Research Approach*

David L. Peterson, Craig D. Allen, Jill S. Baron, Daniel B. Fagre, Donald McKenzie, Nathan L. Stephenson, Andrew G. Fountain, Jeffrey A. Hicke, George P. Malanson, Dennis S. Ojima, Christina L. Tague, and Phillip J. van Mantgem

CONTENTS

INTRODUCTION

Mountains in western North America are beginning to see changes in ecosystem processes primarily from climate-forced changes in water dynamics. With earlier snowmelt and increasing proportions of rain versus snow (Mote 2003; Stewart et al. 2005; Knowles et al. 2006), drought stress is increasing. Cascading effects include increasing vegetation mortality (van Mantgem et al. 2009), dieback of entire forest stands (Allen and Breshears 1998; Breshears et al. 2005), longer and more intense fire seasons (McKenzie et al. 2004; Westerling et al. 2006; Littell et al. 2009a), and increased susceptibility to insects and pathogens (Carroll et al. 2004; Breshears et al. 2005; Hicke et al. 2007). Model projections (Littell et al. 2009) suggest that all these phenomena will become even more pronounced in coming years (IPCC 2008; Milly et al. 2008).

Climate-forced changes in hydrology at high elevations are being caused by temperature-driven changes in winter precipitation form and snowmelt and by direct summer warming (Redmond 2007; Baron et al. 2009). Permanent snow and glaciers are retreating rapidly (Hall and Fagre 2003; Granshaw and Fountain 2006; Hoffman et al. 2007), portending the loss of smaller glaciers in the continental United States within the century (Dyurgerov and Meier 2000), and water chemistry is changing as weathering rates increase and new sediments are exposed (Lafreniere and Sharp 2005; Williams et al. 2007b; Thies et al. 2007; Baron et al. 2009). Increased stream temperature and decreased summer flow will strongly affect mountain aquatic ecosystems (Hauer et al. 1997; Ficke et al. 2007).

These ongoing and projected changes in western mountains are important for both people and ecosystems in the western United States. Western mountains are the water source for agriculture and 60 million people, provide timber and other commodities, and support recreation. National parks and national forests are reservoirs of reasonably intact natural ecosystems that protect critical wildlife habitat and maintain natural ecological processes. The Western Mountain Initiative (WMI) has developed a collaborative, multidisciplinary research program to address how climatic variability and change influence forest processes, disturbance dynamics, mountain hydrologic changes, and ecohydrologic interactions among climate, water, and vegetation. Long-term studies by WMI have identified regional responses in hydrology, disturbance patterns, and vegetation dynamics caused by a combination of regional climatic patterns (van Mantgem and Stephenson 2007; Groisman and Knight 2008; Littell et al. 2009, 2010), topographic complexity (Hicke et al. 2006; Christensen et al. 2008), and landscape controls (Tague et al. 2008).

The WMI research program has addressed four research questions for Western mountains:

1. How are climatic variability and change likely to affect disturbance regimes (particularly fire)?
2. How are changing climate and disturbance regimes likely to affect the composition, structure, and productivity of vegetation (particularly forests)?
3. How will climatic variability and change affect hydrologic processes in the mountainous West?

4. Which mountain resources and ecosystems are likely to be most sensitive to future climatic change, and what are possible management responses?

These research questions address priorities of the US Climate Change Science Program (CCSP), especially Goal 4 of the CCSP Strategic Plan: "Understand the sensitivity and adaptability of different natural and managed ecosystems and human systems to climatic and related global changes," with priority on improving knowledge of "thresholds of change that could lead to discontinuities (sudden changes) in some ecosystems and climate-sensitive resources."

BIOGEOGRAPHIC FOCUS

The WMI research program is based on long-term research and monitoring in five bioregions. The scientific focus for each bioregion is summarized as follows.

Pacific Northwest

Research has been conducted primarily in Olympic National Park (NP) and North Cascades NP (Washington), with supporting field sites and studies on US National Forest lands in the Olympic and Cascade Mountains (Washington and Oregon). This area ranges from marine to continental climates, with striking westside (wet) versus eastside (dry) contrasts. Research has focused on ecosystem productivity and fire disturbance, with empirical studies at a variety of spatial and temporal scales. Simulation modeling was used to integrate responses of natural resources to climatic variability and change, and statistical modeling established the biophysical "niche" of dominant tree species in the region. Growth increases in high-elevation tree species over the past century and widespread establishment of trees in subalpine meadows have been documented, especially for warmer periods with low snowpack. Teleconnections between the Pacific Decadal Oscillation (20- to 30-year cool and warm phases) and temporal variation in (1) regional tree growth, (2) fire, and (3) long-term drought have also been documented. Carbon storage was quantified across the forest ecosystems of the region, thereby improving carbon budgets for mountain forests.

Sierra Nevada

Research has been conducted primarily in Sequoia-Kings Canyon NP and Yosemite NP (California). This area is dominated by Mediterranean climate with cool, wet winters and dry summers. Research has focused primarily on the effects of climatic variability and change on forest ecosystem dynamics and fire regimes, with collaborative studies on forest ecology, paleoecology, and simulation modeling. Significant findings include documentation of major periods of drought prior to 1950, altered demographic patterns as a possible function of climatic variability in mixed-conifer forest, and high-resolution quantification of past fire regimes in cool-wet winters mixed-conifer forest. Investigations of fire effects on forest pattern and dynamics have led to modifications in prescribed fire and timber harvest approaches. Modeled projections of the consequences of natural fire, prescribed fire, and timber harvest provided insight on the effects of potential management options in the context of climatic change.

Northern Rocky Mountains

Research has been conducted mostly in Glacier NP (Montana). An integrated program of mountain ecosystem modeling and empirical studies has quantified ecological processes and biodiversity patterns affected by climatic variability. Ecosystem models have been used to quantify snow distribution, annual watershed discharge, and stream temperature variation under the warmer, wetter climate projected for this region. Interannual variation in snowpack is closely related to the Pacific Decadal Oscillation and modifies fire cycles through its effects on fuels. Glaciers are rapidly disappearing from Glacier NP. In 1850, there were 150 glaciers, but today there are only 37, and simulation models predict that all glaciers in the park will be gone by 2030 at current warming rates. Modeling indicates that in a warmer climate high-elevation conifer productivity will decrease but alpine vegetation productivity will increase. Modeling also suggests that low-elevation forests will be more productive but generate higher fuel loads that contribute to high-intensity wildfires.

Central Rocky Mountains

Research has been conducted mostly in Rocky Mountain NP (Colorado). Paleoecological, experimental, remotely sensed, modeling, and monitoring tools have bounded the range of fire and insect outbreaks over the past 400 years and correlated these with El Niño–Southern Oscillation events. Tree-line expansion into tundra has not occurred since the mid-1800s, although changes in growth form from krummholz to upright trees have occurred. Late-successional coniferous forests are vulnerable to changes in disturbance regimes rather than directly responsive to changing climate. Aspen forests—systems with high plant and insect diversity—are sensitive to fire suppression (which allows conifers to establish) and high-intensity fires (which allow exotic species to establish). Simulations of stream discharge and ecosystem processes show that the timing of snowmelt, but not water volume, is sensitive to warming. Recent observations also show that spring snowmelt is not occurring earlier at high elevations, but summer discharge is increasing as a result of increased summer temperatures and melting glaciers and permafrost. Regional climate has responded to changes in land use that add moisture to the atmosphere near mountains.

Southern Rocky Mountains

Research has been conducted mostly in Bandelier National Monument and adjacent landscapes in northern New Mexico and southern Colorado. Studies have focused on the sensitivity of semiarid forests and woodlands to climatic change by (1) quantifying patterns of drought-related tree mortality and associated ecotone shifts, (2) inferring long-term relationships between climate and severe fire events, and (3) measuring long-term responses of key ecosystem parameters (herbaceous cover, runoff, erosion, tree growth, arthropod populations) to interannual and decadal climatic variation. Documentation of forest die-off following extended drought has shown that drought and heat can exceed critical thresholds of tree mortality at the regional scale. Reconstructed crown fire dates from 12 of the largest aspen stands in the southern Rockies reveal correlations with past droughts and region-wide surface fire years. Sediment cores contain charcoal deposits that record evidence of fires throughout the Holocene and suggest that the post-1900 cessation of widespread fire in the southern Rockies is anomalous on millennial time scales. Ecohydrologic research is revealing the critical importance of climate-sensitive land surface cover characteristics in controlling patterns and rates of soil erosion in semiarid landscapes.

COLLABORATIVE APPROACH

WMI builds on past research in each of the five bioregions. The depth and breadth of place-based knowledge represented by this work strengthens multisite regional comparisons. WMI employs cross-site syntheses of long-term data, modeling, and workshops to convene subject-matter experts from a variety of scientific disciplines. Mechanistic understanding of how climatic variability and change affect ecosystems directly (e.g., by affecting productivity) and indirectly (as mediated through disturbance) has been explored primarily through natural experiments in time, natural experiments in space, and synthetic modeling.

WMI conducts *natural experiments in time* through a variety of paleoecological data and contemporary information sources. At millennial time scales, pollen and charcoal in sediments reveal ecosystem responses to long-term climatic shifts (Whitlock and Anderson 2003). At decadal to millennial time scales, tree rings reveal annual to seasonal changes in climatic and fire regimes (Swetnam and Baisan 2003), and broad demographic trends in forests (Veblen and Lorenz 1991; Swetnam and Betancourt 1998). At time scales of a few years to a century, instrumental records, written records, photographs, and plot data offer fine resolution for mechanistic understanding. Quasi-periodic climatic phenomena such as the Pacific Decadal Oscillation (Mantua et al. 1997) and El Niño–Southern Oscillation (Diaz and Markgraf 2000) offer a context for climatic change, particularly when they drive climatic extremes.

WMI conducts *natural experiments in space* by a subcontinental network of field sites in aquatic and terrestrial systems. This allows WMI to seek generalizations across temperature regimes (continental vs. maritime [longitudinal comparisons], warm vs. cool [latitudinal comparisons]) and precipitation regimes (e.g., mediterranean vs. monsoonal [Sierra Nevada vs. southern Rockies], wet vs. dry [Pacific Northwest to southern Rockies]). At regional scales, steep elevational gradients that exist within each WMI region are paralleled by steep temperature and moisture gradients.

WMI conducts modeling studies to understand ecosystem function and effects of climatic warming. We primarily use RHESSys (Regional Hydro-Ecological Simulation System) (Baron et al. 2000a; Christensen et al. 2008) as a framework to assess effects on aquatic and terrestrial systems. RHESSys allows us to frame hypotheses based on past modeling results, scale up empirical results, and identify sensitivities of specific mountain landscapes to climatic change (Urban 2000).

WMI research is guided by the research questions noted previously and uses the following themes to identify critical properties of ecosystem structure and function: (1) *linkages* (e.g., climatic change affects disturbance regimes, hence vegetation, hence erosion), (2) *sensitivities* (resources, ecosystems, or landscapes susceptible to abrupt or profound change), and (3) *thresholds* (magnitudes of climatic events sufficient to drive abrupt and profound change). Although not all themes are present in all projects, at least one theme is present in each project.

WMI has a strong focus on technology transfer and communication in order to put scientific information in the hands of resource managers, media, and the general public. Primary clients are federal land management agencies, but we also communicate regularly with state resource agencies, counties, municipalities, and nongovernmental organizations. Results are summarized in fact sheets and reports written in common language and distributed in hard copy and on the Web. Resource managers, legislative staff, and educators are invited to workshops at various locations in the west, and resource managers often assist with the design and implementation of WMI studies.

STATUS AND TRENDS IN WESTERN MOUNTAIN ECOSYSTEMS

WMI research has documented important characteristics and trends of the structure and function of Western mountain ecosystems, the effects of exceeding critical thresholds, and the potential for future changes in aquatic and terrestrial systems. Although it is not possible to definitively identify all recent changes as being caused by climatic change, the effects of climatic variability—especially dominant modes of climatic variability, such as the Pacific Decadal Oscillation—can be used with some confidence to infer expected changes in ecosystems in a warmer climate.

The following sections discuss the results of recent WMI research. They are intended to cover a range of aquatic and terrestrial systems across the WMI bioregions, and suggest topics that merit additional research and monitoring.

SNOW, ICE, AND HYDROLOGY

Loss of glacier mass has accelerated in several regions of the world (e.g., Hall and Fagre 2003), with effects on both mountainous regions and downstream resources. Glaciers and snow are the source of water for over half of humanity and provide 85% of the water for agriculture and municipalities in the western United States. Changes in mountain glaciers have social and economic impacts on water supplies and hydropower production, and affect the structure and function of mountain ecosystems and species adapted to cold mountain environments. WMI has addressed the need for a cryospheric assessment of Western mountains with targeted studies of glaciers and snow and mountain catchment hydrology. First, we conducted a Westwide survey of glacier areal extent. Second, we developed time series of glacier recession rates in different mountain areas. Third, we investigated relationships between climate and mass balance of glaciers by intensively monitoring selected glaciers.

A 1:100,000-scale map of all extant glaciers of the American West was digitally compiled from existing maps and aerial photography (Fountain et al. 2007), including 1,523 glaciers ranging from 3,070 m^2 to 13.6 km^2. At this scale some smaller glaciers were not mapped, so a 1:24,000 glacier inventory was created to provide a comprehensive snapshot of glacier cover in the American West. Fountain et al. (2007) report nearly 8,300 "glaciers," which in this inventory includes perennial snow patches and small ice features. The 1:24,000-scale map added 6,770 glaciers to the inventory ranging from 347 m^2 to 10.59 km^2. The glacier inventory (http://www.glaciers.us) contains interactive maps, photographs, and other information, and documents significant reductions in ice and snow since 1900: 66% in Montana, 42% in Wyoming, 40% in Colorado, 56% in California, 30% in Oregon, and 24–46% in various Washington locations. Mapped glaciers that are less than 0.01 km^2 (40% of the total) are especially vulnerable to warming trends.

In Glacier NP, Grinnell Glacier was first explored by George Bird Grinnell in 1887 and described "as 1,000 feet high and several miles across" (Diettert 1992). About 90% of the ice that Grinnell saw is now gone. Repeat photographs (Figure 8.1) show that Grinnell Glacier has retreated substantially, become thinner, separated from what is now Salamander Glacier, formed a lake from its own meltwater, and discharged icebergs into the lake. A time-series map, observations, terminus measurements, and photographs document a reduction from 2.3 km^2 to 0.6 km^2 from 1850 to 2006, with a recent acceleration in glacier recession (Key et al. 2002; Fagre and McKcon 2010).

To achieve a more accurate understanding of glacier responses to climate, four years of seasonal and annual mass balances for Sperry Glacier (Glacier NP) were calculated using direct glaciological methods augmented with data from an adjacent weather station and snow pillow. Net annual balance averaged –1.0-m water equivalent for 2005–2007. Despite significant spatial variability, Sperry Glacier is primarily driven by climatic variation rather than topographic shading, snow redistribution by wind, or avalanche inputs. Mass balance loss from 2005 to 2008 is three to four times higher than during the 1950s, indicating that ice loss at Sperry Glacier has accelerated in recent decades due to warmer temperatures.

Snow is critical to maintaining glacier mass and drives other mountain ecosystem dynamics ranging from avalanche frequency to timing of alpine plant emergence. Snowpacks across Western mountains now contain less moisture (Mote et al. 2005; Mote 2006; Barnett et al. 2008), melt earlier (Regonda et al. 2005; Stewart et al. 2005; McCabe and Clark 2005), and contribute less to streamflow (Hamlet et al. 2007) than they did prior to 1950, with spring and summer temperatures as the primary driver of earlier snowmelt runoff. WMI has examined the following for each of the five WMI bioregions: (1) changes in evolution of snowpack throughout the accumulation and melt season; (2) regional rates of temperature change, especially changes in the 0°C freeze threshold at low- and mid-elevations; (3) changes in the amount and timing of regional streamflows; and (4) relationships between changes in the evolution of regional snowpack and streamflow.

FIGURE 8.1 **(See color insert.)** Grinnell Glacier from the summit of Mt. Gould, Glacier NP. Upper left (1938): Oblique view of Grinnell Glacier shows decreased glacier area and reduced depth of the glacier along the cirque wall where, prior to 1938, the ice-surface elevation was high enough to connect with the upper band of ice (T.J. Hileman, Glacier NP archives). Upper right (1981): (C. Key, U.S. Geological Survey [USGS]). Lower left (1998): (D. Fagre, USGS). Lower right (2006): (K. Holzer, USGS).

An analysis for the northern Rocky Mountains shows that since 1969 maximum snow-water equivalent (SWE) has declined and peaked earlier (14 days) in the water-year, with minimal temporal change between peak and 0 SWE. The number of snow-free days and ablation days indicates temperature-driven loss of snow cover or increasing loss of snow through sublimation. Although accumulation and ablation days did not change significantly, the annual number of snow-free days has increased, resulting in an average loss of 14 days of snow cover between 1969 and 2007 due to higher temperatures. Minimum temperatures have increased more at higher elevations than in valleys, suggesting that altered lapse rates may have a disproportionate warming effect on snowpack retention at higher elevations. This effect on snowpack ablation is compounded by spring nighttime minimums with fewer freeze events since the 1980s, leading to more rapid melting of snow. These results suggest that spring runoff, peak discharge, and other metrics of streamflow would be expected to change, although our analysis found no significant changes in these factors. Increased summer precipitation in the northern Rockies appears to have compensated for earlier snowmelt, although it has also contributed more variability and a "flashier" flow regime.

Monitoring in Rocky Mountain NP since 1982 shows that there was a period of below-average precipitation during 2000–2006 and a steady increase in summer and fall temperatures of 0.12°C per year since 1991. This is causing increased melting of rock glaciers and possibly other cryic features, with a consequent increase in summer stream discharge. Nitrate concentrations, as well as the weathering products calcium and sulfate, were higher in rock glacier meltwater for the period 2000–2006 (Baron et al. 2000b, 2009). Melting ice in glaciers and rock glaciers is exposing sediments from which nitrogen produced by nitrification can be flushed. It appears that a water quality threshold was crossed in Rocky Mountain NP around 2000, and that nitrogen release from ice features such as rock glaciers may occur throughout the West as mountain glaciers retreat.

In cold-water fisheries, regulation of stream temperature, as influenced by snowpack and glaciers, controls the distribution and abundance of invertebrates (Hauer et al. 1997) and fish (Keleher and Rahel 1996; Dunham et al. 2003). The body temperature of salmonids depends on the temperature of their surroundings, with a characteristically narrow range of thermal tolerance. Bull trout (*Salvelinus confluentus*) is a native char to northwestern North America that requires cold, connected, and complex habitats. However, populations have declined over the past century due to habitat degradation, nonnative species introductions, and elevated water temperature (Rieman et al. 1997). A warming climate and higher daily temperature extremes during summer may negatively impact existing trout populations by increasing water temperatures beyond critical physiological thresholds (Selong et al. 2001) in natal habitat area and habitat patches (Rieman et al. 1997).

In conjunction with losing the cooling effect of glaciers and perennial snow and ice, stream temperature will be affected by changes in maximum summer temperatures and minimum winter temperatures (Keleher and Rahel 1996). Using an upper temperature threshold of 22°C as a constraining variable for cold-water fish, Keleher and Rahel (1996) predicted that the length of streams occupied in Wyoming would decrease 8–43% for increases in temperature from 1°C to 5°C. Changes in temperature of this magnitude will further fragment available suitable habitat, increase the risk of severe fire, change the timing and quantity of water from snowpack, increase winter flooding, and create habitat that favors introduced species (Williams et al. 2007a).

Forest Ecology, Demography, and Die-Off

WMI places a strong emphasis on understanding the direct and indirect influences of climatic variability and change on Western forests, and has contributed to recent efforts linking regional warming to chronic and acute increases in tree mortality in the West. Even small changes in background tree mortality rates (rates considered "normal" or not steeply declining), when compounded over time, could substantially alter forest structure, composition, and function. For example, a persistent doubling of mortality rate (such as from 1% to 2% per year) ultimately would cause a greater than 50% reduction in average tree age, and hence a potential reduction in carbon storage. Increasing background mortality rates could indicate forests approaching thresholds for abrupt die-back.

We compiled and analyzed long-term data from unmanaged old-forest plots across the western United States, finding that background tree mortality rates have more than doubled in recent decades (van Mantgem et al. 2009) (Figure 8.2). Mortality rates not only increased in each of three subregions (Pacific Northwest, California, and interior), but also within dominant taxonomic groups (*Abies*, *Pinus*, *Tsuga*, and all other genera combined) and across elevation zones and tree size classes. In sharp contrast, tree recruitment rates remained unchanged. Consistent with what would be expected when mortality increases without compensating increases in recruitment, in recent decades the density and basal area of old-forests in the West have declined slightly (van Mantgem et al. 2009).

We examined two broad classes of possible causes of the observed increases in tree mortality rates: endogenous processes (those inherent to forest dynamics) and exogenous drivers (such as environmental changes). Several endogenous processes—such as increasing competition among trees, successional dynamics, aging cohorts of trees, and effects of fire exclusion—were not considered to be causes of elevated tree mortality. Available evidence was also inconsistent with a significant role for two possible exogenous causes: forest fragmentation and air pollution. For a number of reasons, regional warming appears to be a likely contributor to the widespread increases in tree mortality.

Specifically, from the 1970s to 2006 mean annual temperature of the West increased at a rate of 0.3–0.4°C per decade, even approaching 0.5°C per decade at the higher elevations typically occupied by forests (Diaz and Eischeid 2007). This regional warming has contributed to widespread hydrologic changes such as declining fraction of precipitation falling as snow (Knowles et al. 2006), declining snowpack water content (Mote et al. 2005), earlier spring snowmelt and runoff (Stewart et al. 2005), and a consequent lengthening of the summer drought (Westerling et al. 2006). Thus, both temperature and climatic water deficit (evaporative demand that is not met by available water) increased over the study period, and both were positively correlated with tree mortality rates (van Mantgem et al. 2009).

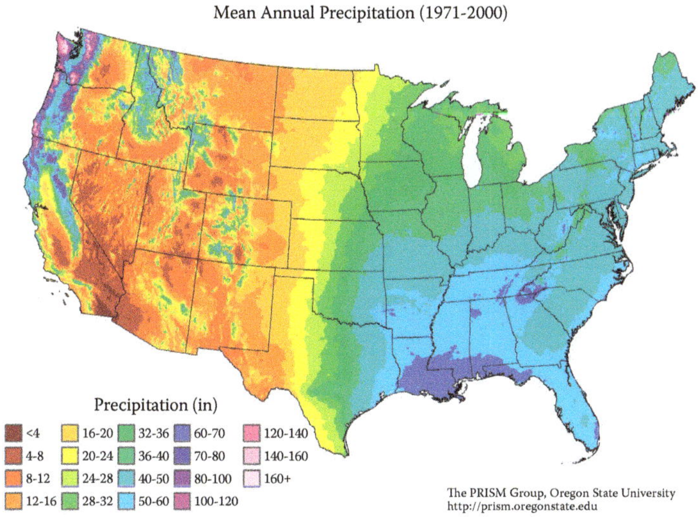

FIGURE 1.1 Mean annual precipitation from PRISM (Daly et al. 2002, 2004, 2008) for the 1971–2000 period of record. Note the sharp gradients in much of the West.

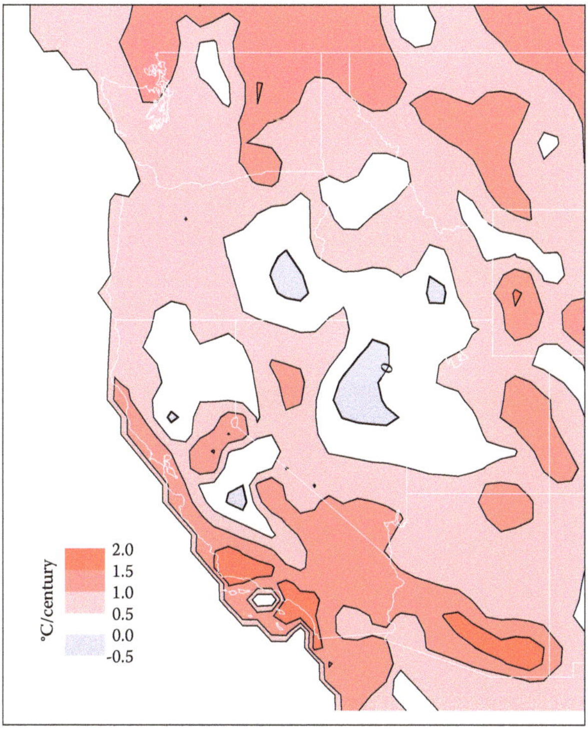

FIGURE 1.2 Linear trends in temperature (°C/century) from the HadCRU 0.5° × 0.5° dataset, evaluated over the 1901–2000 period. The contour interval is 0.5°C per century.

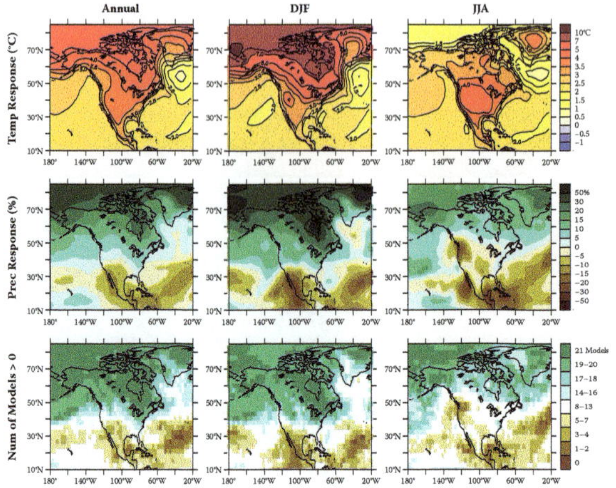

FIGURE 1.5 Maps of changes (departures from 1971–2000 mean) in climate elements for the late-twenty-first century (2080–2099) derived from an ensemble summary of 21 global climate models (Christensen et al. 2007, their Figure 11.12) for emissions scenario A1B. Top row: Temperature departure in °C. Middle row: Precipitation departure in percent, with green indicating wetter and brown indicating drier. Bottom: Number of models agreeing on wetter than average out of 21 models; green indicates agreement on wetter and brown is agreement on drier. Columns—Left: Annual. Middle: Winter (December–February). Right: Summer (June–August).

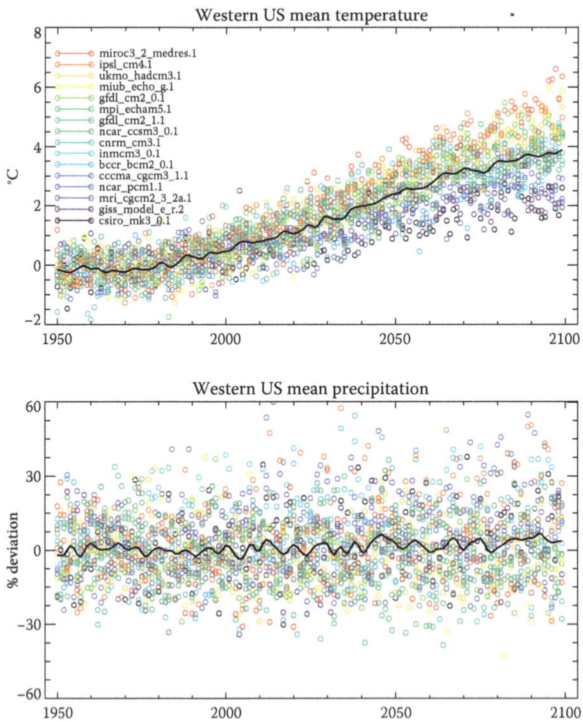

FIGURE 1.6 Western-US annual mean temperature and precipitation for 16 climate models (colored circles, names and runs shown) along with the all-model mean (heavy black curve). Greenhouse gas and aerosol forcing follows the observed for 1950–2000 and then the A1B scenario for 2000–2100.

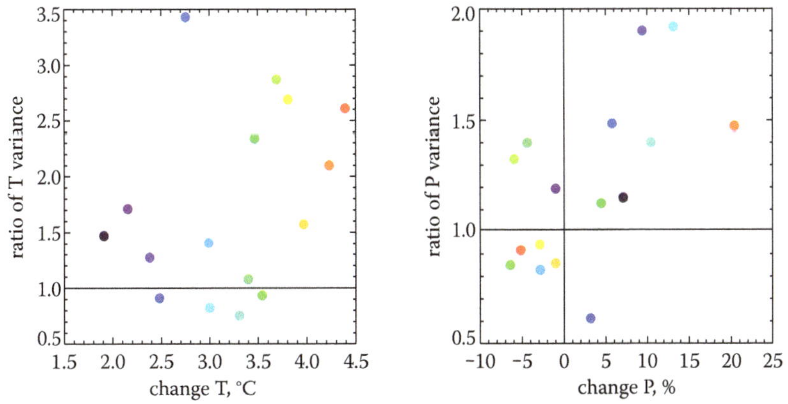

FIGURE 1.7 Scatterplots of change in mean and ratio of change in variance, 1950–2000 versus 2050–2100, for temperature (left) and precipitation (right). Colors indicate the models and are the same as in Figure 1.6.

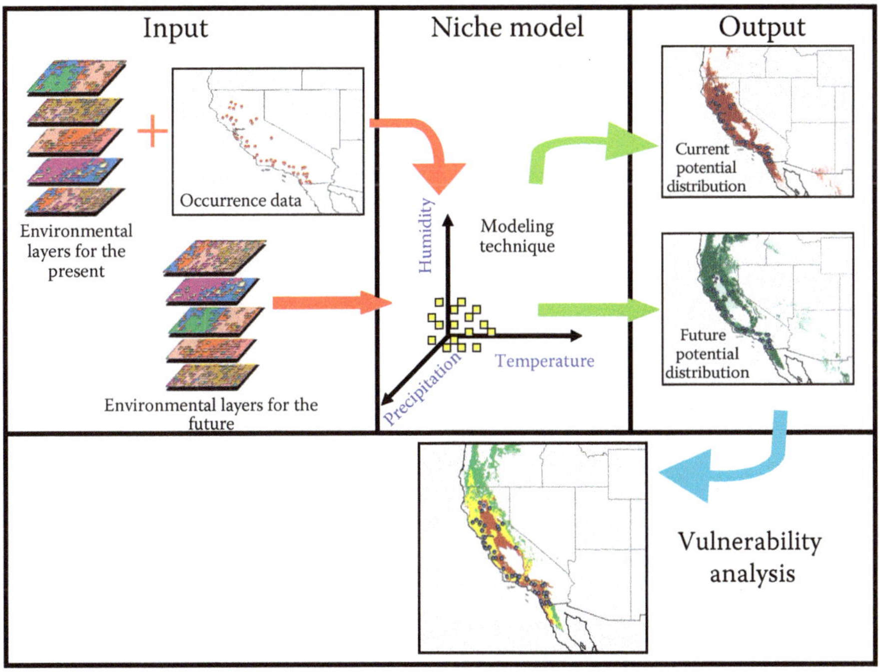

FIGURE 4.1 Diagrammatic representation of the Ecological Niche Modeling process.

FIGURE 8.1 Grinnell Glacier from the summit of Mt. Gould, Glacier NP. Upper left (1938): Oblique view of Grinnell Glacier shows decreased glacier area and reduced depth of the glacier along the cirque wall where, prior to 1938, the ice-surface elevation was high enough to connect with the upper band of ice (T.J. Hileman, Glacier NP archives). Upper right (1981): (C. Key, U.S. Geological Survey [USGS]). Lower left (1998): (D. Fagre, USGS). Lower right (2006): (K. Holzer, USGS).

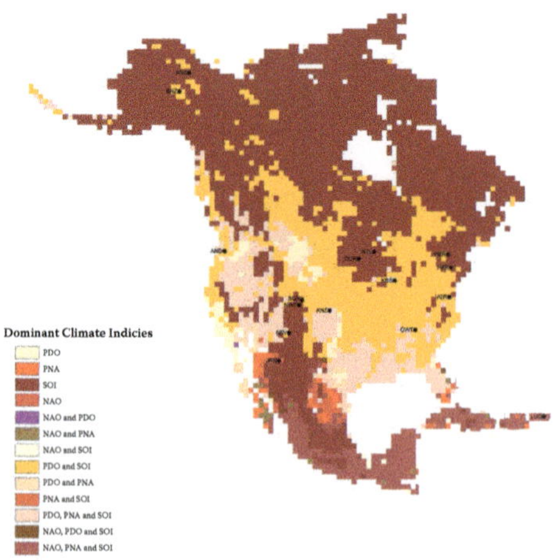

FIGURE 9.3 Dominant climate indices for the gridded precipitation data, determined using Maximum Basic Probability Assignment.

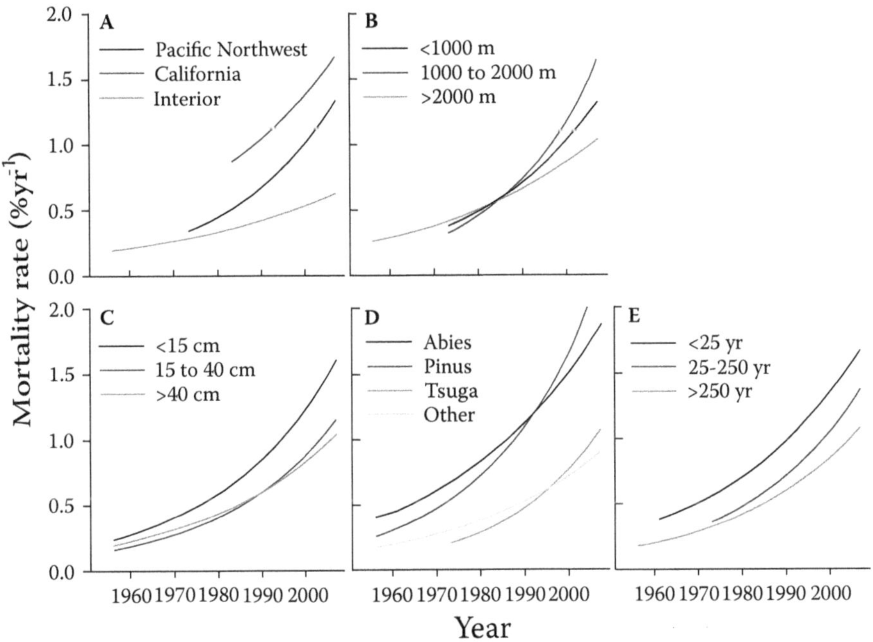

FIGURE 8.2 Modeled trends in tree mortality rates for (A) regions, (B) elevational class, (C) stem diameter class, (D) genus, and (E) historical fire return interval class. (From van Mantgem, P.J., N.L. Stephenson, J.C. Byrne, L.D. Daniels, J.F. Franklin, P.Z. Fulé, M.E. Harmon, J.M. Smith, A.H. Taylor, and T.T. Veblen. 2009. Widespread increase of tree mortality rates in the western United States. *Science* 323:521–24.)

Warming-induced tree mortality is consistent with both the apparent role of warming in episodes of recent forest die-back in western North America (Breshears et al. 2005; Raffa et al. 2008; Adams et al. 2009) and the positive correlation between short-term fluctuations in background tree mortality and water deficits reported in California and Colorado (Bigler et al. 2007; van Mantgem and Stephenson 2007). Over time, an increase in background tree mortality rates could lead to altered forest structure, habitat quality, fire hazard, and carbon storage. Perhaps even more important, in some cases these chronic increases in mortality could be symptomatic of forests that are stressed and vulnerable to abrupt die-off.

WMI is making significant contributions to regional and global syntheses of climate-induced forest mortality through the Cordillera Forest Dynamics Network (CORFOR, North and South American mountains; http://mri.scnatweb.ch/networks/mri-amercian-cordillera/-cordillera-forest-dynamics-networkcorfor.html). In 2007, WMI convened an international workshop on climate-induced forest dieback, leading to subsequent synthesis publications (Allen 2009; Allen et al. 2010). Recent and ongoing WMI research continues to address key uncertainties in forest mortality processes (McDowell et al. 2008; Breshears et al. 2009; McDowell et al. 2010).

Tree-Line Dynamics

The transition from subalpine forest to alpine tundra (alpine forest-tundra ecotone, AFTE) is one of the most distinctive features of mountain environments. Although the ecological dynamics of the AFTE are influenced by climate, it is an imperfect indicator of climatic change. Mechanistic processes that shape the ecotone—seed rain, seed germination, seedling establishment, and subsequent tree growth form—depend on microsite patterns. Although mechanistic processes are similar for all AFTEs in western North America, prior climate, geology and geomorphology, and genetic constraints of tree species create geographic differences in responses of ecotones to climatic change. Thus, significant variation in tree-line spatial composition and history can occur within and among mountain ecosystems. We examined tree-line phenomena and limiting factors (Malanson et al. 2007): in the northern Rockies centered on Glacier NP, the southern Rockies centered on Niwot Ridge and Rocky Mountain NP (with insights from the Medicine Bow Mountains, Wyoming), the Sierra Nevada centered on Sequoia-Kings Canyon NP, and the Pacific Northwest centered on Olympic NP.

Biological mechanisms controlling tree establishment in the AFTE are similar across western North America. Climate, energy, and water directly affect these processes, with differing outcomes due to abiotic differences and biogeographic history, although thresholds for effects on carbon balance of seedlings is unknown (Cairns and Malanson 1998). Climatic change may affect successful dispersal, germination, and survival by modifying soil availability, quality, and moisture; precipitation fraction arriving as snow; snow redistribution and melt; extent of glacial forefields; extent of permanent snowfields; and disturbance regimes.

The three-dimensional pattern at tree line is often patchy, including krummholz and dwarf trees (Allen and Walsh 1996), with expansion often facilitated by other plants (Resler 2006). The formation of vegetation structures that add wind protection, snow collection, and soil development allows subalpine forest species to initiate patches and then to have patches expand in the upper tree line (Smith et al. 2003). These structures can affect the infilling of openings in tree cover in the lower AFTE, sometimes with a linear configuration of trees (Holtmeier 1982).

Climate and variation in geomorphology or geology control the dynamics of tree lines (e.g., Butler and Walsh 1994). For example, the freeze-thaw cycle in the periglacial climate of tree line leads to solifluction, soil creep, landslides, and erosion. Snow avalanches and debris flows cause tree mortality and limit tree-line elevation. Insects and pathogens are spatially diffuse and alter tree line at large spatial scales. Recent examples include outbreaks of mountain pine beetle (MPB, *Dendroctonus ponderosae*) in high-elevation pine species (Logan and Powell 2001) and ongoing infection by white pine blister rust (*Cronartium ribicola*), in five-needle pines (Tomback and Resler 2007).

At regional to continental scales, control of AFTE by temperature is locally modified by moisture (Malanson and Butler 2002). In warmer, wetter conditions significant changes in the structure of tree line in all WMI bioregions are likely. The Pacific Northwest has ample moisture, and increased snow will inhibit the establishment and growth of trees, so longer-lasting snowpack could limit the expansion of krummholz higher in the ecotone and could expand lower meadows. The other bioregions are drier, so increasing moisture would improve conditions for tree establishment and growth. Improvement would be greatest in the southern Rocky Mountains and least in the Sierra Nevada, because of differences in substrate development and water-holding capacity. For example, tree growth increased in Sierra Nevada tree lines for warmer, wetter periods in the twentieth century (Millar et al. 2004). Millar et al. (2004) also found increased tree establishment in subalpine meadows during drier periods when the limitation of deep snow was reduced. At all three sites, the effect is likely to be largest in the lower AFTE where deeper soils and root zones can utilize increased water.

In warmer, drier conditions, the Pacific Northwest will differ from the other sites: less moisture will not reduce tree species establishment and growth in the AFTE, and reduced snow suppression could allow expansion in the upper ecotone and infilling of meadows in the lower ecotone. At the other tree-line sites, warmer and drier climate could reduce tree establishment in the ecotone. While much of the evidence for AFTE response suggests upslope advance during warmer conditions, in many places advance may be moisture limited. For example, Glacier NP experienced upward movement of the ecotone in the nineteenth century (Bekker 2005), while during the twentieth century, advance was limited but density of existing patches increased (Butler et al. 1994; Klasner and Fagre 2002). Tree-line advances in the latter half of the twentieth century coincided with the cool (wet) phase of the Pacific Decadal Oscillation, and advance stopped coincident with the warm (dry) phase of the 1980s and 1990s (Alftine et al. 2003). The effects of general warming on the AFTE must be evaluated in the context of multidecadal modes of climatic variability.

Advance of trees into tundra can alter ecosystem services, such as carbon, water, and nutrient cycles; feedbacks to the climate system; and maintenance of biodiversity. Tree establishment at some locations has increased carbon storage through accretion of living vegetation, and transfer of carbon to dead biomass and soil organic matter results in a potential long-term carbon sink. Advance of trees, or change in pattern in general, also has the potential to alter snow retention and hydrology, with implications for local soil moisture and nutrient transport.

Ecological Disturbance and Interaction of Stressors

WMI disturbance research addresses spatial scales from landscapes to regions. We have used both process-based and empirical models to estimate the extent and magnitude of future disturbances across the West. Data sources for modeling include observational records from recent decades, and in the case of fire, historical tree-ring reconstructions dating back to the sixteenth century and charcoal deposits in sediments extending back as much as 15,000 years before present.

Changing disturbance regimes in a warmer climate are likely to precipitate more rapid changes in ecosystems than the direct effects of warmer temperatures. Particularly in forest ecosystems dominated by long-lived conifer species and subject to stand-replacing disturbances, accelerated species turnover will occur after severe disturbance, because seedlings are less resistant to changing climate than mature individuals (Figure 8.3). In light of this central role of disturbance in Western forests, WMI research has recently emphasized quantifying the effects of climatic

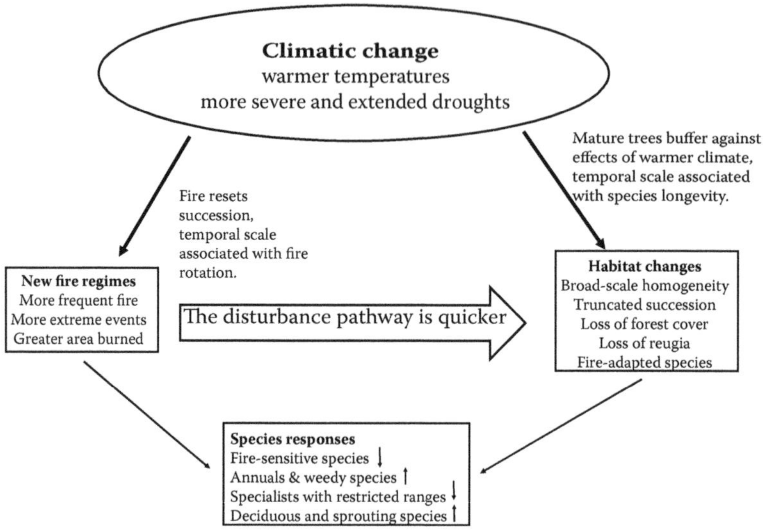

FIGURE 8.3 Conceptual model of the relative time scales for disturbance versus climatic change alone to alter ecosystems. Times are approximate. The focus here is on fire, but much of the same logic applies to insect outbreaks. (Adapted from McKenzie, D., Z. Gedalof, D.L. Peterson, and P. Mote. 2004. Climatic change, wildfire, and conservation. *Conservation Biology* 18:890–902.).

variability and change on the areal extent and broad-scale spatial patterns of two key disturbance processes: wildfire and insect outbreaks, particularly of MPB.

Westwide we have linked wildfire to climatic variability via studies of Holocene charcoal and pollen sediments (Anderson et al. 2008), dendrochronological reconstructions of forest fire histories, stand ages, and climate patterns (e.g., Margolis et al. 2007), and statistical models using twentieth-century instrumental records (McKenzie et al. 2004 and references therein). WMI convened 75 tree-ring and charcoal specialists for a 2005 workshop on Fire History and Climate Synthesis in Western North America, leading to special symposia at two international scientific meetings; subsequent research papers compose a special issue of the *International Journal of Wildland Fire* (Swetnam and Anderson 2008).

WMI research has provided important paleoecological context on late Pleistocene and Holocene fire regimes in relation to both climatic variability and patterns of dominant vegetation (e.g., Anderson et al. 2008; Prichard et al. 2009). Synchronous changes in fire occurrence and species composition suggest the central role of disturbance in structuring forest vegetation. Using historical fire-scar records from sites ranging from northern New Mexico to eastern Washington, we have documented the overarching control of climate, particularly drought, on fire when examined at broad scales (Hessl et al. 2004; Margolis et al. 2007), in contrast with the fine-scale controls of fuels and topography (McKenzie et al. 2006; Kellogg et al. 2008). In the southern Rocky Mountains, we have combined charcoal sediment and tree-ring fire scar data to document the effects of climate and land use histories on overlapping Holocene and historical fire regimes (Allen et al. 2008).

Of particular concern Westwide are increases in fire area in a warming climate and the effects of extreme wildfire events on ecosystems (Gedalof et al. 2005; Lutz 2008; Littell et al. 2009). For example, in 2006, the Tripod Complex Fire in north-central Washington burned over 80,000 ha, much of it higher severity than expected from historical fires. WMI researchers have documented strong climatic controls on area burned by wildfire across the West at the spatial scales of entire states (McKenzie et al. 2004), ecoprovinces (Littell et al. 2009), and sections within ecoprovinces (Littell et al. 2010). In forests across the Northwest, climate during the fire season appears to control area burned, whereas in arid mountains and shrublands, antecedent climate (e.g., wetter and cooler summers or winters preceding the fire season) can increase area burned during the fire season by increasing fuel abundance and continuity (Littell et al. 2009). WMI research has also synthesized sediment yields after wildland fire in different rainfall regimes across the western United States (Moody and Martin 2008), providing key context for potential climate-mediated watershed changes in post-fire runoff and erosion relationships.

MPB infestations have historically occurred frequently and extensively throughout the American West (Logan and Powell 2001). Warming and drought affect development rates of beetle life stages, winter mortality, and host-tree susceptibility (Logan and Powell 2001; Carroll et al. 2004; Oneil 2006). WMI has examined MPB outbreaks Westwide, with focused studies on outbreak potential within Washington State. Across the West, stand structural conditions make host species susceptible to beetle attack (Oneil 2006; Hicke and Jenkins 2008), but we expect host vulnerability to become increasingly linked to drought stress as warming continues (DeLucia et al. 2000; Oneil 2006). MPB outbreaks are facilitated when the insect's reproductive cycle is very close to one year, such that larvae emerge at an optimal time for feeding and dispersal (Logan and Powell 2001). As temperatures increase, the life cycle shortens; therefore, future climatic change is predicted to reduce the area of climatic suitability for the MPB at low elevations, but increase suitability at higher elevations (Hicke et al. 2006; Littell et al. 2010). WMI contributed significantly to a recent synthesis of the causes and ecological consequences of MPB outbreaks in western North America (Bentz et al. 2009).

Although their nature, timing, and effects are only beginning to be understood, synergistic interactions between disturbances are producing larger effects than would occur from either disturbance independently. For example, MPB outbreaks have been linked to the increased likelihood of stand-replacing fire and changes in fire behavior, with the nature of the effect depending on the time since infestation (Lynch et al. 2006; Jenkins et al. 2008), although there is conflicting evidence to support the idea of elevated fire hazard following bark beetle outbreaks (Romme et al. 2006; Kulakowski and Veblen 2007; Bentz et al. 2009). Combined with increasing climatic stress on tree populations and growth, disturbance interactions can alter forest structure and function faster than could be expected from species redistribution or disturbance alone. Simultaneous climatically driven shifts in the locations of species' optima, ecosystem productivity, disturbance regimes, and the interactions between them could reset forest succession over large areas and short time frames.

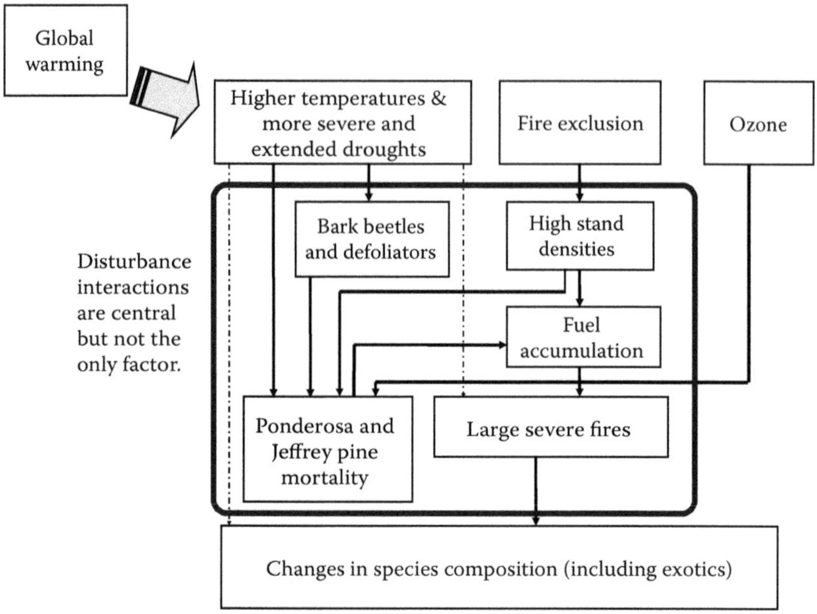

FIGURE 8.4 Stress complex in Sierra Nevada and southern Californian mixed-conifer forests. The effects of disturbance regimes (insects and fire) and fire exclusion are exacerbated by global warming. Stand-replacing fires and drought-induced mortality contribute to changes in species composition including increased abundance of exotic species. (From McKenzie, D., D.L. Peterson, and J.S. Littell. 2009. Global warming and stress complexes in forests of western North America. In *Wildland fires and air pollution*, Eds. A. Bytnerowicz, M. Arbaugh, A. Riebau, and C. Anderson. Amsterdam: Elsevier Science, 319–37.).

WMI researchers have begun to document disturbance interactions, and how these in turn affect ecosystem process, by building conceptual models of "stress complexes" (Allen 2007; McKenzie et al. 2009) and convening a structured workshop to establish a research agenda for this difficult but key problem area. Early results include conceptual frameworks for linking global warming to disturbance interactions in WMI bioregions (Figure 8.4) and guidelines for landscape disturbance modeling under climatic change (Cushman et al. 2007).

Simulating Ecosystem Response to Climatic Change

The response of hydrologic and ecological processes of Western US mountains to historical climatic extremes provides insight on how these systems may vary in their responses to future climates (Baron et al. 2000a). The climates of Western mountain catchments vary by their location in maritime, continental, windward, or leeward locations, as well as by latitude and elevation. We used RHESSys, a spatially distributed, dynamic process model of water, carbon, and nitrogen fluxes, to compare measures of net primary production, evapotranspiration, soil moisture, and stream flow across sites in response to climatic extremes. The five highest and lowest precipitation, snow depth, and temperature years when compared with their 47-year mean values for catchments in North Cascades NP, Glacier NP, Yosemite NP, and the Snake River of Colorado were used to evaluate ecosystem and hydrological responses.

At all sites, precipitation was more variable than temperature from year to year. Wet and dry precipitation years ranged 30–63% higher to 32–51% lower, depending on the site, than the 47-year mean, while annual mean temperature variation for the four catchments ranged 1.1–1.4°C higher to 0.7–1.7°C lower compared with the 47-year mean. The driest years and those with the lowest snow depths influenced ecological and hydrologic responses at all sites. Primary productivity was stimulated by warmer temperatures at the high-elevation, low-temperature site in the Snake River watershed. Productivity at Yosemite NP and Glacier NP was responsive to precipitation and declined in dry years. Primary productivity at North Cascades NP, the warmest and wettest of the four watersheds, was the least responsive of the watersheds to climatic extremes.

Hydrologic processes were proportionally more responsive than ecological processes to climatic extremes, and these responses varied across watersheds. All four watersheds showed statistically significant increases in annual stream flow in extremely wet years and significant decreases in extremely dry years. Between-watershed differences in the relative change in stream flow with change in precipitation should indicate the extent to which evapotranspiration moderates the sensitivity of stream flow to a change in precipitation. North Cascades NP showed the greatest influence of evapotranspiration, consistent with the minimal vegetation responses to climatic extremes.

Stream flow in all catchments responded strongly to snow depth extremes. The North Cascades was least sensitive to extremes, and Glacier stream flow was most responsive. While Glacier discharge responded to both maximum and minimum snow depth, Yosemite stream flow responded significantly to high-snow-depth years only. All four watersheds had significant responses to high- and low-snow years in their fraction of summer flow, but the Snake River watershed had the largest proportional changes, with the North Cascades watershed having the smallest changes.

Our results suggest that historically, hydrological and ecological processes in four Western US mountain watersheds have been more affected by precipitation extremes than temperature extremes. Temperature is not unimportant (and was very important for the cold Snake River catchment), but model analysis suggests that basin-scale ecohydrologic fluxes may be more sensitive to precipitation. The ability of general circulation models to project precipitation patterns with increased greenhouse gas concentrations is poor, reducing our capability to provide meaningful projections of future ecohydrologic responses. The analysis of how mountain watersheds have responded in the past is one way of adding insight into the possible consequences.

In addition to comparing processes across watersheds, RHESSys modeling has elucidated interesting within-watershed patterns in response to climatic variability and change. The response of forests to a warmer climate depends on the direct effects of temperature on ecophysiological function and indirect effects related to a range of biogeophysical processes. In alpine regions, reduced snow accumulation and earlier melt of seasonal snowpacks are expected hydrologic responses to warming. For forests, this leads to earlier soil moisture recharge and increased summer drought stress. At the same time, increased air temperature alters net primary productivity of vegetation. Most models of climatic change effects focus either on hydrologic behavior or ecosystem structure or function. WMI has addressed the interactions between them.

We used a coupled model of ecohydrologic processes to estimate changes in evapotranspiration and vegetation productivity in a warmer climate. Results for Yosemite suggest that, for most snow-dominated elevations, the shift in the timing of recharge is significant and that this shift is likely to lead to declines in productivity and vegetation water use. The strength of this effect, however, depends on interactions between several factors that vary substantially across elevation gradients, including the initial timing of melt relative to the summer growing season, vegetation growth, and the extent to which initial vegetation is limited by water or energy. These climate-driven changes in water use by vegetation also have implications for summer stream flow. Results from this analysis provide a framework that can be used to develop strategic measurements and a context for scaling local measurements of vegetation responses, including sap flow, carbon flux, tree mortality, and tree growth (Tague et al. 2009).

Developing Adaptation Science

Efforts to develop strategies that facilitate adaptation to documented and expected responses of natural resources to climatic change are now beginning in earnest. In the most substantive effort to date, the US Climate Change Science Program has developed Synthesis and Assessment products for federal land management agencies (Baron et al. 2008; Joyce et al. 2008). We initiated science-management collaborations at Olympic National Forest (Washington) in order to develop management options that will facilitate adaptation to climatic change (Littell et al.). This was the first attempt to work with a federal agency to develop specific concepts and applications that could potentially be implemented in management and planning. The focus of this effort was to develop strategies and management options for adapting to climatic change across multiple resources.

Olympic National Forest resource staff developed the following *general adaptation strategies* (Littell et al. 2011):

- *Manage for resilience, decrease vulnerability* – The success of adaptation strategies can be defined by their ability to reduce the vulnerability of resources to a changing climate while attaining specific management goals for the condition of resources and production of ecosystem services.
- *Prioritize climate-smart treatments* – Prioritizing treatments with the greatest likelihood of being effective in the long run recognizes that some treatments may cause short-term detrimental effects but have long-term benefits. For example, fish species may be vulnerable to increased sedimentation associated with flood-driven failures of unmaintained roads, but road rehabilitation may produce temporary sedimentation and may invite invasive weeds. Ideally, triage situations could be avoided, but given limited financial resources, it is necessary to prioritize management actions

with the highest likelihood of success at the expense of those that divert resources and have less certainty of favorable outcomes.

- *Consider trade-offs and conflicts* – Future impacts on ecological and socioeconomic sensitivities can result in potential trade-offs and conflicts for species conservation and other resource values. For example, stress complexes exacerbated by climatic change may cause threatened species to become even more rare, thus undermining the likelihood of successful protection. These trade-offs and conflicts can be considered collectively and incorporated in land management planning.

Olympic National Forest resource staff also developed *specific adaptation options*:

- *Increase landscape diversity* – This option focuses on increasing variety in stand structures and species assemblages over large areas and avoiding "one size fits all" management prescriptions (Millar et al. 2007). This can include applying forest thinning to increase variability in stand structure, increase resilience to stress by increasing tree vigor, and reduce vulnerability to disturbance (Parker et al. 2000; Dale et al. 2001; Spittlehouse and Stewart 2003).
- *Maintain biological diversity* – Appropriate species and genotypes can be planted in anticipation of a warmer climate (Smith and Lenhart 1996; Parker et al. 2000; Spittlehouse and Stewart 2003; Millar et al. 2007), thus allowing resource managers to "hedge their bets" by diversifying the phenotypic and genotypic template on which climate and competition interact, and to avoid widespread mortality at the regeneration stage. For example, nursery stock from warmer, drier locations than what is prescribed in genetic guidelines based on current seed zones can be planted following a crown fire (Spittlehouse and Stewart 2003).
- *Treat large-scale disturbance as a management opportunity* – Large-scale disturbance causes rapid changes in ecosystems, but also provides opportunities to apply adaptation strategies (Dale et al. 2001; Millar et al. 2007). Carefully designed management experiments for adapting to climatic change can be implemented, provided that plans are in place in anticipation of large disturbances. For example, one could experiment with mixed-species tree planting even though the standard prescription might be for a monoculture (Millar et al. 2007).
- *Implement early detection/rapid response for invasive species* – A focus on treating small problems before they become large, unsolvable problems recognizes that proactive management is more effective than delayed implementation (Millar et al. 2007). For example, recently burned areas may be susceptible to invasive species that can be detected by monitoring during the first few years after fire.
- *Match engineering of infrastructure to expected future conditions* – This refers primarily to road and drainage engineering that can accommodate future changes in hydrology (Spittlehouse and Stewart 2003). However,

it might be possible to design road networks to reduce damage caused to valley-bottom roads when flooding occurs.

- *Collaborate with a variety of partners* – Working with a diversity of landowners, agencies, and stakeholders will develop support for and consistency in adaptation options. For example, federal land managers can work with adjacent state forest managers to agree on conservation strategies for anadromous fish species.
- *Promote education and awareness about climatic change* – It is critical that internal and external education on climatic change is scientifically credible and consistent (Spittlehouse and Stewart 2003), with emphasis on the role of active management in adaptation. For example, local residents can be informed that wildfire may be more frequent in a warmer climate, which makes it imperative that they clear brush around homes to reduce fire hazard.

UNDERSTANDING, FORECASTING, AND ADAPTING—THE NEXT PHASE

The next phase of WMI research targets scientific activities that will inform decision support in federal agencies and beyond (Figure 8.5). Although our understanding of the effects of climatic variability and change on Western ecosystems is far from complete, there is increasing demand from resource managers to move forward with implementation of climatic-change science in management and planning and to develop options for adaptation to a warmer climate. Sufficient information is available to make informed decisions, despite some uncertainty, and guidance is needed now to comply with agency mandates to address climatic change. WMI is collaborating with federal agencies in this transition while maintaining scientific studies in key areas.

First, WMI will strive to *understand climate-ecosystem interactions, and detect and attribute change.* Land managers and policy makers need evidence that ecosystems are changing (detection), and that changes are caused by specific agents (attribution), in order to develop and implement adaptive management. They also need a strong scientific basis for all decision making related to climatic change. Proposed studies expand the geographic scope of WMI bioregions and provide empirical data and mechanistic understanding for the modeling effort described next.

Second, WMI will use *ecological forecasting to identify thresholds and vulnerabilities.* We will build on previous modeling efforts and provide forecasts of specific changes across the West, which in turn will provide the scientific basis for developing descriptions of future ecosystem conditions in Western mountains. We will emphasize local and regional vulnerabilities of ecosystem components or processes to climatic variability and change, thresholds at which specific forest or hydrologic ecosystems may experience state changes, and uncertainties associated with the forecasts.

Third, we will collaborate with land managers throughout the western United States to *develop adaptation strategies* to address potential effects on natural resources. Building on our initial efforts described previously, a variety of syntheses,

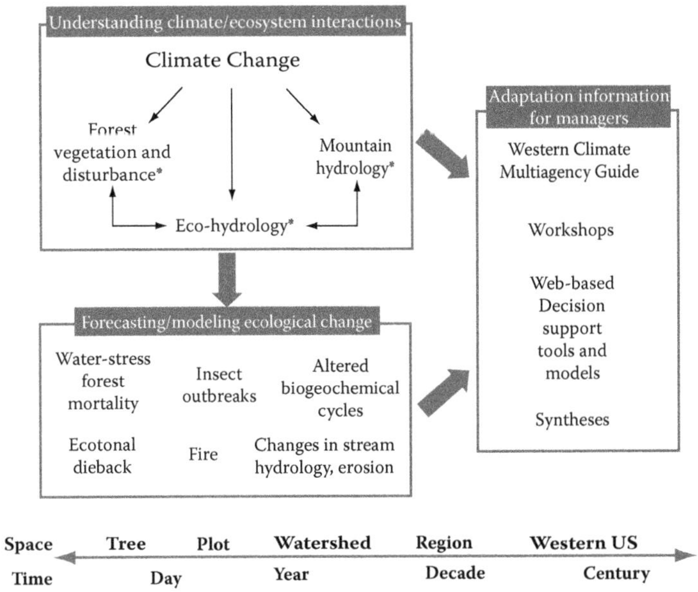

FIGURE 8.5 Diagrammatic summary of interactions and integration of scientific activities by the Western Mountain Initiative during the next phase of research.

tools, and strategies for adaptation will be developed, driven by the goal of maintaining native biodiversity and critical ecosystem structures and functions. All authors of this chapter have participated in writing Synthesis and Assessment Products for the U.S. Climate Change Science Program, which along with initial efforts on adaptation described previously, will provide much of the scientific basis for the development of adaptation options (Baron et al. 2008; Joyce et al. 2008; Fagre et al. 2009).

SUMMARY

The Western Mountain Initiative (WMI), a consortium of research groups in the western United States, focuses on *understanding and predicting responses—emphasizing sensitivities, thresholds, resistance, and resilience—of mountain ecosystems to climatic variability and change*. WMI addresses four key questions: (1) How are climatic variability and change likely to affect disturbance regimes? (2) How are changing climate and disturbance regimes likely to affect the composition, structure, and productivity of vegetation? (3) How will climatic variability and change effect hydrologic processes in the mountainous West? (4) Which mountain resources and ecosystems are likely to be most sensitive to future climatic change, and what are possible management responses? The research framework for WMI is premised on natural experiments in time (paleoecological and long-term studies), natural experiments in space (studies across regions and elevation gradients), and synthetic modeling. Results to date have documented how climatic variability and change effect long-term patterns of snow, glaciers, and water geochemistry; forest productivity, vigor, and demography; and long-term patterns of fire occurrence. Some significant physical and biological thresholds are already being reached (e.g., glacier recession, forest dieback), and empirical and simulation modeling indicates that major changes in hydrologic function and ecological disturbance will occur as climate continues to warm. WMI is working with national parks and national forests to develop science-based management options for adapting to climatic change.

REFERENCES

Adams, H.D., M. Guardiola-Claramonte, G.A. Barron-Gafford, J. Camilo Villegas, D.D. Breshears, C.B. Zou, P.A. Troch, and T.E. Huxman. 2009. Temperature sensitivity of drought-induced tree mortality portends increased regional die-off under global change-type drought. *Proceedings of the National Academy of Sciences, USA* 106:7063–66.

Alftine, K.J., G.P. Malanson, and D.B. Fagre. 2003. Feedback-driven response to multi-decadal climatic variability at an alpine forest-tundra ecotone. *Physical Geography* 24:520–35.

Allen, C.D. 2007. Cross-scale interactions among forest dieback, fire, and erosion in northern New Mexico landscapes. *Ecosystems* 10:797–808.

———. 2009. Climate-induced forest dieback: an escalating global phenomenon? *Unasylva* 231/232:43–49.

Allen, C.D., R.S. Anderson, R.B. Jass, J.L. Toney, and C.H. Baisan. 2008. Paired charcoal and tree-ring records of high-frequency fire from two New Mexico bog sites. *International Journal of Wildland Fire* 17:115–30.

Allen, C.D., and D.D. Breshears. 1998. Drought-induced shift of a forest-woodland ecotone: rapid landscape response to climatic variation. *Proceedings of the National Academy of Sciences, USA* 95:14, 839–42.

Allen, C.D., A.K. Macalady, H. Chenchouni, D. Bachelet, N. McDowell, M. Vennetier, T. Kitzberger, A. Rigling, D.D. Breshears, E.H. Hogg, P. Gonzalez, R. Fensham, Z. Zhang, J.-H. Lim, J. Castro, N. Demidova, G. Allard, and S.W. Running. 2010. A global overview of drought and heat-induced tree mortality reveals emerging climate change risks for forests. *Forest Ecology and Management* 259:660–84.

Allen, T.R., and S.J. Walsh. 1996. Spatial and compositional pattern of alpine treeline, Glacier National Park, Montana. *Photogrammetric Engineering and Remote Sensing* 62:1261–68.

Anderson, R.S., C.D. Allen, J.L. Toney, R.B. Jass, and A.N. Bair. 2008. Holocene vegetation and forest fire regimes in subalpine and mixed conifer forest sites, southern Rocky Mountains, USA. *International Journal of Wildland Fire* 17:96–114.

Barnett, T.P., D.W. Pierce, H.G. Hidalgo, C. Bonfils, B.D. Santer, T. Das, G. Bala, A.W. Wood, T. Nozawa, A.A. Mirin, D.R. Cayan, and M.D. Dettinger. 2008. Human-induced changes in the hydrology of the western United States. *Science* 319:1080–83.

Baron, J.S., C.D. Allen, E. Fleishman, L. Gunderson, D. McKenzie, L. Meyerson, J. Oropeza, and N. Stephenson. 2008. National parks. In *Preliminary review of adaptation options for climate-sensitive ecosystems and resources: A report by the U.S. Climate Change Science Program and the Subcommittee on Global Change Research*, Eds. S.H. Julius and J.M. West. Washington, DC: US Environmental Protection Agency, 4-1–4-68.

Baron, J.S., M. Hartman, L.E. Band, and R. Lammers. 2000a. Sensitivity of a high elevation Rocky Mountain watershed to altered climate and CO_2. *Water Resources Research* 36:89–99.

Baron J.S., H.M. Rueth, A.M. Wolfe, K.R. Nydick, E.J. Allstott, J.T. Minear, and B. Moraska. 2000b. Ecosystem responses to nitrogen deposition in the Colorado Front Range. *Ecosystems* 3:352–68.

Baron, J.S., T.W. Schmidt, and M.D. Hartman. 2009. Climate-induced changes in high elevation dynamics stream nitrate. *Global Change Biology*, 15:1777–1789.

Bekker, M.F. 2005. Positive feedback between tree establishment and patterns of subalpine forest advancement, Glacier National Park, Montana, USA *Arctic, Antarctic, and Alpine Research* 37:97–107.

Bentz, B., C.D. Allen, M. Ayres, E. Berg, A. Carroll, M. Hansen, J. Hicke, L. Joyce, J. Logan, W. MacFarlane, J. MacMahon, S. Munson, J. Negron, T. Paine, J. Powell, K. Raffa, J. Régnière, M. Reid, W. Romme, S. Seybold, D. Six, D. Tomback, J. Vandygriff, T. Veblen, M. White, J. Witcosky, and D. Wood. 2009. *Bark beetle outbreaks in western North America: causes and consequences*. Salt Lake City: University of Utah Press.

Bigler, C., D.G. Gavin, C. Gunning, and T.T. Veblen. 2007. Drought induces lagged tree mortality in a subalpine forest in the Rocky Mountains. *Oikos* 116:1983–94.

Breshears D.D., N.S. Cobb, P.M. Rich, K.P. Price, C.D. Allen, R.G. Balice, W.H. Romme, J.H. Kastens, M.L. Floyd, J. Belnap, J.J. Anderson, O.B. Myers, and C.W. Meyer. 2005. Regional vegetation die-off in response to global-change-type drought. *Proceedings of the National Academy of Science USA* 102:15, 144–48.

Breshears, D.D., O.B. Myers, C.W. Meyer, F.J. Barnes, C.B. Zou, C.D. Allen, N.G. McDowell, and W.T. Pockman. 2009. Tree die-off in response to global-change-type drought: mortality insights from a decade of plant water potential measurements. *Frontiers in Ecology and the Environment* 7:185–89.

Butler, D.R., G.P. Malanson, and D.M. Cairns. 1994. Stability of alpine treeline in northern Montana, USA. *Phytocoenologia* 22:485–500.

Butler, D.R., and S.J. Walsh. 1994. Site characteristics of debris flows and their relationship to alpine treeline in Glacier National Park. *Physical Geography* 15:181–99.

Cairns, D.M., and G.P. Malanson. 1998. Environmental variables influencing carbon balance at the alpine treeline ecotone: a modeling approach. *Journal of Vegetation Science* 9:679–92.

Carroll, A.L., S.W. Taylor, J. Regniere, and L. Safranyik. 2004. Effects of climate change on range expansion by the mountain pine beetle in British Columbia. In *Mountain pine beetle symposium: challenges and solutions*, Eds. T.L. Shore, J.E. Brooks, and J.E. Stone, Information Report BC-X-399. Kelowna, British Columbia: Natural Resources Canada, Canadian Forest Service, Pacific Forestry Centre, 223–32.

Christensen, L., C.L. Tague, and J.S. Baron. 2008. Spatial patterns of transpiration response to climate variability in a snow-dominated mountain ecosystem. *Hydrological Processes* 22:3576–88.

Cushman, S.A., D. McKenzie, D.L. Peterson, J.S. Littell, and K.S. McKelvey. 2007. Research agenda for integrated landscape modeling. USDA Forest Service General Technical Report RMRS-GTR-194. Fort Collins, CO: Rocky Mountain Research Station.

Dale, V.H, L.A. Joyce, S. McNulty, R.P. Neilson, M.P. Ayres, M.D. Flannigan, P.J. Hanson, L.C. Irland, A.E. Lugo, C.J. Peterson, D. Simberloff, F.J. Swanson, B.J. Stocks, and B.M. Wotton. 2001. Climate change and forest disturbances. *BioScience* 51:723–34.

DeLucia, E.H., H. Maherali, and E.V. Carey. 2000. Climate-driven changes in biomass allocation in pines. *Global Change Biology* 6:587–93.

Diaz, H.F., and J.K. Eischeid. 2007. Disappearing "alpine tundra" Köppen climatic type in the western United States. *Geophysical Research Letters* 34:L18707doi:10.1029/2007 GL031253.

Diaz, H.F., and V. Markgraf. 2000. *El Niño and the Southern Oscillation*. New York: Cambridge University Press.

Diettert, G.A. 1992. *Grinnell's glacier: George Bird Grinnell and Glacier National Park*. Missoula, MT: Mountain Press.

Dunham, J., B. Rieman, and G. Chandler. 2003. Influences of temperature and environmental variables on the distribution of bull trout within streams at the southern margin of its range. *North American Journal of Fisheries Management* 23:894–904.

Dyurgerov, M.B., and M.F. Meier. 2000. 20th century climate change: evidence from small glaciers. *Proceedings of the National Academy of Science, USA* 97:1406–11.

Fagre, D.B., C.W. Charles, C.D. Allen, C. Birkeland, F.S. Chapin III, P.M. Groffman, G.R. Guntenspergen, A.K. Knapp, A.D. McGuire, P.J. Mulholland, D.P.C. Peters, D.D. Roby, and G.J. Sugihara. 2009. *Thresholds of climate change in ecosystems: a report by the U.S. Climate Change Science Program and the Subcommittee on Global Change Research.* Washington, DC: US Geological Survey.

Fagre, D.B., and L.A. McKeon. 2010. Documenting disappearing glaciers: repeat photography at Glacier National Park. In *Repeat photography: methods and applications in the natural sciences*, Eds. R.H. Webb, D.E. Boyer, and R.M. Turner. Washington D.C.: Island Press, 77–88.

Ficke, A.D., C.A. Myrick, and L.J. Hansen. 2007. Potential impacts of global climate change on freshwater fishes. *Reviews in Fish Biology and Fisheries* 17:581–613.

Fountain, A., M. Hoffman, K. Jackson, H. Basagic, T. Nylen, and D. Percy. 2007. Digital outlines and topography of the glaciers of the American West. US Geological Survey Open File Report 2006-1340.

Gedalof, Z., D.L. Peterson, and N.J. Mantua. 2005. Atmospheric, climatic, and ecological controls on extreme wildfire years in the northwestern United States. *Ecological Applications* 15:154–74.

Granshaw, F.D., and A.G. Fountain. 2006. Glacier change (1958–1998) in the North Cascades National Park Complex,Washington, USA. *Journal of Glaciology* 52:251–56.

Groisman, P.Y., and R.W. Knight. 2008. Prolonged dry episodes over the conterminous United States: new tendencies. *Journal of Climate* 21:1850–62.

Hall, M.P., and D.B. Fagre. 2003. Modeled climate-induced glacier change in Glacier National Park, 1850–2100. *BioScience* 53:13140.

Hamlet, A.F., P.W. Mote, M.P. Clark, and D.P. Lettenmaier. 2007. Twentieth-century trends in runoff, evapotranspiration, and soil moisture in the western United States. *Journal of Climate* 20:1468–86.

Hauer, R.F., J.S. Baron, D.H. Campbell, K.D. Fausch, S.W. Hostetler, G.H. Leavesley, P.R. Leavitt, D.M. McKnight, and J.A. Stanford. 1997. Assessment of climate change and freshwater ecosystems of the Rocky Mountains, USA and Canada. *Hydrological Processes* 11:903–24.

Hessl, A.E., D. McKenzie, and R. Schellhaas. 2004. Drought and Pacific Decadal Oscillation linked to fire occurrence in the inland Pacific Northwest. *Ecological Applications* 14:425–42.

Hicke, J.A., and J.C. Jenkins. 2008. Mapping lodgepole pine stand structure susceptibility to mountain pine beetle attack across the western United States. *Forest Ecology and Management* 255:1536–47.

Hicke, J.A., J.C. Jenkins, D.S. Ojima, and M. Ducey. 2007. Spatial patterns of forest characteristics in the western United States derived from inventories. *Ecological Applications* 17:2387–2402.

Hicke J.A., J.A. Logan, J.A. Powell, and D.S. Ojima. 2006. Changing temperatures influence suitability for modeled mountain pine beetle (*Dendroctonus ponderosae*) outbreaks in the western United States, *Journal of Geophysical Research B*, 111, G02019, doi: 10.1029/2005JG000101.

Hoffman, M.J., A.G. Fountain, and J.M. Achuff. 2007. Twentieth-century variations in area of cirque glaciers and glacierets, Rocky Mountain National Park, Rocky Mountains, Colorado, USA. *Annals of Glaciology* 46:349–54.

Holtmeier, F.-K. 1982. "Ribbon-forest" and "hedges": Strip-like distribution patterns of trees at the upper timberline in the Rocky Mountains. *Erdkunde* 36:142–53.

IPCC (Intergovernmental Panel on Climate Change). 2008. Climate change 2007: impacts, adaptation, and vulnerability, Working Group II contribution to the Fourth Assessment Report. United Kingdom: Cambridge University Press.

Jenkins, M.J., E. Hebertson, W. Page, and C.A. Jorgensen. 2008. Bark beetles, fuels, fires and implications for forest management in the Intermountain West. *Forest Ecology and Management* 254:16–34.

Joyce, L.A., G.M. Blate, J.S. Littell, S.G. McNulty, C.I. Millar, S.C. Moser, R.P. Neilson, K. O'Halloran, and D.L. Peterson, 2008. National Forests. In *Preliminary review of adaptation options for climate-sensitive ecosystems and resources: a report by the U.S. Climate Change Science Program and the Subcommittee on Global Change Research*, Eds. S.H. Julius and J.M. West. Washington, DC: US Environmental Protection Agency, 3-1–3-127.

Keleher, C.J., and F.J. Rahel. 1996. Thermal limits to salmonid distributions in the Rocky Mountain region and potential habitat loss due to global warming: a geographic information system (GIS) approach. *Transactions of the American Fisheries Society* 125:1–13.

Kellogg, L.-K.B., D. McKenzie, D.L. Peterson, and A.E. Hessl. 2008. Spatial models for inferring topographic controls on low-severity fire in the eastern Cascade Range of Washington, USA. *Landscape Ecology* 23:227–40.

Key, C. H., D.B. Fagre, and R.K. Menicke. 2002. Glacier retreat in Glacier National Park, Montana. In *Satellite image atlas of glaciers of the world, glaciers of North America—glaciers of the Western United States*, Eds. R.S. Williams and J.G. Ferrigno. US Geological Survey Professional Paper 1386-J. Washington, DC: US Government Printing Office, J365–J381.

Klasner, F.L., and D.B. Fagre. 2002. A half century of change in alpine treeline patterns at Glacier National Park, Montana, USA. *Arctic, Antarctic and Alpine Research* 34:49–56.

Knowles, N., M.D. Dettinger, and D.R. Cayan, 2006. Trends in snowfall versus rainfall in the western United States. *Journal of Climate* 19:4545–59.

Kulakowski, D., and T.T. Veblen. 2007. Effect of prior disturbances on the extent and severity of a 2002 wildfire in Colorado subalpine forests. *Ecology* 88:759–69.

Lafreniere, M., and M. Sharp. 2005. A comparison of solute fluxes and sources from glacial and non-glacial catchments over contrasting melt seasons. *Hydrological Processes* 19:2991–3012.

Littell, J.S., McKenzie, D., Peterson, D.L., and Westerling, A.L. 2009a. Climate and wildfire area burned in western U.S. ecoprovinces, 1916–2003. *Ecological Applications* 19:1003–21.

Littell, J.S., E.E. Oneil, D. McKenzie, J.A. Hicke, J.A. Lutz, R.A. Norheim, and M.M. Elsner. 2010. Forest ecosystems, disturbance, and climatic change in Washington State, USA. *Climatic Change* 102:129–158.

Littell, J.S., D.L. Peterson, C.I. Millar, and K.A. O'Halloran. 2011. U.S. national forests adapt to climate change through science-management partnerships.

Logan, J.A., and J.A. Powell. 2001. Ghost forests, global warming and the mountain pine beetle (Coleoptera: Scolytidae). *American Entomologist* 47:160–73.

Lutz, J.A. 2008. Climate, fire, and vegetation change in Yosemite National Park. Ph.D. dissertation. Seattle, WA: University of Washington.

Lynch, H.J., R.A. Renkin, R.L. Crabtree, and P.R. Moorcroft. 2006. The influence of previous mountain pine beetle (*Dendroctonus ponderosae*) activity on the 1988 Yellowstone fires. *Ecosystems* 9:1318–27.

Malanson, G.P., and D.R. Butler. 2002. The western cordillera. In *Physical geography of North America*, Ed. A. Orme. United Kingdom: Oxford University Press, 363–79.

Malanson, G.P., D.R. Butler, D.B. Fagre, S.J. Walsh, D.F. Tomback, L.D. Daniels, L.M. Resler, W.K. Smith, D.J. Weiss, D.L. Peterson, A.G. Bunn, C.A. Hiemstra, D. Liptzin, P.S. Bourgeron, Z. Shen, and C.I. Millar. 2007. Alpine treeline of western North America: linking organism-to-landscape dynamics. *Physical Geography* 28:378–96.

Mantua, N.J., S.R. Hare, Y. Zhang, J.M. Wallace, and R.C. Francis. 1997. A Pacific interdecadal climate oscillation with impacts on salmon production. *Bulletin of the American Meteorological Society* 78:1069–79.

Margolis, E.Q., T.W. Swetnam, and C.D. Allen. 2007. A stand-replacing fire history in upper montane forests of the Southern Rocky Mountains. *Canadian Journal of Forest Research* 37:2227–41.

McCabe, G.J., and M.P. Clark. 2005. Trends and variability in snowmelt runoff in the western United States. *Journal of Hydrometeorology* 6:476–82.

McDowell, N., C.D. Allen, and L. Marshall, L. 2010. Growth, carbon isotope discrimination, and climate-induced mortality across a *Pinus ponderosa* elevation transect. *Global Change Biology* 16:399–415.

McDowell, N., W.T. Pockman, C.D. Allen, D.D. Breshears, N. Cobb, T. Kolb, J. Sperry, A. West, D. Williams, and E.A. Yepez. 2008. Mechanisms of plant survival and mortality during drought: why do some plants survive while others succumb to drought? *New Phytologist* 178:719–39.

McKenzie, D., Z.M. Gedalof, D.L. Peterson, and P. Mote. 2004. Climatic change, wildfire, and conservation. *Conservation Biology* 18:890–902.

McKenzie, D., A.E. Hessl, and L.-K.B. Kellogg. 2006. Using neutral models to identify constraints on low-severity fire regimes. *Landscape Ecology* 21:139–52.

McKenzie, D., D.L. Peterson, and J.S. Littell. 2009. Global warming and stress complexes in forests of western North America. In *Wildland fires and air pollution*, Eds. A. Bytnerowicz, M. Arbaugh, A. Riebau, and C. Anderson. Amsterdam: Elsevier Science, 319–37.

Millar, C.I., N.L. Stephenson, S.L. Stephens. 2007. Climate change and forests of the future: managing in the face of uncertainty. *Ecological Applications* 17:2145–51.

Millar, C.I., R.D. Westfall, D.L. Delany, J.C. King, and L.J. Graumlich. 2004. Response of subalpine conifers in the Sierra Nevada, California, USA, to 20th-century warming and decadal climate variability. *Arctic, Antarctic, and Alpine Research* 36:181–200.

Milly, P.C.D., J. Betancourt, M. Falkenmark, R.M. Hirsch, Z.W. Kundewicz, D.P. Lettenmaier, and R.J. Stouffer. 2008. Stationarity is dead: whither water management? *Science* 319:573–74.

Moody, J.A., and D.A. Martin. 2008. Synthesis of sediment yields after wildland fire in different rainfall regimes in the western United States. *International Journal of Wildland Fire* 18:96–115.

Mote, P.W. 2003. Trends in snow water equivalent in the Pacific Northwest and their climatic causes. *Geophysical Research Letters* 30:1601, DOI:10.1029/2003GL017258.

———. 2006. Climate-driven variability and trends in mountain snowpack in western North America. *Journal of Climate* 19:6209–6220.

Mote, P.W., A.F. Hamlet, M.P. Clark, and D.P. Lettenmaier. 2005. Declining mountain snowpack in western North America. *Bulletin of American Meteorological Society* 86:39–49.

Oneil, E.E. 2006. Developing stand density thresholds to address mountain pine beetle susceptibility in eastern Washington forests. Ph.D. dissertation. Seattle, WA: University of Washington.

Parker, W.C., S.J. Colombo, M.L. Cherry, M.D. Flannigan, S. Greifenhagen, R.S. McAlpine, C. Papadopol, and T. Scarr. 2000. Third millennium forestry: what climate change might mean to forests and forest management in Ontario. *Forestry Chronicle* 76:445–63.

Prichard, S.J., Oswald, W.W., Gedalof, Z., and Peterson, D.L. 2009. Holocene fire and vegetation dynamics in a montane forest, North Cascade Range, Washington, USA. *Quaternary Research* 72:57–67.

Raffa, K.F., B.H. Aukema, B.J. Bentz, A.L. Carroll, J.A. Hicke, M.G. Turner, and W. Romme. 2008. Cross-scale drivers of natural disturbances prone to anthropogenic amplification: dynamics of bark beetle eruptions. *BioScience* 58:501–17.

Redmond, K. 2007. Evaporation and the hydrologic budget of Crater Lake, Oregon. *Hydrobiologia* 574:29–46.

Regonda, S.K., B. Rajagopalan, M. Clark, and J. Pitlick. 2005. Seasonal cycle shifts in hydroclimatology over the western United States. *Journal of Climate* 18:372–84.

Resler, L.M. 2006. Geomorphic controls of spatial pattern and process at alpine treeline. *Professional Geographer* 58:124–38.

Rieman, B.E., D.C. Lee, and R.F. Thurow. 1997. Distribution, status, and likely future trends of bull trout within the Columbia River and Klamath River basins. *North American Journal of Fisheries Management* 17:1111–25.

Romme, W.H., J. Clement, J. Hicke, D. Kulakowski, L.H. MacDonald, T.L. Schoennagel, and T.T. Veblen. 2006. Recent forest insect outbreaks and fire risk in Colorado forests: a brief synthesis of relevant research. Colorado Forest Restoration Institute Report. Fort Collins: University of Colorado.

Selong, J.H., T.E. McMahon, A.V. Zale, and F.T. Barrows. 2001. Effect of temperature on growth and survival of bull trout, with application of an improved method for determining thermal tolerances in fishes. *Transactions of the American Fisheries Society* 130:1026–37.

Smith, J.B., and S.S. Lenhart. 1996. Climate change adaptation policy options. *Climate Research* 6:193–201.

Smith, W.K., M.J. Germino, T.E. Hancock, and D.M. Johnson. 2003. Another perspective on altitudinal limits of alpine timberlines. *Tree Physiology* 23:1101–12.

Spittlehouse, D.L., and R.B. Stewart. 2003. Adaptation to climate change in forest management. *BC Journal of Ecosystems and Management* 4:1–11.

Stewart, I.T., D.R. Cayan, and M.D. Dettinger. 2005. Changes toward earlier streamflow timing across western North America. *Journal of Climatology* 18:1136–55.

Swetnam, T.W., and R.S. Anderson. 2008. Fire climatology in the western United States: introduction to special issue. *International Journal of Wildland Fire* 17:1–7.

Swetnam, T.W., and C.H. Baisan. 2003. Tree-ring reconstructions of fire and climate history in the Sierra Nevada and southwestern United States. In *Fire and climatic change in temperate ecosystems of the western Americas*, Eds. T.T. Veblen, W.L. Baker, G. Montenegro, and T.W. Swetnam. New York: Springer, 158–95.

Swetnam, T.W., and J.L. Betancourt. 1998. Mesoscale disturbance and ecological response to decadal climatic variability in the American Southwest. *Journal of Climate* 11:3128–47.

Tague, C., G. Grant, M. Farrell, J. Choate, and A. Jefferson. 2008. Deep groundwater mediates streamflow response to climate warming in the Oregon Cascades. *Climatic Change* 86:189–210.

Tague, C., K. Heyn, and L. Christensen. 2009. Topographic controls on spatial patterns of conifer transpiration and net primary productivity under climate warming in mountain ecosystems. *Ecohydrology* 2:541–54.

Thies, H., U. Nickus, V. Mair, R. Tessadri, D. Tait, B. Thaller, and R. Psenner. 2007. Unexpected response of high alpine lake waters to climate warming. *Environmental Science and Technology* 10.1021/es0708060.

Tomback, D.F., and L.M. Resler. 2007. Invasive pathogens at alpine treeline: consequences for treeline dynamics. *Physical Geography* 28:397–418.

Urban, D.L. 2000. Using model analysis to design monitoring programs for landscape management and impact assessment. *Ecological Applications* 10:1820–32.

van Mantgem, P.J., and N.L. Stephenson. 2007. Apparent climatically induced increase of tree mortality rates in a temperate forest. *Ecology Letters* 10:909–16.

van Mantgem, P.J., N.L. Stephenson, J.C. Byrne, L.D. Daniels, J.F. Franklin, P.Z. Fulé, M.E. Harmon, J.M. Smith, A.H. Taylor, and T.T. Veblen. 2009. Widespread increase of tree mortality rates in the western United States. *Science* 323:521–24.

Veblen, T.T., and D.C. Lorenz. 1991. *The Colorado Front Range: a century of change.* Salt Lake City: University of Utah Press.

Westerling, A.L., H.G. Hidalgo, D.R. Cayan, and T.W. Swetnam. 2006. Warming and earlier spring increase western U.S. forest wildfire activity. *Science* 313:940–43.

Whitlock, C., and R.S. Anderson. 2003. Fire history reconstructions based on sediment records from lakes and wetlands. In *Fire and climatic change in temperate ecosystems of the western Americas*, Eds. T.T. Veblen, W.L. Baker, G. Montenegro, and T.W. Swetnam. New York: Springer, 3–31.

Williams, J.E., A.L. Haak, H.M. Neville, W.T. Colyer, and N.G. Gillespie. 2007a. Climate change and western trout: strategies for restoring resistance and resilience in native populations. In *Sustaining wild trout in a changing world: proceedings of Wild Trout IX symposium*, Eds. R.F. Carline and C. LoSapio, 236–46. http:// www.wildtroutsymposium.com/proceedings-9.pdf. (Accessed April 15, 2011.)

Williams, M.W., M. Knauf, R. Cory, N. Caine, and F. Liu. 2007b. Nitrate content and potential microbial signature of rock glacier outflow, Colorado Front Range. *Earth Surface Processes and Landforms* DOI:10.1002/esp.1455.

9 Precipitation Climatology at Selected LTER Sites: *Regionalization and Dominant Circulation Patterns*

Douglas G. Goodin

CONTENTS

INTRODUCTION

There is now abundant evidence that the Earth's climate is changing at a rapid rate (IPCC 2007). Although the role of humans in this change is controversial, the consensus opinion among the majority of the climate research community is that anthropogenic emissions of radiatively active greenhouse gases are a major driver of the observed climatic changes. Global climate models have consistently predicted increases in mean global temperature with increased atmospheric greenhouse gas concentrations (Cox et al. 2000). These predictions have been supported by direct observational evidence, for example, the association between increased atmospheric CO_2 and warming in the latter half of the twentieth century (Mann et al. 2008). Additional support of the enhanced greenhouse effect hypothesis has been provided by a number of proxy indicators of temperature change, among them glacial dynamics (Oerlemans 2001), reduction in ice sheet volume in both the northern (Clark et al. 1999) and southern (Rignot and Thomas 2002) hemispheres, and global phenology changes (Badeck et al. 2004), as well as other changes in the Earth's biota (Root et al. 2003).

Regardless of the reasons for climatic change, the reality of such changes and the need to understand their effect on the Earth system is one of the most pressing challenges facing the scientific community. The potential impacts of climate change will manifest in both the biotic and abiotic components of the Earth system, in ways that may not be readily predictable based on past analogs (Williams and Jackson 2007). In this context, even consideration of the terminology used to describe the problem is illustrative. Although frequently framed in both the scientific and popular media as "global warming," the actual impacts of climate change are likely to be more complex and geographically heterogeneous than the simple hypothesis of temperature increase would suggest. Results for similar emissions scenarios simulated using different climate models show widely varying patterns of temperature and precipitation in a greenhouse-enhanced world (Watson et al. 1997). Geographic and temporal variations in the distribution of precipitation are expected. Some areas might have little or no change in total interannual precipitation, but with the timing altered to include less frequent but larger events. Other areas could see changes in both the amount and timing of precipitation, as well as changes in the relative amounts of rain and snow. These global alterations in temperature and hydrology represent changes in the distribution of energy through the surface-atmospheric system, which in turn may feed back to force the system in unforeseen ways.

Changes in the distribution of precipitation and temperature are important in their own right, but also because of their role in regulating the Earth's biosphere. Across spatial scale, climate (operationally defined here as the mean and variability of temperature and precipitation) is a major ecological control of terrestrial biota. At larger spatial scales, aggregate patterns of climate control the location and distribution of major terrestrial biomes (Holdridge 1967). Vegetation assemblies and species distribution can often be related to climatic conditions, even at scales as fine as a few square centimeters. Understanding the interaction of climate and the biosphere is a major theme in ecological and biogeographic research (Stenseth et al. 2003). Relating them remains the subject of intense investigation, made all the more significant because of the potential need to understand climate–biosphere interactions under changing conditions.

Climate and Ecology Research in the LTER Program

As with many problems in ecology, comprehensive understanding of climate–biosphere interactions requires systematic observation at time scales appropriate to the problem. The Long Term Ecological Research (LTER) program was established in 1980 in order to address these needs within the ecological research community. LTER consists of a network of 26 research sites distributed throughout North America, Antarctica, and the Pacific Islands. The sites were chosen to represent a wide variety of ecosystem types, including agricultural, forest, grassland, arctic and alpine tundra, hot and cold desert, estuarine, urban, and marine. Since its inception, LTER research has concentrated on five core areas: primary productivity (patterns and control), trophic structures, organic matter accumulation, nutrient cycling, and disturbance. The configuration of the network as a series of permanent research sites allows for observation of both the spatial and temporal patterns of these ecosystem properties. Clearly, climate (as previously defined) is a major control over several of these core ecological variables, especially net primary productivity. For this reason, climate observations have been among the fundamental data sets collected at each LTER site since the inception of the program.

There have been a number of research projects aimed at relating climate and ecological response within LTER, including the Climate Variability and Ecosystem Response (CVER) project, a comprehensive, network-wide, cross-site program (Greenland et al. 2003). The CVER program was organized into a framework of time scales. Within the CVER project, investigations between climate drivers and various ecological responses were conducted at a number of LTER sites, at temporal scales ranging from less than a year to centuries/millennia (using modeling approaches and proxy measures). At each of these temporal scales, it was found that the ecosystems responded to climatic events or episodes that exceeded a certain threshold and that ecosystems to a certain extent seemed to buffer the effects of climatic variability (Greenland and Kittel 2002). The studies also found that at many LTER sites, ecosystem responses were related to a number of well-known climate/circulation patterns, including the El Niño/Southern Oscillation (ENSO), North Atlantic Oscillation (NAO), Pacific North American pressure cycle (PNA), and the Pacific Decadal Oscillation (PDO). Linking site-based ecosystem response variables to global circulation modes is significant, because it shows how large-scale climate events "connect" with localized responses. The results of these studies also clearly demonstrated the importance of long-term, site-based science for understanding and monitoring the linkages between climate and ecosystem response.

Although the CVER program provided information and comparative analysis of the LTER sites, it did not consider sites within their regional context. Hence, a question arises: How well do LTER sites represent the region in which they are located? The question is important, because an explicit goal of climate/environment research at sites such as those that comprise the LTER network is to understand how the biotic component of a particular ecosystem may respond to its environment. Because the type of intensive ecosystem monitoring done at individual sites is not practical at the biome scale, it becomes necessary to extrapolate based on site-specific results. In order for these site-specific results to be valid representations, the site itself must represent its region well. From a climatic perspective, this problem is complicated by two problems arising from the concept of a region. First, climate regions often do not coincide exactly with ecoregions. Second, climatic regionalizations are typically based on measurable climatic state variables (i.e., temperature, precipitation) and not on the dynamic processes that shape the climate of a region. Clearly, a better understanding of how these dynamic processes couple to the ecological system would enable ecologists to better understand how an ecosystem is linked to the larger global system. In this chapter, I explore this idea by (1) using correlation and multivariate analysis to link a climatic state variable (precipitation) to several modes of global circulation, in order to establish a geographic pattern of these linkages; and (2) determine how well a selected subset of LTER sites "represent" the circulation region in which they are located.

DATA

There are two components to this analysis: (1) regionalization of North American precipitation based on correlation with global circulation modes, and (2) evaluation of the linkages of precipitation climatology to the same set of circulation modes for a subset of LTER sites. The data used for these two components derives from three sources, each of which is discussed below.

Gridded Precipitation Data

For the climate regionalization component, data from the CRU05 data set were used. This data set was obtained from the Climate Research Unit of the University of East Anglia (www.cru.uea.ac.uk/cru/data/hrg.htm). The CRU05 data set consists of gridded mean monthly values of nine climate variables for the terrestrial surface of the Earth (excluding Antarctica) for the 30-year period from 1961 to 1990. For this analysis, only the precipitation data were used. The precipitation component of the data set was constructed from 19,295 global stations, interpolated to 0.5° × 0.5° resolution using thin-plate splines. The entire global database was subdivided into a series of geographic tiles, then each tile was interpolated using a separate spline function. Interpolation accuracy was assessed via generalized cross validation (GCV), where each interpolated value is removed from the data set and then the summed square of the differences between the omitted point and all other points is calculated (see New et al. 1999 for details of the data set). The values were also validated by comparison to other climatologies.

LTER Precipitation Data

Three criteria were used to select sites for this analysis. These were (1) located in the Northern Hemisphere, (2) not primarily a marine site, and (3) an available climate data series that includes the time period 1961–1990. Based on these conditions, 16 sites were chosen for analysis (Figure 9.1, Table 9.1). Precipitation data were obtained from the LTER CLIMDB database (http://www.fsl.orst.edu/climhy/). Because the LTER program was not initiated until 1981, some sites do not have climate data sets prior to this (some LTERs were established at existing research sites with operating climate stations). For these sites, observations from nearby recording stations were used. Regardless of the actual location of the observation sites, all measurements were collected using methods consistent with US Weather Service observational standards. For precipitation, this entails use of the Weather Bureau standard 8-inch cylindrical rain gauge.

Atmospheric Circulation Mode Data

Four atmospheric circulation modes were chosen for this study. These are described in the following, along with their data source. Together, these four modes provide a comprehensive set of potential climate drivers for the North American LTER sites.

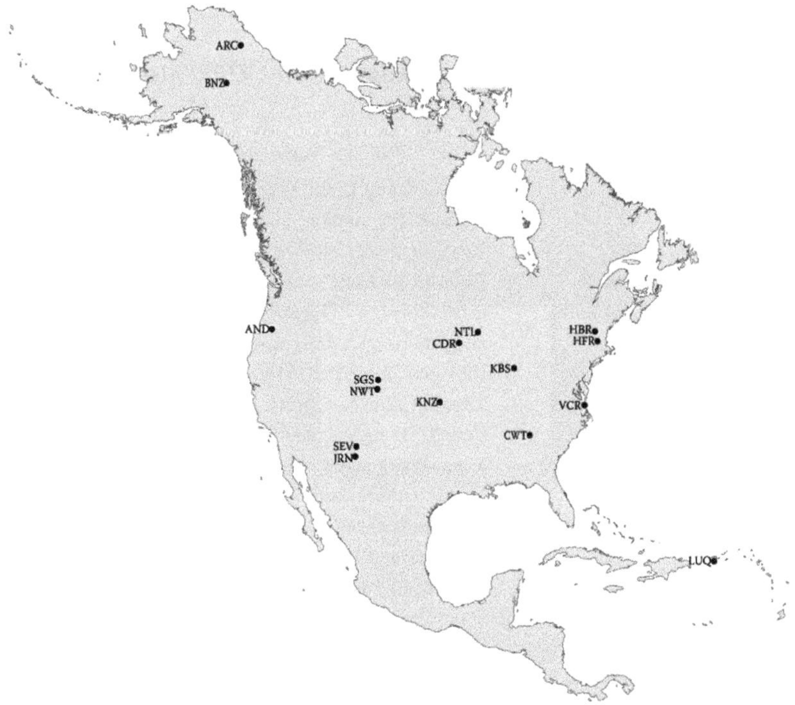

FIGURE 9.1 Location of LTER sites used in this chapter. See Table 9.1 for full site names and guide to abbreviations.

1. *Southern Oscillation Index (SOI).* The SOI is calculated from the difference in barometric pressure between Tahiti, in the central Pacific Ocean, and Darwin, in the Australian Northern Territory. These measurements are used to construct a monthly index that is widely used to diagnose the presence of the ENSO phenomenon, one of the best-known and widely studied of all atmospheric teleconnections. ENSO is a pattern of pressure change across the Pacific Ocean, producing altered sea surface temperatures and changes in the strength and persistence of the southern Pacific trade winds. ENSO is correlated with precipitation and temperature patterns in central North America (Greenland 1999). ENSO events are quasi-periodic, varying at time scales ranging from 4 to 7 years (Kaplan et al. 1998). SOI values used here were calculated from monthly values (Trenberth 1984) and were obtained from the National Weather Service Climate Prediction Center (http://www.cpc.ncep.noaa.gov/data/indices/).
2. *North Atlantic Oscillation (NAO).* The NAO is an oscillation in atmospheric mass and pressure across the North Atlantic Ocean. NAO effects on climate are manifest in a number of ways. Low-frequency oscillation between a negative and positive phase are associated with long-term temperature and precipitation trends in temperature and precipitation in North America. The NAO also manifests interseasonal and interannual variability (Hurrell

TABLE 9.1
Full Site Names and Abbreviations for the LTER Sites Used in This Chapter

Site Abbreviation	Full Site Name and Location
AND	Andrew Forest LTER, Oregon
ARC	Arctic LTER, Alaska
BNZ	Bonanza Creek LTER, Alaska
CDR	Cedar Creek LTER, Minnesota
CWT	Coweeta Forest LTER, North Carolina
HBR	Hubbard Brook LTER, New Hampshire
HFR	Harvard Forest LTER, Massachusetts
JRN	Jornada Basin, New Mexico
KBS	Kellogg Biological Station LTER, Michigan
KNZ	Konza Prairie LTER, Kansas
LUQ	Luquillo LTER, Puerto Rico
NTL	North Temperate Lakes LTER, Wisconsin
NWT	Niwot Ridge LTER, Colorado
SEV	Sevilleta LTER, New Mexico
SGS	Short Grass Steppe LTER, Colorado
VCR	Virginia Coast Reserve LTER, Virginia

1995). It is indexed by the difference in sea-level pressure between stations at Gibraltar and Iceland. Physically, it represents east-west oscillations in the position of the Icelandic Low and Bermuda High semipermanent pressure centers. Fluctuations in the NAO area are associated with changes in the position and strength of the upper-air westerly flow in the Northern Hemisphere, which in turn influence patterns of temperature and pressure over the major northern land masses. Although the effects of the NAO are most prominent in Europe, NAO can affect climate in the North American interior, producing warmer winters when the NAO index is high (van Loon and Rogers 1978). Amplitude and phase of the NAO are highly variable, with variations at both interannual and interdecadal time scales (Hurrell 1995). NAO index values were obtained from the NWS Climate Prediction Center (http://www.cpc.noaa.gov/products/precip/CWlink/pna/nao.shtml).

3. *Pacific North American Pattern (PNA).* The PNA is a teleconnection pattern characterized by low-frequency (≈30 year) variability in extratropical Northern Hemisphere pressure. PNA indices are calculated from a weighted combination of normalized differences of 500 hPa geopotential height values at 4 latitudes (Wallace and Gutzler 1981). The PNA is closely associated with upper air-flow over North America; positive values of the PNA index indicate zonal flow in the upper air westerlies, whereas negative (or reverse) values indicate meridional flow. These alterations in the upper-level flow translate into variations in temperature and precipitation at the surface. During the positive or warm PNA phase, precipitation shifts

south and the central\northern areas tend to be dry. The opposite is true during the negative or warm phase. PNA values were obtained from the Joint Institute for the study of the Atmosphere and Ocean (JISAO) of the University of Washington (http://jisao.washington.edu/data/pna/).

4. *Pacific Decadal Oscillation (PDO).* The Pacific Decadal Oscillation is a teleconnection pattern that resembles El Niño in its effect, but operates on a much longer time scale (15–25 years). It is indexed by the first principal component of North Pacific sea surface temperature variability (Mantua et al. 1997). Of the circulation modes used here, its causes and effects are probably the least well understood (Latif and Barnett 1994). Warm-phase PDO is associated with above-average precipitation in the southwestern United States and northern Mexico, and below average precipitation in northwestern North America. Cool-phase PDO effects are essentially the opposite. PDO index values were obtained from the JISAO of the University of Washington (http://jisao.washington.edu/pdo/).

METHODS

Methods used for analyzing the gridded and LTER site data were similar. For the gridded precipitation data, a spatial subset of the global data large enough to encompass all of the 16 LTER sites used in this analysis was extracted from the complete CRU05 data set. These data were further subset temporally to extract only those monthly values between 1961 and 1990. These values were converted to anomalies using the mean and standard deviation values for each grid cell. The gridded values were then correlated to each of the four circulation indices using Spearman's Rank Order Correlation (ρ). Rank order correlation was used due to concerns about the linearity of the relationship between precipitation and correlation indices. Significance of the ρ-values was determined using a permutation approach. Note that the purpose of this analysis was not to predict precipitation using the circulation indices, nor was it to find the optimal timing of the correlation between these two time series. Instead, the goal was to understand the geographical pattern of the correlations and thus the pattern of influence by each circulation mode. Therefore, no attempt was made to lag the cross correlation of the time series in order to improve the overall fit. All correlations were between values of corresponding months. The results of these correlations were a series of gridded maps with each grid cell containing an ρ-value (Figure 9.2). Although the four circulation indices operate on different time scales, it cannot be assumed that they are independent of each other. Evaluating their spatial effects therefore requires a multivariate approach capable of evaluating the most significantly correlated circulation modes for each grid cell, assuming that more than one mode might be significant. Maximum basic probability assignment (MBPA) was used for this classification (Denœux 2008). MPBA is a form of Bayesian probability classification that uses the Dempster-Shafer weight-of-evidence approach to assign to the most important classification variables to each cell. The resulting map shows geographical patterns of association between grid cells and circulation modes (Figure 9.3). MPBA not only classifies data according to a set of input variables, it also discriminates between significant and insignificant variables

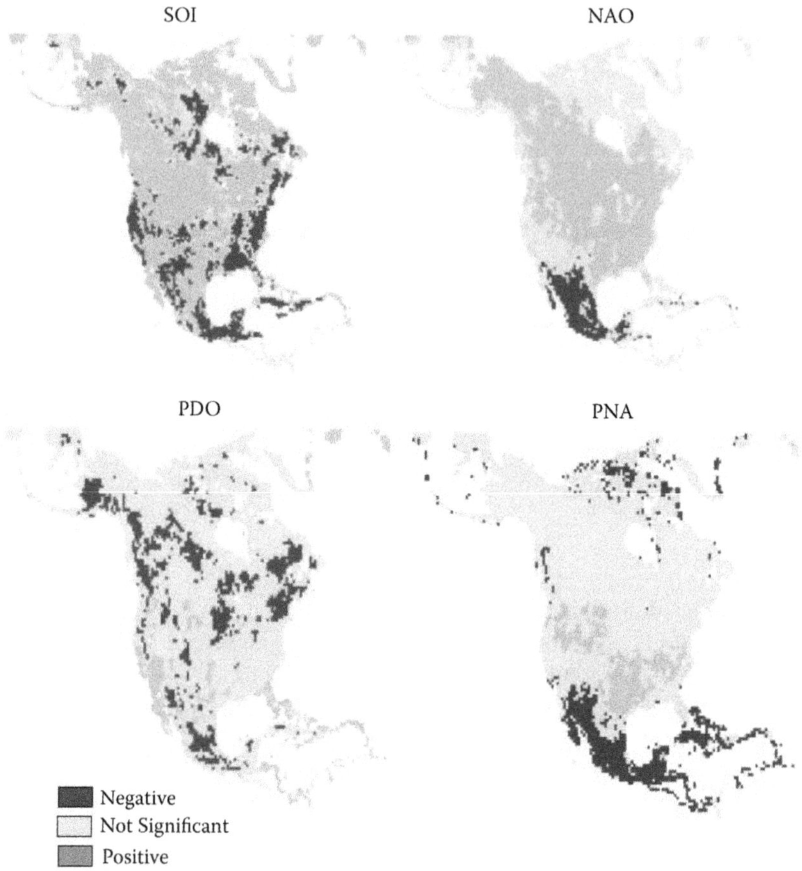

FIGURE 9.2 Geography of positive, negative, and insignificant correlation between the four climate indices and precipitation, using the gridded precipitation data. See Figure 10.1 for locations of LTER sites.

for determining this classification and, if more than one input is significant, ranks them in order of importance.

For the LTER station data, a similar procedure was used, but with station values instead of gridded values. Monthly values for precipitation (converted to anomalies using the mean and standard deviation for each station) at each station were correlated to the same index values that were used for the gridded data (Table 9.2). These correlation values were then subjected to MBPA. A summary of the results are given in Table 9.3.

RESULTS

Correlation with Gridded Precipitation Data

Spatial patterns of correlation between the four atmospheric circulation modes and gridded precipitation data are summarized in Figure 9.2. Correlation values in

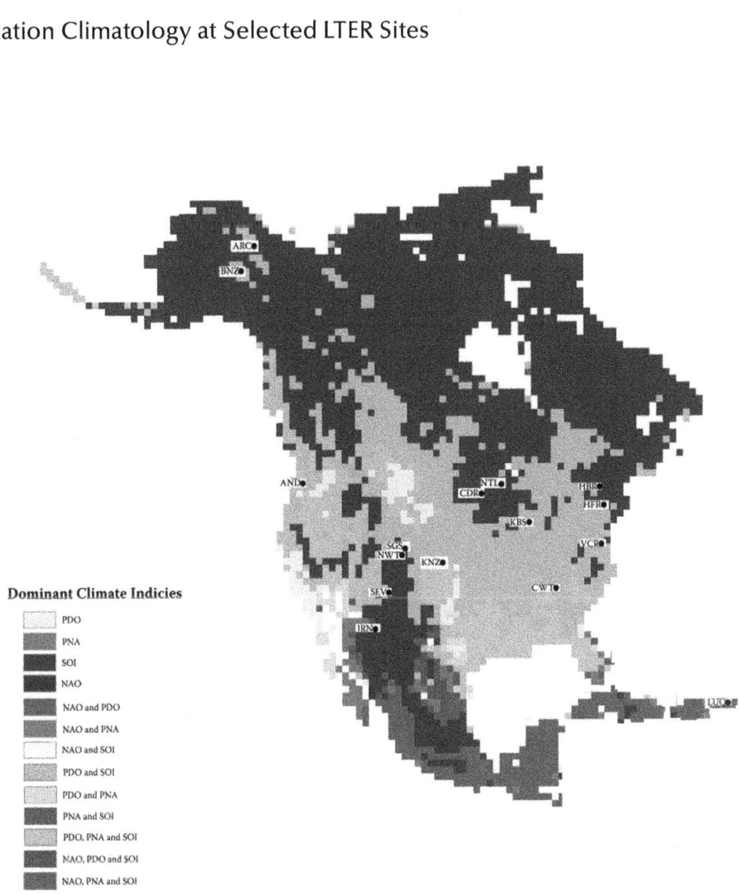

FIGURE 9.3 **(See color insert.)** Dominant climate indices for the gridded precipitation data, determined using Maximum Basic Probability Assignment.

Figure 9.2 have been categorized into areas of positive, negative, and insignificant correlation. In most cases, significant correlation varied from absolute values of about 0.20 to 0.40. Insignificant values fell between –0.20 and 0.20. Although not high in the absolute sense, these values are similar to other values reported in the literature (van Oldenborgh and Burgers 2005). It should also be noted that some of the highest ρ-values are negative—this does not necessarily indicate a negative relationship between the index and precipitation. All of the circulation indices used here have both negative and positive values depending on the phase of the climate mode. Thus, a negative correlation value might mean that precipitation amounts are greater when a climate mode is in its negative phase. The meaning of the phase varies from depending on the mode in question, thus it is useful to consider the absolute value of correlation when evaluating the overall importance of a circulation index.

Correlation between gridded precipitation and atmospheric circulation modes show some spatial patterns of dominance; however, there were also large areas of insignificant correlation associated with each circulation index (Figure 9.2). Significant correlations between SOI and precipitation were found over most of

TABLE 9.2
Spearmann Rank Correlation (ρ) Values between Monthly Precipitation Values and Atmospheric Circulation Modes

Site	NAO*	PDO*	PNA*	SOI*
AND	−0.09	**−0.15**	−0.02	**−0.1**
ARC	−0.01	−0.03	0.04	**−0.16**
BNZ	0.05	**−0.11**	**−0.18**	0.06
CDR	0.01	0.03	**−0.14**	0.03
CWT	−0.01	**−0.13**	−0.05	0.06
HBR	−0.01	−0.02	−0.35	0.16
HFR	0.09	0.09	0.04	0.01
JRN	0.03	0.06	0.04	0.02
KBS	0.03	−0.08	−0.07	**0.1**
KNZ	0.08	0.03	0.06	0.01
LUQ	**0.17**	0.09	0.04	0.05
NTL	0.05	−0.01	0.06	−0.01
NWT	0.02	0.1	0.06	−0.01
SEV	**0.14**	0.15	0.01	−0.03
SGS	0.09	0.07	**−0.11**	−0.04
VCR	0.04	0.03	0.05	−0.03

Notes: Significant values ($P < 0.05$) are shown in bold. Index abbreviations are as follows: NAO, North Atlantic Oscillation; PDO, Pacific Decadal Oscillation; PNA, Pacific North American Index; SOI, Southern Oscillation Index. See Table 10.1 for site abbreviations.

North America; however, the spatial patterning of these values is complex. Positive ρ-values dominated in Alaska, the Canadian interior, the US Great Plains, and the Intermountain Basin. Negative ρ-values were found along the northwest Pacific Coast, the southern Rocky Mountains and interior arid areas (southwestern United States, northern Mexico, southern Mexico), and the Atlantic and Gulf coasts. Smaller areas of negative correlations were found in the Canadian Arctic.

Patterns of correlation for the NAO appear to be somewhat more spatially coherent than those associated with the SOI. Significant positive values occur throughout most of North America, with the exception of the Pacific Coast and interior arid-desert region. Precipitation over most of Mexico, with the exception of the arid and desert north, is negatively associated with the NAO. Precipitation in the extreme north and interior arid-desert regions are not significantly correlated with the NAO, nor is precipitation throughout most of the Caribbean islands.

The remaining two circulation indices (PDO, PNA) are, for the most part, not significantly correlated with precipitation throughout most of the North American continent. The PDO shows scattered areas of positive and negative association, most notably along the Pacific coast of California (a region of positive association) and extending upward into the Pacific Northwest (where the association is significantly negative). Florida and the Caribbean islands are also positively correlated with PDO.

TABLE 9.3
A Summary and Comparison of the Principal Climate Modes Associated with the Various LTER Sites, as Determined from Gridded and Station Precipitation Data

Site	Principal Climate Mode(s) Gridded Data[a]	Principal Climate Mode(s) Station Data[a]
AND	PDO PNA SOI	PNA SOI
ARC	SOI	SOI
BNZ	PDO SOI	SOI
CDR	SOI	SOI
CWT	PDO SOI	SOI PNA
HFR	SOI	PDO PNA
HBR	SOI	SOI
JRN	SOI	PNA SOI
KBS	PDO PNA SOI	PNA SOI
KNZ	PDO PNA SOI	PNA SOI
LUQ	NAO PNA SOI	NAO SOI
NWT	SOI	SOI PNA
NTL	SOI	SOI
SEV	PDO SOI	PDO
SGS	SOI	SOI PDO PNA
VCR	PDO SOI	PNA SOI

Notes: Circulation modes are listed in order of importance. See Tables 10.1 and 10.2 for header abbreviations.

[a] Classified using MBPA.

Correlation values for the PNA are negative throughout most of Mexico and are insignificant elsewhere, with the exception of a region of positive ρ-values in the southern Great Plains.

Correlation with LTER Site Data

Correlation analysis between the precipitation values at the selected LTER sites and the circulation indices is summarized in Table 9.2. In general, the ρ-values for the site precipitation values were less than those for the gridded data, even when compared to the ρ-value for the grid cell containing the LTER site. Perhaps the most notable result of the site data analysis is the fact that, for 7 of the 16 sites, there were no significant correlations between precipitation and any of the four circulation indices. For the remaining sites (i.e., those with at least one significant correlation), each of the four circulation modes correlates with at least one site. Only the Bonanza Creek site (located near Fairbanks, AK) was significantly correlated with more than one index.

Analysis of MBPA

Results of the MBPA analysis were, in some ways, more revealing than those from the simple correlation analysis. As previously noted, MBPA is a technique both for classifying a data set and for determining the most important variables for class membership. When interpreting these results, however, it is important to remember that Spearman rank values are used as classification variables regardless of their significance. Thus, circulations that were not significant according to univariate correlation may be significant contributors to the multivariate regionalization process.

Classification and regionalization of the climate mode correlations (Figure 9.3) show a complex pattern across North America, which can roughly be divided into four regions, each determined by a different pattern of dominant indices. The most spatially extensive of these categories is the area dominated by correlation between precipitation and the SOI. This class occurs in two large regions: one generally in the northern latitudes and extending into the Great Lakes region, and the other extending from central Mexico into the United States along the cordillera. These two regions correspond to areas known to be affected by ENSO events (Hoerling et al. 1997) and also match areas of positive SOI correlation (Figure 9.2).

The next largest class consists of areas dominated by the combined effects of the PDO and SOI, with the PDO being the more significant of the two. This class is found mainly in one large, roughly contiguous region dominating the Atlantic coast, Midwest, and the northern Great Plains in the United States and extending north onto the Canadian Shield. A second region lies along the Pacific Coast of Oregon, Washington, and British Columbia. A few smaller, outlying regions occur throughout the interior of Alaska and the northern Canadian provinces.

The third region is determined by correlation with three circulation modes: the PDO, PNA, and SOI. This region is found along the Gulf Coast, extending northward into Texas, the southern Great Plains, and the Mississippi River valley. A second, smaller contiguous area dominates the Columbian plateau. Like the third class, the fourth circulation class is also determined by correlation with three climate modes, but the effects of the NAO replace the PDO as the most important determinant. This class is found almost exclusively in the subtropics, dominating the Caribbean islands and the Gulf Coast of Mexico, but with an extension into the Pacific coast and Baja areas, as well. Numerous other grid cells were classified based on other combinations of circulation indices, but the four classes summarized here were the most important and spatially extensive results of the regionalization.

The results of MBPA classification of the station data are summarized in Table 9.3, along with the classification for the precipitation grid containing each of the LTER sites used in this study (extracted from Figure 9.3). Presentation of both results side by side facilitates comparison of the two results and determination of whether the relationship between circulation indices and precipitation for the station data resembles that of the region in which the station is embedded.

Examination of the results presented in Table 9.3 shows that, in general, the dominant climate indices for precipitation at the LTER stations are similar to those of the surrounding areas, but often with a smaller set of significant correlates. Several of the sites among the ARC, CDR, HBR, and NTL sites are dominated exclusively by the

SOI in both analyses. Examination of the location of these sites (Figure 9.3) shows that each of these sites is located well within a region dominated by correlation with the SOI. That is, they do not lie near a regional boundary. None of the other categories exactly match for both classifications. At the AND site, the station data fell into a class dominated by the PNA and SOI patterns, whereas the region includes both of those modes, plus the PDO. This precipitation grid, in which the AND station is adjacent to a small region, is dominated exclusively by the PDO. The disagreement between the two classifications may be an artifact of comparing a gridded data set to a point observation. The AND station may therefore be representative of a somewhat larger region that receives some influence of all three circulation modes. Although located in a different climatic setting (midlatitude mesic versus coastal), the KNZ and KBS sites show a similar pattern of association to AND. Again, both sites are located in the boundaries between classified regions, and thus may be responding to fine-scale variations in climate influences that are lost when the precipitation values within a 0.5° × 0.5° region are gridded.

Among the four circulation patterns, the PDO occurred much more commonly as a significant correlate in the gridded data, compared to the station data. For a number of sites, including BNZ, KNZ, KBS, and VCR, the PDO was the most significant classification variable in the gridded data set, but did not contribute to classifying the actual station data. Again, examination of the location of these sites shows that each lies nearby a different class, suggesting that scale effects might be affecting the classification. In particular, the BNZ site lies within a small region of dominance by PDO/SOI, which is embedded in a larger homogeneous region dominated by SOI. Only SOI emerged as a determinant of class membership in the station classification. Only one site (HFR) had no common significant classifiers between the two analyses.

SUMMARY

The major goal of this chapter was to determine how geographically representative the precipitation climatology of each LTER site is, using four well-known circulation indices as the basis for regionalizing climate. The results of the analysis are not simple, but in general suggest that while the precipitation regime for the grid cell in which most LTER sites are located responds to climate circulation modes in a way that is consistent with the region in which they are located, the values for the station data associated with each site present a much murkier picture. Of the four indices used, the SOI has the largest overall influence. This reflects the fact that, with the exception of the NAO, the other circulation modes are thermodynamically linked to it.

In evaluating these results, it is important to keep in mind that, as in any correlative study, it cannot be said whether a particular circulation pattern determines the response at a place, or whether both the pattern and the response are related to some other process. In climate research, teleconnections between circulation modes and climatic responses are widely studied, and modeling has provided theoretical grounding for the empirical relationships between circulation modes and some climatic state variables (Lau 1985). It is also important to consider the nature of the comparison being made. Recall that the cells for the gridded data are 0.5° × 0.5°, which is much larger than the area associated with any of the LTER sites. The value for each grid cell is determined by interpolation from among several measurement stations, each of which is essentially a point measurement. In many cases, the LTER or its surrogate station was used in the gridding process, along with other first- or second-order observation stations. Measurements of climate variables, including precipitation, are often spatially variable; even nearby stations can be quite different in the magnitude of their measurements. This is often due to gauge location, localized topography, urban effects, presence of water bodies, or other factors not directly linked to large-scale atmospheric circulation (McHugh and Goodin 2003). For precipitation the nature of the causal mechanisms themselves must also be considered, especially in areas where convectional processes dominate and precipitation events tend to be intense, short-lived, and limited in their geographical extent. Selecting one gauge from within a region (which is essentially what is done when a single station is used) may therefore not be entirely representative of the entire grid cell.

One noteworthy conclusion that can be drawn from the results shown here is that the pattern of influence by teleconnection patterns with circulation modes is much more geographically heterogeneous than expected. The broad patterns discussed in Results are largely consistent with previous understanding of the spatial effects of various teleconnection patterns. For example, the pattern of response of precipitation to the SOI is consistent with previous research showing the influence of ENSO at LTER sites (Greenland 1999). The influence of the Pacific North American pattern interior continental sites is also consistent with previously observed effects of this pattern, which indexes the position and direction of the upper air westerlies and is therefore closely tied to the formation and movement of precipitation-producing low pressure systems. However, the mapped relationships between circulation modes and precipitation patterns often show small areas that deviate from the broad patterns. An example of this can be seen in the Alaskan interior, where a small region of influence by multiple circulation modes lies within an area that is dominated by the SOI. It is unclear why this particular region occurs, because it seems not to be associated with large-scale topography or other factors that might influence precipitation patterns. Because this region contains the Bonanza Creek LTER, it also illustrates how temporal patterns in ecological response at the BNZ site might not be entirely representative of the surrounding region. Of course, it might also represent some anomaly in the data gridding process (see previous discussion)—a possibility that should be considered when evaluating any of the results presented here.

The results presented in this chapter also highlight some of the difficulties with point- or station-based climatic research. The LTER network was established to understand the long-term ecology at sites that were representative of a variety of ecosystems. The sites were never intended to be an exhaustive sampling of regions or biomes, however, and the nonsystematic selection of sites is apparent in their spatial distribution (Figure 9.1). The pattern of sampling favors coastal and western sites, leaving a noticeable "gap" in the central plains east of the Mississippi, where few sites are located. This undersampling makes understanding coherent climate patterns using LTER data difficult, at best. However, the efficacy of this current study lies not in producing information about the spatial distribution of climate, but in providing a better framework for understanding the effect of climatic input (i.e., precipitation) into ecosystem functioning and in regionalizing those findings. The patterns of circulation influence presented here (see Figure 9.3) support the idea that temporal patterns of precipitation at many LTER sites (and by extension, across much of the North American continent) are influenced by several climatic circulation modes, often acting in complex ways that can either positively or negatively reinforce each other. Goodin et al. (2003) analyzed the influences and interactions between temperature, precipitation, and various climate indices in detail for one LTER site (Konza Prairie), finding that more than one index affected climatic response at this site, and that the effects of different indices often manifested themselves seasonally. Similar analysis at other LTER sites where multiple climate indices dominated the precipitation pattern would likely reveal similar, complex relationships between circulation indices and response.

From an ecological perspective, the findings summarized in this chapter are useful in two ways. First, they can inform research on the ecosystem responses to climate change. As noted in the introduction, global climatic change is one of the major issues confronting the earth system science community, and the ecological responses to these changes are largely unknown. By observing how ecosystems responded past climatic fluctuations, it is possible to gain insight into how systems may respond under future conditions. Linking ecological response to climate is complex at best, but the use of indices such as the ones in this study provides valuable simplification of the process (Stenseth et al. 2003), especially because in some cases, ecological variables actually correlate to climate indices better than do weather variables themselves (Stenseth et al. 2002). The regionalization results presented here suggest that in some cases, LTER sites provide adequate analogs for their regional setting, but caution should be exercised in generalizing the results.

The second use of these results is as a way to disentangle the ecological signals of climate change versus climate fluctuation. Recognition of the role of various climate indices in site-specific ecological process is important because it shows that "normal" climate conditions can vary over a range of time scales. Without long-term observation, it is difficult to determine what observed effects are due to climatic fluctuation, and which might be the result of (or evidence for) climate change. By understanding what processes affect a particular site or region, ecological researchers will be in a better position to identify change.

REFERENCES

Badeck FW, Bondeau A, Böttcher K, Doktor D, Lucht W, et al. 2004. Responses of spring phenology to climate change. *New Phytologist* 162:295–309.

Clark PU, Alley RB, Pollard D. 1999. Northern Hemisphere ice-sheet influences on global climate change. *Science* 286:1104–11.

Cox PM, Betts RA, Jones CD, Spall SA, Totterdell IJ. 2000. Acceleration of global warming due to carbon-cycle feedbacks in a coupled climate model. *Nature* 408:184–87.

Denœux T. 2008. A k-nearest neighbor classification rule based on Dempster-Shafer theory. In *Classic Works of the Dempster-Shafer Theory of Belief Functions*, pp. 737–60. Berlini Springer-Verlag.

Goodin DG, Fay PA, McHugh MJ. 2003. Climate variability in tallgrass prairie at multiple time scales: Konza Prairie Biological Station. In *Climate Variability and Ecosystem Response*, Eds. D Greenland, D Goodin, R Smith. Chapel Hill, NC: Oxford USA Press.

Greenland D. 1999. ENSO-related phenomena at long-term ecological research sites. *Physical Geography* 20:491–507.

Greenland D, Goodin DG, Smith RC, Eds. 2003. *Climate Variability and Ecosystem Response at Long-Term Ecological Research Sites*. New York: Oxford University Press.

Greenland DE, Kittel TGF. 2002. Temporal variability of climate at the US Long-Term, Ecological Research (LTER) sites. *Climate Research* 19:213–31.

Hoerling MP, Kumar A, Zhong M. 1997. El Niño, La Niña, and the nonlinearity of their teleconnections. *Journal of Climate* 10:1769–86.

Holdridge LR. 1967. *Life Zone Ecology*. San Jose: Tropical Science Center.

Hurrell JW. 1995. Decadal trends in the North Atlantic Oscillation: regional temperatures and precipitation. *Science* 269:676–79.

IPCC (Intergovernmental Panel on Climate Change). 2007. Climate Change 2007: The physical science basis. WMO/UNEP. Geneva, Switzerland.

Kaplan AM, Cane M, Kushnir Y, Clement A, Blumenthal M, Rajagopalan B. 1998. Analysis of global sea surface temperature. *Journal of Geophysical Research* 103:18567–89.

Latif M, Barnett TP. 1994. Causes of decadal climate variability of the North Pacific and North America. *Science* 266:634–37.

Lau N-C. 1985. Modeling the seasonal dependence of the atmospheric response to observed El Niños in 1972–1976. *Monthly Weather Review* 113:1970–96.

Mann ME, Zhang Z, Hughes MK, Bradley RS, Miller SK, et al. 2008. Proxy-based reconstructions of hemispheric and global surface temperature variations over the past two millennia. *Proceedings of the National Academy of Sciences* 105:13252–57.

Mantua NJ, Hare SR, Zhang Y, Wallace JM, Francis RC. 1997. A Pacific interdecadal climate oscillation with impacts on salmon production. *Bulleting of the American Meteorological Society* 78:1069–79.

McHugh MJ, Goodin DG. 2003. Interdecadal-scale variability: an assessment of LTER climate data. In *Climate Variability and Ecosystem Response*, Eds. D Greenland, D Goodin, R Smith. Chapel Hill, NC: Oxford USA Press.

New M, Hulme M, Jones P. 1999. Representing twentieth-century space-time climate variability. Part I: development of a 1961–90 mean monthly terrestrial climatology. *Journal of Climate* 12:829–56.

Oerlemans J. 2001. *Glaciers and Climate Change*. Utrecht: Balkema.

Rignot E, Thomas RH. 2002. Mass balance of polar ice sheets. *Science* 297:1502–1506.

Root TL, Price JT, Hall KR, Schneider SH, Rosenzweig C, Pounds JA. 2003. Fingerprints of global warming on wild animals and plants. *Nature* 421:57–60.

Stenseth N, Myerstud C, Ottersen G, Hurrell JW. 2002. Ecological effects of climatic fluctuations. *Science* 297:1292–96.

Stenseth NC, Ottersen G, Hurrell JW, Mysterud A, Lima M, et al. 2003. Study the effects on ecology through the use of climate indices: the North Atlantic Oscillation, El Niño Southern Oscillation and beyond. *Proceedings of the Royal Society of London, Series B* 270:2087–96.

Trenberth KE. 1984. Signal versus noise in the Southern Oscillation. *Monthly Weather Review* 112:326–32.

van Loon H, Rogers JC. 1978. The seesaw in winter temperatures between Greenland and northern Europe. Part I: General Description. *Monthly Weather Review* 106:296–310.

van Oldenborgh GJ, Burgers G. 2005. Searching for decadal variations in ENSO precipitation teleconnections. *Geophysical Research Letters* 32:L15701.

Wallace JM, Gutzler DS. 1981. Teleconnections in the geopotential height field during the northern hemisphere winter. *Monthly Weather Review* 114:784–812.

Watson RT, Zinyowera MC, Moss RH, Eds. 1997. *The Regional Impacts of Climate Change: An Assessment of Vulnerability*. United Kingdom: Cambridge University Press.

Williams JW, Jackson ST. 2007. Novel climates, no-analog communities, and ecological surprises. *Frontiers in Ecology and the Environment* 5:475–82.

10 Dealing with Uncertainty: *Managing and Monitoring Canada's Northern National Parks in a Rapidly Changing World**

Donald McLennan

CONTENTS

INTRODUCTION

National parks located in arctic and subarctic biomes ("northern parks") of Canada protect some of the world's most intact ecosystems. Although presently in an excellent state of ecological integrity Canada National Parks Act (2000), the combination

* Disclaimer: The opinions expressed in this chapter are those of the author and not Parks Canada Agency.

of rapidly changing climate and associated increasing pressure from industrial and community developments is threatening this integrity. This emerging reality represents a serious challenge for protected areas managers in the north, where climate has been warming about twice as fast as in southern latitudes (ACIA 2005; ICARP II 2005; Anisimov et al. 2007), and where projections for future climate change are predicted to be more immediate as well (Callaghan et al. 2005; Chapin et al. 2006).

The potential impacts of a changing climate on Canada's national parks were first described 20 years ago by Lopoukhine (1990, 1991) and Rowe (1989). In 1997, nine of 39 parks identified climate change as a major stressor to park ecosystems (PCA 1997a). Observations, analyses, and recommendations in Scott and Suffling (2000), Scott and Lemieux (2003), Jones et al. (2003), Lemieux and Scott (2005), and Scott and Lemieux (2007), and an increasing awareness by park managers, have instigated the development of a Parks Canada climate-change strategy that is presently being developed.

This chapter builds on previous work at Parks Canada—especially the recommendations for adopting an adaptive-management approach and linking monitoring more directly to park decision making. Although recommended for many years (Lemieux and Scott 2005; Welch 2005, 2008; Heller and Zavaleta 2009; Lawler et al. 2009b), the application of adaptive management strategies in protected areas has seldom been implemented.

This chapter summarizes the potential for ongoing and future climate change to impact the ecological integrity (EI) of northern national parks in Canada, and presents an approach for navigating this inevitable change through adaptation strategies aimed at minimizing impacts on park biodiversity. The development and implementation of proactive adaptive management is put forward here as a required park information system that includes two key elements: inventory and monitoring to measure and understand ecological change as it occurs, and research and modeling to attempt to predict the nature and rate of future change. The system should be designed to provide park managers with key information aimed at reducing uncertainty to inform and support park management through this challenging period of changing climates and ecosystems.

CANADA'S NORTHERN NATIONAL PARKS

Canada's arctic and subarctic national parks (Figure 10.1) protect some of the world's most outstanding examples of wild nature, where predator-prey systems are still largely intact, ecosystem processes are acting without constraint, and the role of indigenous people continues to be an important component of park ecosystems. These parks are large and remote, with an average area of about 15,000 km^2, and covering a total area of 166,000 km^2.

National parks in northern Canada represent a wide ecological sample of arctic and subarctic environments that provide ecological representation in most northern ecozones (Figure 10.1). Torngat Mountains, Auyuittuq, Sirmilik, and Quttinirpaaq National Parks (NPs) are situated along a latitudinal and climatic gradient from tree line to the high arctic in the eastern arctic, and Vuntut, Ivvavik, Aulavik, and Quttinirpaaq NPs make up a similar gradient in the west. Wapusk, Ukkusiksalik,

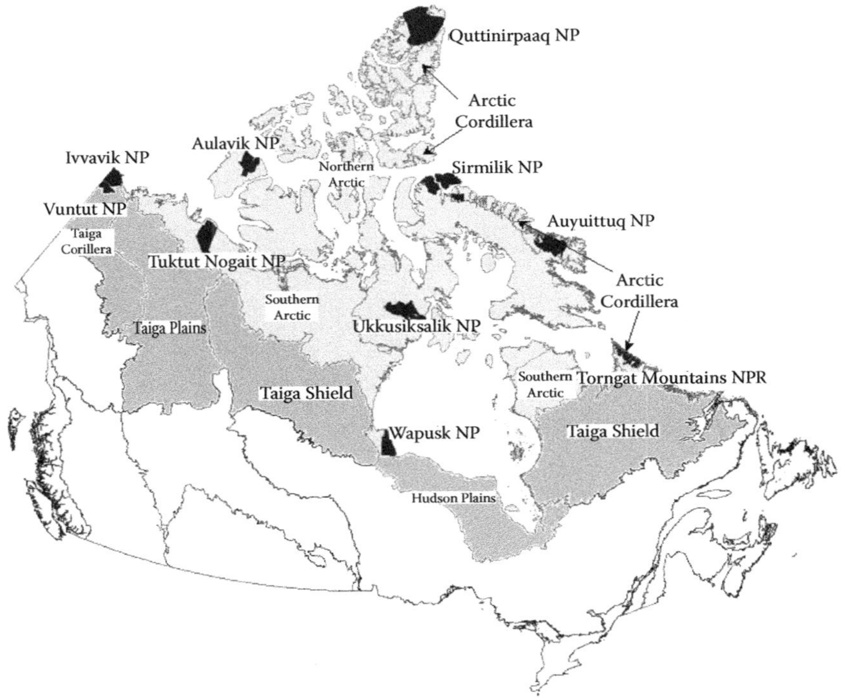

FIGURE 10.1 Locations and ecological representation by ecozone for arctic (light grey) and subarctic (dark grey) national parks in Canada. (From Marshall, I.B. and P.H. Schut. 1999. A National Ecological Framework for Canada. Overview. Ecosystems Science Directorate, Environment Canada, and Agriculture and Agri-Food Canada. Ottawa (http://sis.agr.gc.ca/cansis/nsdb/ecostrat/intro.html).

and Tuktut Nogait NPs complete this latitudinal ecological representation in the central areas of the Canadian North.

The comprehensive ecological representation of northern national parks has been the goal of the Parks Canada System Plan (PCA 1997b), and 5 of the 10 parks shown in Figure 10.1 have been established since 2001. Recent park establishment has been the result of consultations with different indigenous groups across the Canadian North, and follow successful land claims settlements (see, for example, www.gov.nu.ca/hr/site/doc/NLCA.pdf). As a result, traditional cultural activities are maintained, and parks are cooperatively manages with different Inuit, Gwitch'in, and Metis groups through comanagement boards that make all decisions regarding park management.

The broad ecological representation of Canada's northern national parks means that effective and systematic monitoring and reporting of ecological change in national parks will not only inform park management, but will also make a comprehensive ecological statement of the condition of Canada's North. For this reason, and because of the size, cost, and complexity of the monitoring task, Parks Canada strives to work with other northern governments, academics, aboriginal peoples and community partners to share expertise and pool resources to help design and implement monitoring and research within northern parks and across the North.

PREDICTED ECOLOGICAL EFFECTS OF CLIMATE CHANGE

There is a general scientific consensus that the next 50 years will see historically unprecedented rates of change in temperature and precipitation parameters across arctic and subarctic landscapes in northern North America (Sturm and Tape 2001; Lloyd and Fastie 2002, 2003; Jia et al. 2003; Stow et al. 2003; Hinzman et al. 2005; Olthof et al. 2008; Pouliot et al. 2009), and that these changes will have profound implications for the condition of park ecosystems and the successful management of national parks (Jones et al. 2003; Lemieux and Scott 2005; Scott and Lemieux 2003, 2005; Scott and Suffling 2000). By the year 2050, it is predicted (Scott and Suffling 2000) that all northern national parks will experience significant increases in mean summer (2.6–5.0°C) and mean winter (4.3–8.2°C) temperatures, as well as significant but highly variable responses in winter and summer precipitation (Table 10.1). These climate factors are key drivers of the composition and structure of the ecological communities that characterize the biomes that national parks were established to protect, and the magnitude of predicted changes are sufficient to radically alter these characteristic communities. To emphasize the potential magnitude of this biotic change, Lawler et al. (2009a) assessed the central Canadian Arctic as a global hotspot of species change, predicting an overturn of 70–80% in bird and mammal species composition over the next 100 years.

Table 10.2 uses conclusions from the Arctic Climate Impact Assessment (ACIA 2005) and the most recent IPCC publications (Anisimov et al. 2007), as well as a number of reports commissioned specifically on assessment of climate-change impacts for Canadian national parks (Scott and Suffling 2000; Jones et al. 2003; Scott and Lemieux 2003), to provide a summary of potential changes to park ecosystems,

TABLE 10.1
Predicted 2050 Winter and Summer, Temperature and Precipitation Changes for Northern National Parks from 1961 to 1990 Averages Canadian Climate Center General Circulation Model I (CCCma-CGCMI)

	Temperature Change (°C)		Precipitation Change (%)	
Park/Reserve	**Winter**	**Summer**	**Winter**	**Summer**
Aulavik	+5.5	+5.0	−18	+10
Auyuittuq	+5.2	+2.6	−4	−4
Quttinirpaaq	+5.9	+4.0	+23	+20
Ivvavik	+5.9	+4.2	−3	+25
Sirmilik	No data	No data	No data	No data
Torngat Mountains	No data	No data	No data	No data
Tuktut Nogait	+4.3	+4.4	−24	0
Vuntut	+5.9	+4.2	−3	+25
Ukkasiksalik	No data	No data	No data	No data
Wapusk	+8.2	+3.8	+16	+9

Source: Scott, D., and R. Suffling 2000. *Climate Change and Canada's National Parks*. Environment Canada, Toronto, Ontario.

organized by major park ecosystem. Park managers can expect that all northern national parks will experience major shifts in species and community composition, peri-glacial processes, and ecological drivers over the next 35–40 years. Potential effects will vary from park to park, depending on the park's location in relation to change gradients, the physiographic characteristics of the park, and local biological factors. For example, about 95% of the area of Wapusk NP is comprised of wetland ecosystems dotted by thousands of small lakes with little relief, where hydrologic changes will be prominent. By contrast, Torngat Mountains NP is a dramatic, mountainous landscape with few lakes, and where the ecological effects of climate change will be primarily terrestrial. For this reason, potential ecological changes will have to be assessed separately for each park, and this assessment should guide the design of park monitoring and reporting systems.

KEY UNCERTAINTIES

The direction and magnitude of climatic change and potential ecological effects in northern national parks were discussed previously, and these anticipated effects provide a general guide to the kinds of changes park monitoring programs will need to be designed to track and report, to provide the most relevant and timely information required by park managers. What is also clear from the discussion around arctic climate change is that there is a considerable degree of uncertainty in two key components of the scenarios discussed:

TABLE 10.2
Summary of Potential Climate-Related Ecological Changes for EI Indicators in Northern National Parks in Canada

EI Indicator	Key Ecological Issues
Tundra	Regional shifts in plant and animal populations; expanded distribution and increased density of invasives
	Expansion of erect shrubs and grasses; reduced cover of prostrate shrubs and lichens
	Reduction in area of continuous permafrost; increase in active-layer depths and soil temperature
	Moister soils; reduction in periglacial processes (cryoturbation, stone sorting, solifluction)
	Habitat change and direct weather effects (icing events, snow depth and persistence, increased insect pests) on major ungulate populations (caribou, muskoxen, lemmings, Dall's sheep, moose) and their predators (grizzly bear, wolf, red/arctic fox)
	Loss of sea ice connectivity and increased genetic isolation of island-based populations (e.g., Peary caribou, wolves)
	Increased fire frequency, intensity, and area burned
Wetlands	Regional shifts in plant and animal populations; invasives
	Alterations to wetland hydrologic regimes and vegetation composition and structure
	Habitat change effects on wetland requiring species (e.g., grizzly and moose foraging, shorebird nesting, wetland songbirds)
	Changes in wetland periglacial effects (ice wedge polygons, peat palsas and plateaus, thermokarst pools)
Subarctic Forest	Regional shifts in plant and animal populations; invasives
	Increased fire frequency, intensity, and burned area; increased forest insects and pathogens all affecting age class distribution and landscape connectivity
	Increased forest productivity and expansion of forest area
	Change in stand dynamics (seed production, recruitment, mortality, growth) leading to increased stand densities and habitat effects on forest-dependent species (songbirds and raptors, small mammals)
	Change in permafrost and soil temperature and moisture regimes
Freshwater—Streams and Rivers	Regional shifts in plant and animal populations; invasives
	Changes in stream flow (earlier break up and later freeze up, altered peak flows, higher winter flows, glacier melting) and higher air and water temperatures leading to increased stream bank erosion, higher suspended load and nutrient levels, with complex effects on periphyton, invertebrates, and spawning and overwintering of arctic char and other species
	Upslope permafrost melting leading to increased slope instability (mudflows, slumping, solifluction) increasing stream suspended load with possible stream damming events

(*continued*)

TABLE 10.2 (continued)
Summary of Potential Climate-Related Ecological Changes for EI Indicators in Northern National Parks in Canada

EI Indicator	Key Ecological Issues
Freshwater—Lakes	Regional shifts in plant and animal populations; invasives
	Reduced lake-ice duration and warmer air temperatures will affect stratification and increase water temperature and primary productivity (phytoplankton, macrophytes, zooplankton, juvenile fish) above the thermocline; possible reduced area for mature fish such as lake trout and grayling
	Warmer water and a longer ice-free season, coupled with warmer soils and deeper active layers, will increase rates of thermokarst erosion, with related effects on water chemistry (POC, DOC, sedimentation) and lake productivity and biota
Coastal-Marine	Regional shifts in plant and animal populations; invasives
	Changes in sea-ice extent, character, and phenology, with effects on ice-dependent biota (polar bears, ringed seals, walrus)
	Change in the rate of coastal erosion and dynamics of coastal landform processes; submergence of coastal wetlands and estuaries; changes in coastal habitats and plant phenology and foraging opportunities for migratory birds
	Changes in physical oceanographic processes (freshwater inputs, currents, mixing, temperature, deposition) and effects on marine biota and coastal seabird colonies
Glaciers	Localized changes in precipitation (snow accumulation) and air temperature conditions leading to changes decreases in glacier area, and mass balance

Sources: Scott, D., and R. Suffling 2000. *Climate Change and Canada's National Parks*. Environment Canada, Toronto, Ontario; ACIA. 2005. *Impacts of a Warming Arctic: Arctic Climate Impacts Assessment.* Cambridge University Press, Cambridge, 1042 pp; Anisimov, O.A., D.G. Vaughan, T.V. Callaghan, C. Furgal, H. Marchant, T.D. Prowse, H. Vilhalmsson, and J.E. Walsh. 2007. *Polar Regions (Arctic and Antarctic) Climate Changes 2007. Impacts, Adaptation and Vulnerability. Contribution of Working Group II to the Fourth Assessment Report of the International Panel on Climate Change*. Eds. M.L. Parry, O.F. Canziani, J.P. Palutikof, P.J. van der Linden, and C.E. Hanson, 655-685. Cambridge University Publications, Cambridge.

1. Problems with regional- to local-scale predictions and applications of the climate models themselves, and
2. Key uncertainties around the myriad ways in which complex park ecological systems will respond to these climate changes.

With regard to uncertainties around published climate change models for the north, first, it is well acknowledged that there are problems in modeling arctic climate systems that result from our limited understanding of the interaction of climate with key surface processes (e.g., feedback effects from albedo and/or surface roughness from vegetation changes) that could significantly affect the accuracy of scenarios (ACIA 2005; Chapin et al. 2005). Second, there is the problem of separating relatively large natural climatic variability from that induced by greenhouse gas additions, so that projected "differences" are difficult to quantify (ACIA 2005). A third uncertainty is the poor accounting for extreme-weather events that can seriously affect biotas directly (e.g., cold snaps, icing events, or floods) or indirectly, by creating a spate of mass-wasting events or catastrophic coastal slumping. The final concern with climate scenarios is that of scale. Climate modeling averages potential changes over broad areas so that it will be difficult to make accurate predictions that apply directly to the regional and local landscapes in which Canadian parks function, and that we care about as park managers. This lack of spatial resolution means, for example, that the important effects of elevation, relief, and aspect on bioclimate are poorly accounted for in mountainous parks, and similar issues exist for coastal areas of parks due to extra-local scale land-sea interactions. These factors are partially responsible for the relatively low confidence modelers place on precipitation scenarios. Thus, for example, orographic factors that may triple precipitation amounts cannot be reliably predicted due to the poor horizontal resolution of present models (ACIA 2005).

All of these factors emphasize the central problem for park managers, which is uncertainty around the ways in which park biotas will respond to rapid climate change. Climate is a key driver of all ecosystems at global, regional, and local scales. Climate affects biota directly through its effects on physiological tolerances and the rate of cellular metabolism, as well as indirectly by influencing the ecological and physiographic processes that drive ecosystem composition, process, and function (Callaghan et al. 2005). Climate also interacts with many other physical factors such as photoperiod to determine ecosystem phenological processes (e.g., the timing of leaf-out) or with soil physical parameters that control the availability of soil moisture and nutrients, which are key determinants of ecosystem productivity. It is the interaction of all of these physical and biotic components of arctic ecosystems at a range of scales that complicates a simple prediction of potential ecological change for a given park (Callaghan et al. 2005; Anisimov et al. 2007).

As a result of these uncertainties around climate change at regional and local scales, and the complexities of potential ecosystem responses, well-designed park ecological integrity monitoring programs linked to focused research will be the key to reducing uncertainty by providing park managers with timely information on ecological changes as they evolve in northern national parks. As an important caveat, it remains to be seen whether or not PCA can design and afford to implement monitoring that will be sufficient to anticipate and capture the most important ecological changes.

PARKS CANADA MANAGEMENT CONTEXT AND MONITORING APPROACH

Parks Canada Agency (PCA) is responsible for the protection and presentation of significant examples of Canada's natural and cultural heritage, including national parks, national marine conservation areas, and national historic sites (PCA 2008c). More specifically in the context of climate change, PCA's Corporate Plan (PCA 2008a) identifies that a key strategic outcome for managers of national parks is "the maintenance or restoration of overall ecological integrity of all national parks from 2008–2013."

In Section 2 (1) of the Canada National Parks Act (2000) *ecological integrity* is defined as

> … a condition that is determined to be characteristic of its natural region and likely to persist, including abiotic components and the composition and abundance of native species and biological communities, rates of change, and supporting processes.

As summarized in Tables 10.1 and 10.2, it is clear that over the next 35–40 years park biota will change significantly in response to warming climatic regimes, and that all of the elements listed as characteristic of park EI are not "likely to persist." This presents a legislative challenge that park policy staff will need to address in the not-too-distant future. In the shorter term, northern national parks are already changing, and a number of important questions require answers so that managers of northern parks can anticipate the nature, rate, and scale of this ecological change.

To meet its obligation to report to Canadians on the condition of national parks, Parks Canada Agency has invested significant time and human resources in the development of the PCA EI Monitoring and Reporting Program (PCA 2005). Monitoring of ecological integrity and other aspects of the PCA mandate is the "knowledge engine" that drives the management cycle within the Agency (Figure 10.2). Most national-park field units have a designated monitoring ecologist who is supported by monitoring ecologists in park service centers, and in the national office. The program is also supported by database and geomatics specialists, and program data are coordinated within the Information Center for Ecosystems (ICE) data-management program. Park monitoring programs are expected to meet a national standard, as outlined in PCA (2005).

The following are the main components of park EI condition monitoring outlined in PCA (2005):

1. All EI condition monitoring should be implemented to answer the general question, "What is the state of park ecological integrity, and how is it changing?"
2. EI condition is reported through four to eight EI Indicators, which represent major park ecosystems; for northern national parks the EI Indicators are Tundra, (Subarctic) Forest, Wetlands, Freshwater, Glaciers, and Coastal/Marine Ecosystems.
3. EI Indicators are composite indices that are developed based on an assessment of the condition of a carefully selected suite of EI Measures (referred to as "biological indicators" in most monitoring literature).
4. To meet criteria for program ecological comprehensiveness, the suite of EI Measures for an EI Indicator need to
   a. Measure and assess biodiversity, ecological processes, and the stressors/drivers in all EI Indicators in a park; the level of effort for an EI Indicator will reflect its conservation importance to the park.

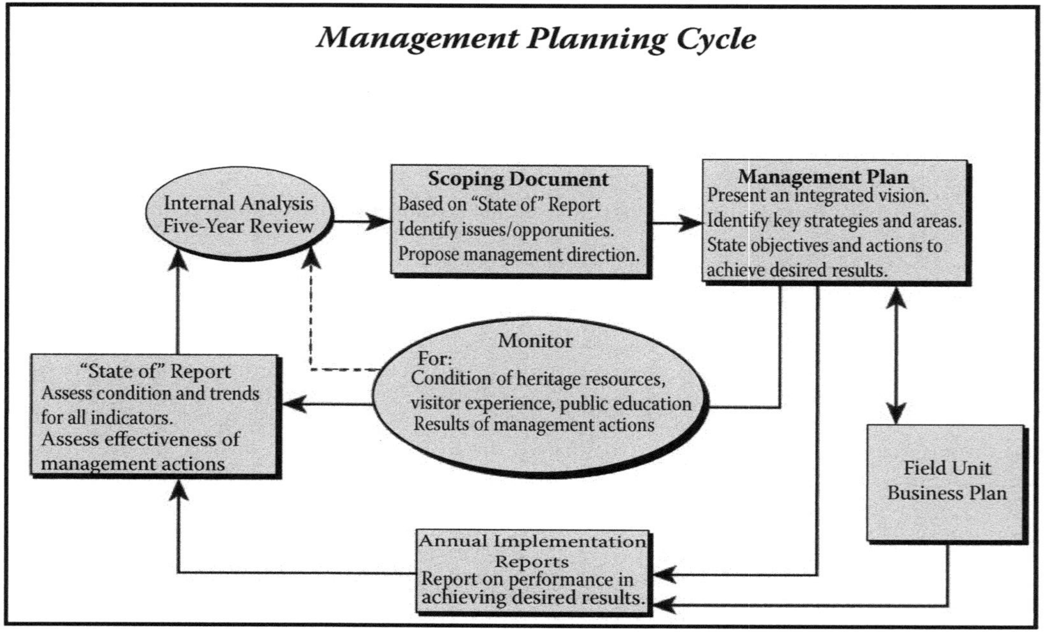

FIGURE 10.2 Parks Canada Agency management planning cycle. Monitoring and reporting is central to the five-year management process. (From PCA 2008. Parks Canada Guide to Management Planning. Parks Canada, Ottawa.)

b. Measure ecological change at two scales: local (ground based) and landscape (remotely sensed).
    i. *Long-term sample site measures*: Co-locate a small suite of conceptually inter-related ground measures in long-term sample areas (e.g., plots, transects) situated and replicated in "focal ecosystems" according to a robust sample design that maximizes inference to the whole park (and accounting for logistic issues) and that can be used to verify or calibrate remote-sensing measures.
    ii. *Other ground-based measures*: Establish various measures focusing on populations of focal park species and important ecological processes.
    iii. *Remote sensing-based measures*: Utilize remote sensing to capture landscape-scale change, such as changes in the area of park ecotypes, ecosystem productivity, and snow and ice phenology.

5. All EI Measures will have biologically based thresholds that provide criteria for assessing Good, Fair, and Poor EI condition and trend. Thresholds for most EI Measures will evolve from those defined by statistical change into biologically based thresholds as monitoring and research evolve. EI Measures are rolled up within EI Indicators to provide a simple and repeatable assessment of the EI condition of each EI Indicator. The assessments of EI condition and trend for each EI Indicator form the basis for reporting park EI condition and trend in State of the Park reports.
6. EI condition is reported by each park every five years in State of the Park reports, which are plain-language documents that communicate significant ecological change to a broad audience. State of the Park reports are supported by a Technical Compendium that presents all rationale, analyses, and data used to make the assessments.

In that all park ecosystems cannot be monitored using long-term ground sampling, the approach proposed is to concentrate local-scale ground sampling in carefully selected "focal ecosystems," and support these observations through monitoring of focal populations and ecological processes, and through remote sensing to reach out to the whole park. Focal ecosystems are terrestrial ecotypes or other logical, local-scale ecosystems (e.g., uniform stream segments or "reaches") that will be selected by each park depending on local priorities. For example, they could be selected to monitor change in those ecosystems most important to park fauna or in ecosystems most responsive to predicted ecological change. Ideally, we will work cooperatively with other agencies to develop a small set of "core measures" (presently undefined) for each EI Indicator/major ecosystems that will be common to all northern national parks, and could also be used by other agencies and community-based monitoring efforts across the North.

A KNOWLEDGE SYSTEM FOR REDUCING UNCERTAINTY

Proactive Adaptive Management

Given the potential complexity and unpredictability of ecological change in Canada's northern national parks, and the need for local information, it is imperative that we develop a knowledge system for collecting, interpreting, and communicating this change at the scale of individual parks. Park managers need to understand as early as possible how park ecosystems are changing, and require clear and timely information on the nature, magnitude, and scale of ecological change in order to make informed and proactive decisions. What, where, and how are park ecosystems changing? How rapidly are these changes occurring? What environmental and biological factors are controlling these changes? What are the implications for subsistence lifestyles and cultural practices of indigenous people, and for visitor experience and education, other park values, park infrastructure, and public safety? Are there cost-effective actions that can be taken to stop, slow, or divert these changes? What are the consequences of taking no action?

To provide this information, a proactive adaptive management approach is proposed (Figure 10.3) that incorporates updated and useful ecosystem inventories, well-designed condition and effectiveness monitoring, management-directed scientific research, forecasting models to estimate rates of change, and clear communication of science results to park managers. Such an approach is not new and has been put forward as the key to science-based ecosystem management for many years (Holling 1978; Gunderson et al. 1995). The fundamental purpose of the approach is to systematically and strategically record and interpret ongoing ecological change and to provide park managers with a range of potential responses, with associated risks of acting or not acting in a given situation.

Two of the foundational steps to develop locally based knowledge systems (Figure 10.3) are presently being developed by Parks Canada: useful and updated dynamic ecosystem inventories for all northern national parks, and effective park EI monitoring programs. These developing initiatives and other components of the proposed knowledge system are described in the following.

Process-Based Ecological Inventories

A first step for understanding, describing, and communicating the ecological integrity of a protected area is an ecological inventory that captures key elements of park biodiversity in the context of the ecological processes that control the distribution, productivity, and character of park ecosystems. The completion of such an inventory for all northern national parks is fundamental to establishing useful and effective park-monitoring programs, and for many other management applications. Through the Canadian contribution to the International Polar Year Program (http://www.ipycanada.ca/), Parks Canada is developing a cost-effective terrestrial ecological inventory method (Fraser et al., submitted) that is based on the ecotype as the fundamental mapping unit and uses mapping concepts and methods related to ecological inventory systems well established in southern Canada (RIC 1998, 2000; Sims et al. 1996).

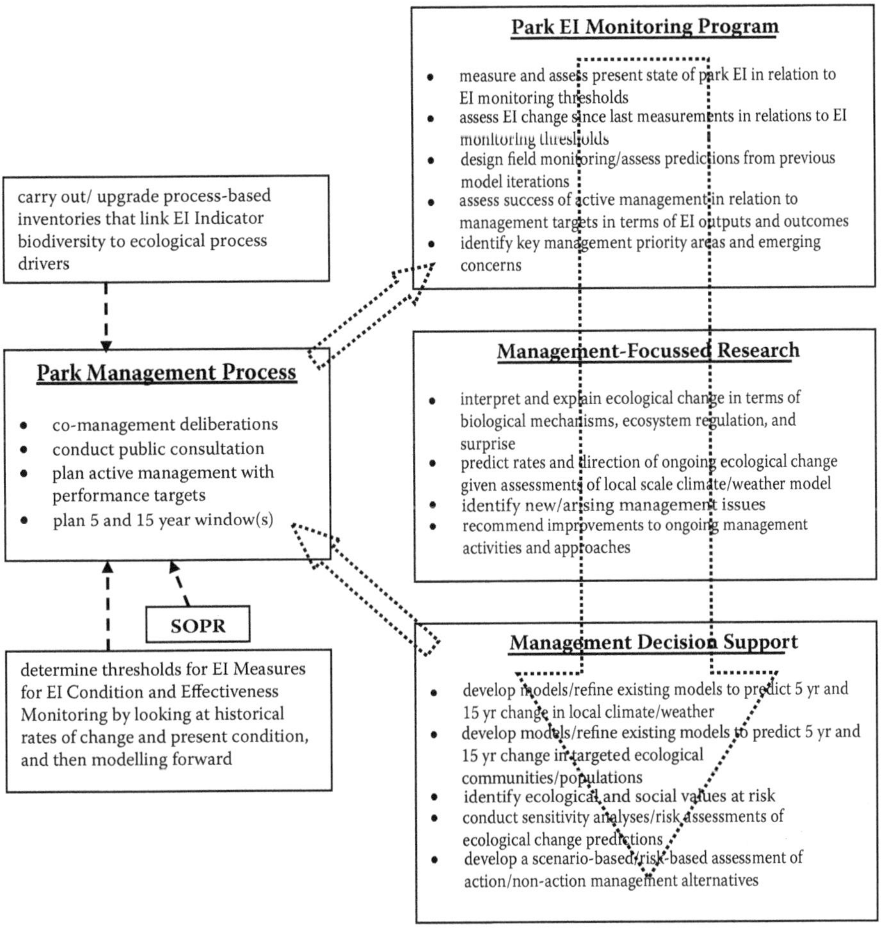

Figure 3: Stages for conducting proactive adaptive management of northern national parks under conditions of evolving uncertainty. The process cycles every five years with management planning and the State of the Park report (SOPR) reporting.

FIGURE 10.3 Stages for conducting proactive adaptive management of northern national parks under conditions of evolving uncertainty. The process cycles every 5 years with management planning and the State of the Park Report (SOPR) reporting.

The following are the key objectives for park terrestrial ecosystem inventories:

- Identify, describe, and classify the park landscape into ecologically equivalent landscape segments (ecotypes).
    - Describe ecotype vascular and nonvascular species, and classify into plant communities following methods outlined in the Canadian System of Vegetation Classification (CNVC 2009).

    - Describe and classify ecotype soils (CSSC 2006), humus forms, and landforms, and relate the structure, composition, and function of ecotype vegetation communities to relevant soil and landform processes (periglacial effects, seepage, inundation, colluvial effects, nivation effects, microclimate effects); identify soil-moisture and soil-nutrient regime; identify characteristic physiographic conditions (slope, aspect, elevation, meso-slope position).
    - Develop a park ecosystem classification based on vegetation and environmental information.
- Develop and map a regional/park bioclimate classification using the distribution of plant communities on zonal sites and linked to the Circum-arctic Vegetation Mapping project (CAVM Team 2003); this will be especially important in mountainous parks or parks located at the boundaries of major climatic zones.

Taken together, the knowledge of ecotypes and their spatial distribution as shown through the mapping provide a logic system for linking the distribution of park terrestrial community diversity to the ecological drivers that control that distribution. Such an inventory provides a basis for establishing and implementing park monitoring by providing the fundamental knowledge to rationalize, locate, and monitor focal ecosystems, given predicted changes in climate and other ecological drivers for the park. The inventory also provides an ecological frame for establishing and conducting research, and for modeling potential ecological change under a range of potential scenarios.

It is understood that such derived inventories contain classification errors (accuracy for the Ivvavik inventory is overall 78% at this time Fraser et al. submitted). The inventories should be considered as useful models for visualizing and understanding community composition, the spatial distribution of classified ecotypes, and the processes that control that condition and distribution. Where the inventories are used locally, for example, to develop ground-based monitoring, map-polygon attributes can be updated based on further field observations. In this way, we expect map accuracy to improve over the years as monitoring and other field programs progress.

The vegetation structure and composition of the ecotypes, soil and site characteristics, as well as their position in the landscape largely determine their value as animal habitat, so ecotypes (along with other factors) can also be interpreted for their suitability for sustaining key wildlife species. By measuring change in ecological drivers and/or predicting how drivers may change in relation to predicted climate change, we can then predict changes in the distributions of ecotypes and interpret potential consequences for wildlife species important to park managers.

Finally, the ecotype maps provide a "baseline" for assessing future changes in the areal extent, productivity, composition, structure, and distribution of park ecological communities. In that northern ecosystems are already changing, the term "baseline" is used in the sense of a starting point for measurement, not a reference condition for the desired distribution and areal extent of park ecosystems. Management thresholds for assessing change in park ecotypes will need to be set in the context of a desired condition for park ecosystems, as part of the evolving park monitoring program.

To date, ecological inventories are in progress under the PCA International Polar Year Program for Ivvavik, Wapusk, and Torngat Mountains NPs. The objective is to complete similar ecosystem inventories for all 10 northern national parks shown in Figure 10.1 as part of the ParkSPACE program to provide the basis for designing and implementing park monitoring programs.

Establishing EI Monitoring in Northern Parks

The ten northern national parks shown in Figure 10.1 are in the process of establishing their park EI monitoring and reporting programs, and there are a number of important issues around northern parks that increase the difficulty of designing and implementing EI monitoring, compared to southern parks. Northern national parks are very large in area and remotely located, and thus are very difficult and expensive to visit and sample. Although Parks Canada does maintain an administrative and public-safety presence in all parks, ambitious ground-based sampling that requires annual or multiannual visitation will be difficult and expensive to sustain in the long term. Given the size of the task and inevitable staff turnover, in order to sustain long-term monitoring, science and technical staff in northern parks will need to be supported internally and by long-term cooperative arrangements with other government agencies and academic institutions. Incorporation of the needs and knowledge of indigenous partners is another important challenge for northern parks, and for the design and sustainability of park monitoring programs. Northern parks are working through their comanagement boards to ensure the inclusion of indigenous knowledge, perspectives, and concerns in the development of park monitoring programs.

Establishing Condition Monitoring

The general model for assessing overall condition utilized across the Parks Canada system is to establish a series of long-term monitoring sites in selected focal ecosystem types (e.g., mesic tundra, shrub tundra at sites where expansion is occurring, estuarine salt marsh, mesic subarctic forest, or targeted wetland ecotypes), coupled with measures of focal park populations and important ecological processes, and supported by parkwide remote-sensing measures, to develop an EI assessment for the entire park. Given the size and remoteness of northern national parks in Canada, park monitoring will develop a relatively heavy reliance on remote-sensing measures to assess ecological change. Within the bounds of logistical feasibility, long-term monitoring sites will be located across the park according to experimental designs that permit statistical inference to the focal watershed, and to the entire park.

At each long-term monitoring site, a small suite of colocated and conceptually inter-related EI measures are established that, taken together, can provide a synopsis of ecological change in that focal ecosystem. Typical EI measures for the Tundra EI Indicator may include

- Changes in vegetation productivity, species composition, and relative species abundance
- Active-layer depth and soil-temperature regime

- Snow regime (depth and duration)
- Local measures of songbirds and small mammals

Some of these measures will be annual (such as the active-layer monitoring, soil temperature, and snow regime) and can be conducted somewhat by park technical staff, and others will be measured on a 5-year basis (vegetation change, songbirds, and small mammals) and will require professional staff to conduct the sampling. Analyzed together, these inter-related measures are linked to conceptual ecosystem models and are designed to "tell a story" of ecological change in the targeted focal ecotypes, should change be occurring. A similar suite of measures would be monitored in other EI Indicators. For example, stream discharge, water quality, benthic invertebrates, and fish community could be monitored at long-term sample locations along targeted stream reaches.

This system of intensively sampled, long-term sample sites can be supported by extensive sampling using photographic techniques, and by observational data designed to observe and record important ecological changes occurring across the park landscape, such as the introduction of invasive species, plant die-back or disease, fire occurrence, mass-wasting events, or presence of new animal species. Parks staff on patrols, indigenous people on the land, and possibly park visitors could be trained to record these observations, either through directed systematic surveys or during the course of their travels through the park. This approach would extend ground-based ecological surveillance to a wide area of the park, and important observations would be investigated and potentially become the focus of more-structured monitoring.

A second important component of park monitoring programs will be EI measures of population trends in focal species important to park EI, including caribou, moose, polar bears and other carnivores, songbirds, colonial seabirds, muskrat, and migratory shore birds and waterfowl. Other supporting measures will follow trends in important ecological processes such as rates of coastal erosion, stream discharge, and sea ice. Much of this work is ongoing with federal and territorial partners and will be included in park monitoring programs as they are developed. Where appropriate, focal ecosystems selected for long-term monitoring of ecosystem productivity and plant species change can be linked to key habitat components of these species to provide a more comprehensive picture of ecological change for focal species.

To support the ground observations and extend monitoring over the entire area of the park, a range of common remote-sensing-based EI measures are being developed for northern national parks through the ParkSPACE program. This is a cooperative program funded by the Canadian Space Agency, where Parks Canada ecologists are working with remote-sensing scientists from the Canadian Centre for Remote Sensing to develop protocols for remote sensing-based EI measures, and to implement an operational system for delivering these EI measures for northern national parks. Remotely sensed EI measures being developed include changes in the area of tundra, forest, wetland, and coastal ecotypes; changes in lake surface area; changes in ecosystem productivity and growing season; changes in active-layer depth and permafrost; changes in the phenology of lake and river ice; and changes in glacier area. As additional components of ParkSPACE, baseline satellite data are being interpreted to support models of specific wildlife species or for general assessments of landscape connectivity. Archival remote-sensing data for all northern parks are also being collected and used to back-cast some of these measures (e.g., shrub cover change, productivity measures, and habitat change) to assess historic trends back to about 1980. This work will support the development of thresholds for establishing park monitoring programs and assessing and reporting trends. As much as possible, ground measurement monitoring will be designed to support remote-sensing measures. For example, ground measures of permafrost depth, soil temperature, soil texture, and soil organic content can be used to parameterize the active-layer model (Zhang et al. 2007) used to predict landscape scale change in active-layer and permafrost conditions.

The coordinated monitoring approach proposed through the Parks Canada Monitoring and Reporting Program is to develop a common suite of ground-based and remotely-sensed core measures in all ten northern national parks, so that a consistent set of observations are made that represent a wide range of arctic and subarctic environmental variability. Along with other monitoring focused on local species and issues relevant to each park (mentioned previously), data from these core variables will make a strong contribution to measuring and reporting ecological change across the Canadian North. Also, by measuring similar ecological elements across such a broad area, regional assessments of patterns and progress of ongoing ecological change can be evaluated and will provide an "early-warning system" of approaching change for park managers. Finally, PCA is also working internationally to coordinate protected areas monitoring in the circumpolar Arctic, in coordination with the Circumpolar Biodiversity Monitoring Program (CBMP 2009).

The Role of Effectiveness Monitoring

Effectiveness monitoring will provide more-specific information on the ecological outcomes of management interventions that may be implemented to adapt to park ecological changes. Effectiveness monitoring in southern parks focuses for the most part on measuring ecological improvements resulting from management activities aimed at restoring EI in degraded areas of a park. Northern parks are largely intact at this time and potential future management interventions will need to respond to situations as they arise. As climate change plays out, however, management actions such as prescribed fire to reduce shrub expansion, species translocations, river-mouth dredging to permit fish access, or revegetation of mass-wasting scars on slopes with accelerated permafrost melting may need to be undertaken to adapt to ongoing ecological change and mitigate effects on park ecological integrity. Effectiveness monitoring will be implemented with these projects to measure and report the success of management investments, and to learn from our experience as projects proceed.

Effectiveness monitoring can also be employed to evaluate the accuracy of model predictions, by using monitoring results to refine models and provide more accurate assessments of ecological change for subsequent model iterations. These measurements may be a component of condition monitoring or effectiveness monitoring, depending on the model scale and application. For example, it is well documented in Alaska that tundra shrubs are presently expanding at the expense of herb and prostrate shrub-dominated communities (Tape et al. 2006). So PCA may want to monitor the spread of tundra shrubs in neighboring Ivvavik NP or other parks as one in a suite of EI measures for assessing the Tundra EI Indicator. The recent terrestrial ecological inventory can be used to identify where shrub communities presently occur in the park. Knowledge of the ecological factors that determine the different types of shrub ecotypes can help focus the analysis to those tundra-shrub ecotypes most vulnerable to spreading out. This analysis can be supported by ground measures to confirm the spreading, and by back-casting aerial imagery or remote-sensing data to understand the scope and scale of the issue.

Once we have identified the target (focal) ecotypes where shrub change is occurring, we can conduct monitoring at two scales: local ground-based measures of rate of spread, and associated ecological effects on snow accumulation, soil temperature, active-layer depth and species effects. We can also link these ground measures to a broader landscape scale using remote-sensing tools. Five-year resampling using high-resolution satellites would provide a cost-effective method at a useful scale to follow the potential spread of tundra shrubs. The tundra-shrub monitoring results would then be used to refine the shrub-spread model and provide more precise estimates for subsequent model iterations. Such ecological models focused on key management issues will be the key to proactively managing park ecosystems and anticipating ecological change.

Science Knowledge and Management-Focused Research

The key objectives for management-focused research will be to measure, interpret, and communicate the biological mechanisms and environmental control factors that determine the most important aspects of ecological change in the park, as identified through EI monitoring. Models will need to be developed that relate climate drivers to the environmental control factors that determine the rate and direction of ecological change.

Using the spread of erect tundra shrubs as an example, the rate of spread can be predicted based on the autecological characteristics of the shrub species (e.g., seed-dispersal potential, germination requirements, time to sexual maturity, potential for vegetative propagation) and the environmental factors that limit the rate of spread such as exposure to winter wind, soil temperature and nutrient levels, and snow accumulations. Given that shrubs are already expanding, we can use historical rates of spread to test the models developed. There is also potential for synergistic interactions, so that more erect shrubs will trap more snow, possibly resulting in warmer ground temperatures and higher germination and rates of growth (Sturm et al. 2005). More winter snowfall with warming climates could also add to this synergy and further increase the rate of spread. These factors can be modeled and a rate of spread predicted. As described previously, the actual rate of spread is then measured through the EI monitoring program, and the model is refined within the 5-year iterations of the park planning cycle. Other important areas of climate-driven ecological change that may be modeled to inform management include changes in the interactions among active-layer depth, annual soil temperature, and soil decomposition (ACIA 2005, Callaghan et al. 2005), with resulting changes in vegetation productivity, composition, and structure, and population models for keystone park species.

Engaging a team of multidisciplinary scientists in park monitoring and management is a very important component of the knowledge system required to successfully guide park decision making to meet the challenges of accommodating, mitigating, and responding to climate change in northern parks. The key to optimizing this relationship is to ensure that research is focused on identifying and elucidating the most important ecological issues the park is facing, and in providing clearly communicated information in a timely manner and in a context useful for informing management.

Management Decision Support

The science and monitoring we can bring to bear on a given ecological issue will only be useful for park management if science knowledge is clearly communicated in plain language, and within a clear management context. Park managers need to know the most critical ecological change that is occurring, why and how rapidly it is happening, and what the implications are for ecological integrity, indigenous lifestyles, and other park-management objectives. Based on this information there may or may not be active management actions that park managers can take that will mitigate or ameliorate the observed and predicted ecological change.

The objective of the management decision support component is to develop tools that present science knowledge in a way that will prove useful for park managers to understand ecological issues and take decisions on active management. Using the tundra shrub example, science advice should attempt to provide answers to questions such as "how will spreading shrubs affect habitat values for caribou, muskoxen, moose, and grizzly bear, and over what time frame? If negatively impacted by these changes, what management options are there for impacted species? Is prescribed fire a viable option? Is there habitat farther north or at higher elevations? Do these changes increase the risk of tundra fire? Are potential actions feasible given financial and human resources? What are the risks of acting or not acting?"

One set of tools to support decision making will come from issue-specific or multiapplication models that may include

1. Models to predict 5-year and 15-year change in local (scaled down) climate, and the potential for critical weather events such as strong winds, heavy precipitation, or tundra icings; refine in 5-year iterations based on assessment of local climate monitoring data; such down-scaled climate models could be used to drive other process based models
2. Models to predict 5-year and 15-year change in targeted ecological communities/populations (e.g., tundra-shrub expansion, tree-line movement upslope, caribou calving production, char population sustenance); refine in 5-year iterations based on assessment of monitoring data
3. Sensitivity analyses/risk assessments of ecological change in relation to proposed management actions (e.g., prescribed burning); test action versus no-action options

4. Identifying and clearly communicating based on model analysis ecological and social values at risk and make recommendations for initiating, improving, or ceasing ongoing management actions
5. Introducing and explaining emerging management issues

It is this modeling component of the knowledge system that provides the "proactive" component of the adaptive management system proposed in this chapter. Through modeling, we attempt to structure and test our understanding of ecological changes relevant for park management by developing models that predict future rates of change based on this understanding, and past monitoring observations or back-casting. Following up monitoring within the 5-year framework of State of the Park reporting tests the accuracy of our predictions; models are subsequently refined to improve predictions. Within a few 5-year model iterations, parks should develop a reasonable understanding of future states and the ecological implications of key issues. The effects of management actions such as prescribed fire or species translocations can also be modeled, and success assessed through follow-up effectiveness monitoring.

Park Management Process

The process of park-management planning is well defined at Parks Canada (PCA 2008b), and the intention of this proposed knowledge system is that decisions can be made based on well-communicated science advice that reduces the uncertainty and describes the implications of ongoing ecological change in a given park. The other key knowledge to be provided to park managers is the assessment and relative risks of actively managing a problem, versus letting an anticipated change unfold without intervention.

The State of the Park Report (SOPR) is an important document that managers can use, and ecological assessments based on the modeling tools could be presented in this report, along with other monitoring results and assessments. The SOPR summarizes changes in the overall EI of the park and reports on the success of EI investments made through active management. Ecological issues raised through the SOPR are vetted through the Scoping Document, which park managers use to set the goals and objectives to be included in the new iteration of the Park Management Plan (refer to Figure 10.2).

The key difference between managing protected areas in a world assumed to be stable and one characterized by accelerating directional ecological change is that managers will need to much more frequently assess and renegotiate ongoing processes and programs in light of considerable uncertainty. Hopefully, as time passes, park management teams will develop a general understanding of the nature and rates of change, and what they mean for core values of the park through focused modeling with follow-up monitoring. However, it is also clear that our understanding of this ongoing change will eventually be usurped by ecological surprise (e.g., ungulate disease, epidemic insect outbreaks, the emergence of increased tundra fire as a disturbance agent), and we will need to understand and respond to these events as well. For all of these challenges, an operational knowledge system like the one described here will allow us to learn by doing, and, over time, will greatly refine our understanding of park ecosystems and the efficacy of our management decisions and actions.

DISCUSSION

Most of the ideas presented in this chapter are not new to the conservation community. The need for an adaptive-management approach to adapt supported by strong EI monitoring has been recommended as the basis for an adaptation strategy at Parks Canada by Lemieux and Scott (2005), Scott and Suffling (2000), Scott and Lemieux (2005), Staple and Wall (1996), and Welch (2005, 2008), and more broadly for protected areas adaptation in general by Heller and Zavaleta (2009), Lawler (2009), Lawler et al. (2009b), Root and Schneider (1995), and Hulme (2005). However, although widely recommended for many years, the direct operational application of ongoing and focused adaptive management approaches in protected areas management is rare. Given the scale and intensity of predicted shifts in arctic and subarctic biotas, and in the ecological drivers and processes that determine their distribution and vigor, the adoption of a proactive adaptation system by park management teams will be fundamental to adapting to these inevitable changes.

Through the PCA commitment to establishing ecological integrity monitoring in all national parks, and through the impetus provided through the International Polar Year and ParkSPACE projects, the development of the crucial inventory and monitoring components of the knowledge system proposed in this chapter is well under way in northern national parks in Canada. The ability of park inventory and monitoring programs to provide the level of information required to measure and assess the most important elements of ecological change will depend on the development of effective program designs, especially EI measure selection, and ongoing park investments in monitoring. The Agency is not yet at the point where management needs drive park science, although progress is being made as a result of the relatively recent and evolving need to report every 5 years on the state of the park. Scaled-down climate modeling and predictive models of ecological change are also undeveloped at this time. A key to successfully implementing useful knowledge systems that will permit adaptation to climate change effects will be greatly facilitated by the development of long-term partnerships with other government agencies, academia, and local communities.

This chapter focused on an approach for proactively managing ecological change within northern national parks, but clearly even the most-informed and effective park-management system will not by itself be successful. In spite of the large size of many of our northern national parks, the scale of predicted ecological change will require regional and national, as well as continental and circumpolar international cooperation and focus for effective adaptation. A key task for this cooperation at a regional scale is the development of multistakeholder land-use plans that ensure the functional connectedness and facilitate adequate protection of lands in the matrix between protected areas. In the Canadian North, matrix (i.e., nonpark) lands are generally in a very healthy state, but this is largely due to low populations and lack of economic activity, not to effective land-use policy. This overall healthy situation is rapidly changing as climate warms, ecological change accelerates, and economic and community development opportunities increase. The decreasing usefulness of winter roads and the inevitable pressure to develop all-weather roads is an important example of an impending development that could significantly alter the condition of matrix lands in the Canadian North.

Finally, international circumpolar cooperation and coordination is required to optimize scientific and governmental resources, and to focus international political attention on key issues as they evolve. If these resources can be coordinated and focused to identify and react to the most important emerging ecological issues, and using management tools such as the proactive approach described here coupled with effective land-use frameworks, then potential impacts on biodiversity can be minimized as the effects of climate change evolve at northern latitudes around the globe.

SUMMARY

Canada's northern national parks have been established to protect a wide representative sample of terrestrial, freshwater, and marine ecosystems across the arctic and subarctic landscapes of Canada. Ongoing and predicted climate change will fundamentally alter the composition, structure, and function of park ecosystems, and some of these ecological changes are ongoing and have already been recorded. This ecological change has important implications for park ecological integrity as defined in the Canada National Parks Act (2000) and will compromise the ability of national park managers to meet corporate expectations to "... maintain or restore the overall ecological integrity of all national parks." According to published climate-change models, important ecological change will begin to affect management objectives within the 15-year time frame of management plans being developed over the next 5 years, a period when Parks Canada is striving to develop park management plans for all northern national parks.

The principal idea presented in this chapter is that the most important factor in mitigating and adapting to the effects of climate change on the ecological integrity of northern national parks is the operational implementation of a proactive adaptive management approach that would reduce uncertainty in two areas:

1. Refinement of broad-scale climate models to better predict potential change at a scale more relevant to park managers
2. Monitoring and research to measure, interpret, and predict the rate and nature of ecological change within northern national parks

The development of scaled-down climate models can occur through partnerships with federal agencies or academic partners, and could be coordinated with regional land managers with similar interests in this scale of climate change information. Information on how park ecosystems are responding to evolving climate change can be provided through park monitoring programs linked to research focused on identifying and interpreting those issues most central and immediate for park managers. A key to effective proactive management is the modeling forward of the most important ecological changes, through 5-year iterations of these models with each State of the Park report, to foster ongoing learning and model refinement. Comanagement agreements in northern national parks will facilitate inclusion of the needs and knowledge of indigenous people in this evolving situation that is certain to directly impact land-based lifestyles that rely on access to abundant and healthy wildlife populations.

REFERENCES

ACIA. 2005. *Impacts of a Warming Arctic: Arctic Climate Impacts Assessment.* Cambridge University Press, Cambridge, 1042 pp.

Anisimov, O.A., D.G. Vaughan, T.V. Callaghan, C. Furgal, H. Marchant, T.D. Prowse, H. Vilhalmsson, and J.E. Walsh. 2007. *Polar Regions (Arctic and Antarctic) Climate Changes 2007. Impacts, Adaptation and Vulnerability. Contribution of Working Group II to the Fourth Assessment Report of the International Panel on Climate Change.* Eds. M.L. Parry, O.F. Canziani, J.P. Palutikof, P.J. van der Linden, and C.E. Hanson, 655-685. Cambridge University Publications, Cambridge.

Callaghan, T.V., L.O. Björn, Y.I. Chernov, F.S. Chapin III, T.R. Christensen, B. Huntley, R. Ims, M. Johansson, D. Jolly, N.V. Matveyeva, N. Panikov, W.C. Oechel, and G.R. Shaver. 2005. Arctic tundra and polar ecosystems. In: *Arctic Climate Impact Assessment (ACIA)*, Eds. C. Symon, L. Arris, and B. Heal, 243-351. Cambridge University Press, Cambridge.

Canada National Parks Act. 2000. http://laws.justice.gc.ca/en/N-14.01/

CAVM Team. 2003. Circum-Arctic Vegetation Map. 2003. Scale 1:7,500,000. Conservation of Arctic Flora and Fauna (CAFF). Map No. 1. U.S. Fish and Wildlife Service, Anchorage, Alaska.

CBMP. 2009. The Circumpolar Biodiversity Monitoring Program. http://cbmp.arcticportal.org/

Chapin, F.S. III, M. Berman, T.V. Callaghan, P. Convey, A.-S. Crepin, K. Danell, H. Ducklow, B. Forbes, G. Kofinas, A.D. McGuire, M. Nuttall, R. Virginia, O.R. Young, and S. Zimov. 2006. Polar systems. In: *Millennium Ecosystem Assessment: Current State and Trends*, Eds. B Fitzharris and K. Shrestha, 719-743. Island Press. Washington D.C.

Chapin, F.S. III, M. Sturm, M.C. Serreze, J.P. McFadden, J.R. Key, A.H. Lloyd, A.D. McGuire, T.S. Rupp, A.H. Lynch, J.P. Shimel, J. Beringer, W.L. Chapman, H.E. Epstein, E.S. Euskirchen, L.D. Hinzman, G. Jia, C.-L. Ping, K.D. Tape, C.D.C. Thompson, D.A. Walker, and J.M. Walker. 2005. Role of land surface changes in Arctic summer warming. *Science* 310 (5748): 657-660.

CNVC. 2009. The Canadian National Vegetation Classification. http://www.natureserve-canada.ca/en/cnvc.htm

CSSC. 2006. *The Canadian System of Soil Classification*, 3rd edition. Agriculture and Agri-food Canada, Ottawa (http://sis.agr.gc.ca/cansis/references/1998sc_a.html).

Fraser, R., D. McLennan, S. Ponomanko, and I. Olthaf. (Submitted.) IPEM-based predictive ecosystem mapping in Canadian National Parks. Submitted to *International Journal of Applied Earth Observation and Geoinformation.*

Gunderson, L.H., C.S. Holling, and S.S. Light (Eds.). 1995. *Barriers and Bridges to the Renewal of Ecosystems and Institutions.* Columbia University Press, New York. ISBN 0231101023.

Heller, N.E., and E.S. Zavaleta. 2009. Biodiversity management in the face of climate change: A review of 22 years of recommendations. *Biological Conservation* 142: 14-32.

Hinzman, L., N.D. Bettez, W.R. Bolton, F.S. Chapin, M.B. Dyurgerov, C.L. Fastie, B. Griffith, R.B. Hollister, A. Hope, H.P. Huntington, A.M. Jensen, G.J. Gia, T. Jorgenson, D.L. Kane, D.R. Klein, G. Kofinas, A.H. Lynch, A.H. Lloyd, A.D. McGuire, F.E. Nelson, W.C Oechel, T.E. Osterkamp, C.H. Racine, V.E. Romanovsky, R.S. Stone, D.A. Stow, M. Sturm, C.E. Tweedie, G.L. Vourlitis, M.D. Walker, D.A. Walker, P.J. Webber, J.M. Welker, K.S. Winker, and K. Yoshikawa. 2005. Evidence and implications of recent climate change in northern Alaska and other Arctic regions. *Climate Change* 72: 251-298.

Holling, C.S. (Ed.). 1978. *Adaptive Environmental Assessment and Management.* Wiley, Chichester. ISBN 0471996327.

Hulme, P.E. 2005. Adapting to climate change: Is there scope for ecological management in the face of a global threat? *Journal of Applied Ecology* 42: 784-794.

ICARP II. 2005. A research plan for the study of rapid change, resilience and vulnerability in socio-ecological systems of the Arctic. Science Plan 10, Second International Conference on Arctic Research Planning, Copenhagen, Denmark (www.icarp.dk).

IPCC. 2007. *Climate Change 2007—The Physical Science Basis.* Contribution of Working Group 1 to the Fourth Assessment Report of the Intergovernmental Panel on Climate Change, Eds. S. Solomon, D. Qin, M. Manning, M. Marquis, K. Avery, M. Tignor, H.L. Miller Jr., and Z. Chen. Cambridge University Press.

Jia, G.J., H.E. Epstein, and D.A. Walker. 2003. Greening of arctic Alaska—1981-2001. *Geophysical Research Letters* 30(20): 2067.

Jones, B., D. Scott, E. Barrow, and N. Wun. 2003. Climate Change and Canada's National Parks—A User's Manual. Report to Parks Canada by Adaptation and Impacts Research Group, Environment Canada and the Faculty of Environmental Studies, University of Waterloo. Parks Canada Agency, Ottawa.

Lawler, J.J. 2009. Climate change adaptation strategies for resource management and conservation planning. The year in ecology and conservation biology. *Annals of the NY Academy of Science* 1162: 79-98.

Lawler, J.J., S.L. Shafer, D. White, P. Kareiva, E.P. Maurer, A.R. Blaustein, and P.J. Bartlein. 2009a. Projected climate-induced faunal change in the Western Hemisphere. *Ecology* 90(3): 588-597.

Lawler, J.J., T.H. Tear, C. Pyke, M.R. Shaw, P. Gonzalez, P. Kareiva, L. Hansen, L. Hannah, K. Klausmeyer, A. Aldous, C. Bienz, and S. Pearsall. 2009b. Resource management in a changing and uncertain climate. *Frontiers in Ecology and the Environment* 7. DOI: 10.1890/070146 (www.frontiersinecology.org).

Lemieux, C., and D. Scott. 2005. Climate change, biodiversity conservation, and protected area planning in Canada. *The Canadian Geographer* 49(4): 384-399.

Lloyd, A.H., and C.L. Fastie. 2002. Spatial and temporal variability in the growth and climate response of treeline trees in Alaska. *Climate Change* 52: 481-509.

———. 2003. Recent changes in forest treeline distribution and structure in interior Alaska. *Ecoscience* 10(2): 176-185.

Lopoukhine, N. 1990. National parks, ecological integrity and climate change. In: *Symposium on the Impacts of Climate Change and Variability on the Great Plains.* Dept. of Geography Publication Series, Occasional Paper No. 12, Ed. G. Wall. University of Waterloo.

———. 1991. The role of national parks in a changed climate. In: *Symposium on the Impacts of Climate Change and Variability on the Great Plains.* Dept. of Geography Publication Series, Occasional Paper No. 12, Ed. G. Wall. University of Waterlo.

Marshall, I.B., and P.H. Schut. 1999. A National Ecological Framework for Canada. Overview. Ecosystems Science Directorate, Environment Canada, and Agriculture and Agri-Food Canada. Ottawa (http://sis.agr.gc.ca/cansis/nsdb/ecostrat/intro.html).

Olthof, I., D. Pouliot, R. Latifovic, and W. Chen. 2008. Recent (1986-2006) vegetation-specific NDVI trends in Northern Canada from satellite data. *Arctic* 61(4): 381-394.

PCA. 1997a. State of the Parks Report. Parks Canada, Ottawa.

———. 1997b. *The National Parks System Plan.* 3rd edition. Parks Canada, Ottawa

———. 2005. Monitoring and Reporting Ecological Integrity in Canada's National Park. Internal Draft. Parks Canada, Ottawa.

———. 2008a. Parks Canada Corporate Plan—2008-2013. Parks Canada, Ottawa.

———. 2008b. Parks Canada Guide to Management Planning. Parks Canada, Ottawa.

———. 2008c. The Parks Canada Charter. Parks Canada, Ottawa. www.pc.gc.ca

Pouliot, D., R. Latifovic, and I. Olthof. (2009). Long-term changes in vegetation from 1-km AVHRR NDVI data over Canada for the period 1985-2006. *International Journal of Remote Sensing* 30(1): 149-168.

RIC. 1998. Standard for Terrestrial Ecosystem Mapping in British Columbia. Resources Inventory Committee. Province of British Columbia, Victoria, B.C. http://ilmbwww.gov.bc.ca/risc/pubs/teecolo/tem/indextem.htm

———. 2000. Standard for Predictive Ecosystem Mapping in British Columbia. Resources Inventory Committee. Province of British Columbia, Victoria, BC. http://ilmbwww.gov.bc.ca/risc/pubs/teecolo/pem/indextem.htm

Root, T.L., and S.H. Schneider. 1995. Ecology and climate: Research strategies and implications. *Science* 269: 334-341.

Rowe, S. 1989. National Parks and Climate Change: Notes for a Talk Given to Canadian Parks Service Personnel in Ottawa, Occasional Paper No. 4, National Parks Branch, Canadian Parks Service, Environment Canada, Ottawa.

Scott, D., and C. Lemieux. 2003. Vegetation Response to Climate Change: Implications for Canada's Conservation Lands. Report to Parks Canada by Adaptation and Impacts Research Group, Environment Canada and the Faculty of Environmental Studies, University of Waterloo. Parks Canada Agency, Ottawa.

———. 2005. Climate change and protected area policy and planning. *The Forestry Chronicle* 81(5): 696-703.

Scott, D.J., and C.J. Lemieux. 2007. Climate change and management of protected areas in the boreal forest. *The Forestry Chronicle* 83(3): 347–357.

Scott, D., and R. Suffling 2000. *Climate Change and Canada's National Parks*. Environment Canada, Toronto, Ontario.

Sims, R., I. Corns, and K. Klinka. 1996. Introduction—Global to local ecological land classification. *Environmental Monitoring and Assessment* 39: 1-10.

Staple, T., and G. Wall. 1996. Climate change and recreation in Nahanni National Park Reserve. *Canadian Geographer* 40: 109-120.

Stow, D., S. Daeschner, A. Hope, D. Douglas, A. Peterson, R. Myneni, L. Zhou, and W. Oechel. 2003. Variability of seasonally integrated normalized difference vegetation index across the North Slope of Alaska in the 1990s. *International Journal of Remote Sensing* 24(5): 1111-1117.

Sturm, M., C. Racine, and K. Tape. 2001. Increasing shrub abundance in the Arctic. *Nature* 411: 546-547.

Sturm, M., J. Schimel, G. Michaelson, J. Welker, S. Oberbauer, G. Liston, Fahnestock, and V. Romanovsky. 2005. Winter biological proceses could help convert Arctic tundra to shrubland. *Bioscience* 55(1): 17–26.

Tape, K., M. Sturm, and C. Racine. 2006. The evidence for shrub expansion in northern Alaska and the Pan-Arctic. *Global Change Biology* 12: 686-702.

Welch, D. 2005. What should protected areas managers do in the face of climate change? *George Wright Forum* 22(1): 75-93.

———. 2008. What should protected area managers do to preserve biodiversity in the face of climate change? *Biodiversity* 9(3/4): 84-88.

Zhang, Y., W. Chen, and D.W. Riseborough. 2007. Transient projections of permafrost distribution in Canada during the 21st century under scenarios of climate change. *Global and Planetary Change* 60 (3/4): 443-456.

Section V

Conservation Efforts in the Face of Rapid Climate Change

11 Ensuring That Protected Areas Play an Effective Role in Mitigating Climate Change

Nigel Dudley, Linda Krueger, Kathy MacKinnon, and Sue Stolton

CONTENTS

Protected area systems can be effective tools for optimizing the contribution of natural ecosystems in climate change response strategies. However, their potential will only be realized if protected areas themselves remain robust in the face of rapidly changing environmental conditions, which in turn implies changes in the way that both individual protected areas are managed and protected area systems are planned. We outline some steps needed to make the best use of protected areas within national strategies to address climate change. We then look briefly at the financial implications and discuss whether protected areas could be suitable for some new carbon-related funding streams emerging from the United Nations Framework Convention on Climate Change (UNFCCC) and voluntary efforts to reduce carbon emissions.

THE ROLE OF PROTECTED AREAS IN CLIMATE RESPONSE STRATEGIES

Protected areas can contribute to climate response strategies through both mitigation and adaptation by addressing the causes of climate change and helping to respond to the changes that are occurring. *Mitigation* includes storage, which includes preventing the loss of carbon that is already present in vegetation and soils, and also capturing additional carbon by sequestering carbon dioxide from the atmosphere in natural ecosystems. *Adaptation* is provided by protection of ecosystem integrity, buffering local climate, and reducing risks and impacts from extreme events such as storms, droughts, and sea-level rise, and provision of the essential ecosystem services that help people cope with climate-related changes in water supplies, fish stocks, and other wild foods, diseases, and agricultural productivity. Some of these opportunities are summarized in Table 11.1.

Although any natural ecosystem can perform these functions, protected area systems have additional advantages in that they are established as efficient, successful, and cost-effective tools for ecosystem management, with laws and policies, management and governance institutions, knowledge, staff, and capacity. The advantages and limitations of protected areas are discussed in more detail in the following.

There is increasing evidence that protected areas work more effectively than other approaches in maintaining vegetation and carbon. Most research looks at deforestation, but some considers degradation. Conservation International assessed threats facing 92 protected areas in 22 tropical countries and concluded that most protected areas are successful in protecting ecosystems (Bruner et al. 2001). The WWF International and the World Bank undertook a survey of 330 protected areas worldwide using a consistent methodology and found biodiversity condition consistently scoring high (Dudley et al. 2007). The University of Queensland is coordinating a global study that has so far analyzed evaluations from over 4,000 protected areas and found that only 13% were "clearly inadequate" in meeting their own criteria for good management (Leverington et al. 2008). For example, research carried out jointly by Indiana University and Ashoka Trust for Research in Ecology and the Environment found lower rates of land-clearing in protected areas compared to surrounding areas (Nagendra 2008). Duke University assessed natural vegetation changes in four tropical areas and

TABLE 11.1
Role of Natural Ecosystems in Mitigating and Adapting to Climate Change

Ecosystem	Role in Climate Response Strategies	Examples and Notes
Mitigation: Preliminary research shows that protected areas contain a huge carbon store, conservatively judged to be at least 15% of terrestrial carbon stock (UNEP-WCMC 2008).		
Forests	Forests are the world's largest terrestrial carbon stock (including soil carbon underneath forests) and continue to sequester in old-growth phases (e.g., Baker et al. 2004).	Tropical forests in protected areas in Bolivia store an estimated 745 million t C, worth US$3.7–14.9 billion at international carbon prices (Emerton and Pabon-Zamora 2009).
Inland wetland and peat	Inland wetlands, particularly peat, contain huge carbon stores. They can be net sources or sinks, depending on conditions and management. Intact peat contains up to 1,300 t of C/ha (Pena 2008) or 550 Gt globally (Sabine et al 2004).	In Belarus 40,000 ha of degraded peat has been restored and 150,000 ha more is planned. Annual greenhouse gas emissions from fires and mineralization were reduced by 448,000 t CO_2 (Rakovich and Bambalov in press).
Marine and coastal systems	Coastal and marine areas also contain carbon; in coastal zones capture is equivalent to 0.2 Gt/year. Salt marshes, mangroves, and sea grass beds all have potential to sequester carbon (Laffoley and Grimsditch 2009).	Analysis of 154 studies of carbon sequestration in mangroves and salt marshes (Chmura et al. 2003) found an average value of around 18.4 Tg C per year assuming a global area of 160,000 km^2.
Grasslands	Grasslands hold 10–30% of global soil carbon (Schuman et al. 2002) but can be a source or sink depending on management, precipitation, and CO_2 levels; loss and degradation currently release large amounts of carbon.	A meta-analysis found careful management increased soil carbon content in 74% of cases, especially conversion from cultivation, introduction of earthworms, and irrigation (Conant et al. 2001).
Soil	Soils are thought to be the largest carbon reservoir of the terrestrial cycle, holding more than the atmosphere and vegetation combined (Lal 2004a), although estimates vary widely.	Low-tillage farming can build soil carbon (Barker et al. 2007), although sequestration from different methods ranges from 50 to 1000 kg/ha/year (Lal 2004b).

(continued)

TABLE 11.1 (continued)
Role of Natural Ecosystems in Mitigating and Adapting to Climate Change

Ecosystem	Role in Climate Response Strategies	Examples and Notes
Adaptation: Some protected areas are established primarily for their wider ecosystem services, although there is still much to be learned about integrating these into national and local adaptation strategies.		
Adaptation	**Role of Protected Areas**	**Examples and Notes**
Disaster mitigation	Protected, well-managed ecosystems including forests and wetlands can buffer against many flood and tidal events, landslides, and storms (Stolton et al. 2008b).	Switzerland manages 17% of forests to prevent rock fall, landslides, and avalanches (Brändli and Gerold 2001); services estimated at US$2 billion and $3.5 billion per year (ISDR 2004).
Safeguarding water supply	Some natural forests (particularly tropical montane cloud forests and some older forests) increase total water flow, so that protecting these forests can maintain water supplies (Hamilton et al. 1994).	90% of Melbourne's water is from forested catchments. Almost half are protected and much of the rest is managed for water. Studies show that yield from forested catchments is related to forest age (Kuczera 1987).
Maintaining water quality	Natural forests usually provide higher quality water, with less sediment and pollutants than other sources (Aylward 2000), and wetlands can also reduce nutrient levels; some water plants concentrate toxic materials, thus purifying the water in which they grow (Jeng and Hong 2005).	Around one-third (33 out of 105) of the world's largest cities obtain a significant proportion of their drinking water directly from protected areas, including Mumbai, Jakarta, Sofia, Bogotá, Dar es Salaam, and Sydney (Dudley and Stolton 2003).
Supporting fisheries	Marine and freshwater protected areas provide safe havens for breeding to rebuild fish populations following over-fishing or coral bleaching (Roberts and Hawkins 2000).	A review of 112 studies in 80 marine protected areas found population densities were 91% higher, biomass 192% higher, and average organism size and diversity 20–30% higher, usually after only 1–3 years (Halpern 2003).
Safeguarding agricultural diversity	Protecting areas can be a way to conserve in situ populations of important crop wild relatives (Stolton et al. 2008a), needed to help crop breeders adapt food species to changing conditions.	In 1982, Erebuni State Reserve in Armenia was protected for its wild wheat diversity (*Triticum* spp.). Species include *T. urartu,* which was discovered in the area in 1935, *T. boeoticum*, *T. araraticum*, and *Aegilops* spp. (Damania 1996).

(continued)

TABLE 11.1 (continued)
Role of Natural Ecosystems in Mitigating and Adapting to Climate Change

Ecosystem	Role in Climate Response Strategies	Examples and Notes
Protecting health	Intact forests are correlated with reduced infection from malaria, leishmaniasis, yellow fever, etc. Protected areas are genetic sources for herbal medicines and pharmaceuticals to help society cope with new disease outbreaks (Stolton and Dudley 2010).	Researchers in Ruteng Park on Flores, Indonesia, found significant correlation between forest protection and a fall in childhood malaria (Pattanayak et al. 2003). Similarly, in the Peruvian Amazon the primary malaria mosquito had a biting rate 278 times higher in deforested than in forested areas (Vittor et al. 2006).

found that, overall, forest cover was often "strikingly higher" in protected areas than in surrounding areas (Joppa et al. 2008). A recent global review by the World Bank also shows that tropical protected areas lose less forest than other management systems, especially those conserved by indigenous peoples (Nelson and Chomitz 2009). This is confirmed by the United Nations Environmental Program (UNEP) World Conservation Monitoring Center, with analysis that shows forests in protected areas accounted for just 3% of tropical forest losses from 2000 to 2005 in countries studied, which is far better than average (Campbell et al. 2008). Finally, Stanford University and partners have also shown that protected areas, indigenous reserves, and national forests in the Amazon generally, but not invariably, afforded protection against logging (Asner et al. 2005).

Country studies confirm this trend. For example, research has estimated the amount of carbon stored in Canada's 39 national parks, which currently occupy about 2.25% of its land mass. In total, these parks store approximately 4,362 million tons (t), of which about 47% is in the soils, another 8% in the plant biomass, and the remaining 45% in peat. Overall, boreal areas in Canada store the largest amount of carbon. The study looked at the costs of replacing this carbon, using two scenarios. The costs of replacing carbon through reforestation of protected areas and afforesting marginal agricultural lands were estimated to be US$15.7 and US$16.9 per ton, respectively, at 2000 prices. Using these prices as proxy values, the value of national parks for carbon sequestration was estimated at between US$70.0 and US$75.4 billion (Kulshreshtha et al. 2000).

Consolidating, expanding, and improving the protected area system is a logical response to climate change that meets many aims of proposed mitigation strategies, particularly those to reduce deforestation and the loss of other ecosystems with large carbon reservoirs (World Bank 2009). There are existing legal and policy initiatives and tools to accelerate this process, so that many of the initial steps needed to implement these responses have already been taken.

MANAGING THE GLOBAL PROTECTED AREAS ESTATE IN THE FACE OF CLIMATE CHANGE

Most protected areas (national parks, wilderness areas, nature reserves, etc.) were established in the second half of the twentieth century, creating what is almost certainly the largest and fastest conscious change in land management in history. These protected areas now cover over 10% of the world's land surface and a small but rapidly increasing proportion of marine and coastal habitats. They exist under a number of management regimes, ranging from strict no-access areas to protected landscapes that include human settlements and cultural management (Dudley 2008). Although most are primarily established to protect biodiversity, their potential role in addressing climate change is being recognized (Dudley et al. 2009). Yet protected areas are themselves under threat from climate change, and risk losing the values that they were originally established to conserve (Hannah et al. 2002). Protected area planners therefore need to look simultaneously at options for maintaining protected area integrity and values under rapidly evolving environmental conditions and for maximizing the climate benefits from protected areas—two new major areas of policy and practice that remain poorly understood. There appear to be seven options available for increasing the role of protected area systems in contributing to climate change response strategies (each of which is discussed in greater detail in the following). The first three focus mainly on the need to expand areas under protection, and the last four on management responses within individual protected areas:

1. **More and larger protected areas and buffers**: to improve ecosystem resilience particularly where stored carbon is likely to be lost without protection or where important ecosystem services are under threat, such as in tropical forests, peat, mangroves, freshwater and coastal marshes, and sea grass.
2. **Connecting protected areas within landscapes/seascapes**: using ecosystem management outside protected areas. This can include buffer zones, biological corridors, and ecological stepping-stones, to build connectivity both to increase resilience to climate change at the landscape/seascape scale and to increase the total amount of habitat under some form of protection.
3. **Recognition and implementation of the full range of protected area governance types**: to encourage more stakeholders to become involved in declaring and managing protected areas as part of community climate response strategies.
4. **Increasing the level of protection within protected areas**: by recognizing protection and management aimed at specific features that have high carbon storage values, for example, to maintain old-growth forest, avoid ground disturbance or drying out of peat, and also restore degraded ecosystems.
5. **Improving management within protected areas**: to ensure that protected ecosystems and the services they provide are not degraded or lost through illegal use or unwise management decisions such as illegal logging and

conversion, other forms of poaching, impacts from invasive species, and poor fire management.

6. **Focusing some management specifically on mitigation and adaptation needs**: including modification of management plans, site selection tools, and management approaches as necessary.
7. **Introducing restoration strategies**: in protected areas where both carbon and biodiversity values could be enhanced by targeted action

More and Larger Protected Areas

Increasing the number of protected areas, and particularly of large protected areas, will become important for maintaining ecosystem integrity and for maximizing ecosystem resilience under conditions of climate change. Steps can include expanding borders of individual protected areas and linking protected areas including across national or regional borders. Appropriate social safeguards are needed to address the needs of local communities living within or adjacent to these areas.

Many governments are currently expanding and consolidating their protected area systems, in line with the commitments made in the Convention of Biological Diversity's (CBD) Program of Work on Protected Areas (SCBD 2004), the main objective of which is to complete ecologically representative, well-managed protected area networks. The Program has agreed actions, a timetable, and political support; in countries such as Mexico and Colombia it has resulted in concrete actions to identify and gazette new protected areas. The CBD already has available a range of tools to help identify likely areas for inclusion within national protected area systems, including a gap analysis methodology that can locate the most suitable areas of land and water (Dudley and Parrish 2006).

Connecting Protected Areas within Landscapes/Seascapes

Protected areas do not exist in isolation but function as part of a larger landscape or seascape—connectivity is therefore increasingly recognized as a critical value (Worboys et al. 2010) and is giving rise to some ambitious projects such as the Yellowstone to Yukon corridor in North America. Given the complexity of issues involved in the establishment and management of such projects, the proportion of territory under protection has to remain flexible to local conditions. A mixture of protection, management, and usually also restoration is required in what has become known as a *landscape approach*, appropriate to particular locations and circumstances (such approaches are also relevant in seascapes). Implicit within this are the twin concepts of increasing ecological connectivity with a view to increasing resilience and thinking constructively about other management systems that can contribute to broader-scale conservation aims. The approach does not imply that there is one "ideal" mosaic that, once achieved, will remain static indefinitely, but rather that there are a range of possible mosaics, which if implemented can help to make a landscape or seascape resilient to environmental change. Any conservation vision will exist alongside other, actually or potentially, competing visions (economic development, sustainable development, cultural values) and planned or unplanned social and political upheavals (Maginnis et al. 2004), making adaptive management an essential component of any such approach. Successful broad-scale conservation programs have therefore built partnerships with governments, private sectors, and local communities.

Recognition and Implementation of the Full Range of Protected Area Governance Types

A major expansion of protected areas driven entirely by the state is a limited and probably unachievable target in many countries. New protected area initiatives are more effective if a broad range of stakeholders are involved. Governments are recognizing this; for example, the report, *Australia's Biodiversity and Climate Change* (Steffen et al. 2009), stresses the need for new governance approaches within protection, including the rapidly expanding Indigenous Protected Areas network.

This also means accepting and welcoming new concepts of protection that contribute to viable climate response strategies, including new governance types such as indigenous and community conserved areas, private protected areas, and comanagement regimes (Borrini-Feyerabend et al. 2004). It will often involve negotiating forms of protection with many stakeholders, accepting different management models, taking risks, and including other peoples' priorities in planning processes. Increasingly, as climate change becomes a reality, local communities are themselves taking the initiative and recognizing the importance of natural ecosystems, sometimes moving faster than the government. Some "bottom-up" responses compiled by the World Resources Institute, for example, include participatory reforestation of Rio de Janeiro's hillside *favelas* to combat flood-induced landslides, reinstating pastoral networks in Mongolia, and reviving traditional enclosures to encourage regeneration in Tanzania (McGray et al. 2007).

Increasing the Level of Protection for Carbon Stores within Protected Areas

Increased levels of protection may be justified to maximize carbon storage in some protected areas. This might involve modifying management aims, for instance, forming strict protection zones in protected areas that have previously allowed some utilization within their borders (in other words, moving from an International Union for Conservation of Nature [IUCN] category V or VI protected area to category Ia, Ib, or II). Alternatively, efforts might be focused on vegetation restoration or changes in patterns of fire management or water flow. Carbon storage and sequestration needs to be measured and planned at a landscape scale and will be subject to some trade-offs, particularly in fire-prone ecosystems. Prescribed burning to reduce fuel load will, for example, release carbon but may prevent future more catastrophic losses. Natural disturbance patterns need to be factored into efforts to increase sequestration.

Improving Management Effectiveness within Protected Areas

Protected areas usually exist in the presence of a range of cross-cutting pressures; if not addressed these can further weaken an area's ability to withstand climate change. Once pressures have been identified and assessed, it is important to build strategies that address both immediate threats, such as poaching, encroachment, forest fires, illegal logging, climate change, and conversion, and underlying causes such as poor governance, poverty, perverse subsidies, trade barriers, and investment flows. Strategic interventions will range from site-based actions to those at landscape, national, ecoregional, and international levels. Approaches to understanding protected area management effectiveness are well developed (Hockings et al. 2006) and assessment tools are increasingly widely applied; for example, there have been protected-area-system-wide assessments of Finland, South Korea, and Colombia in recent years. Some assessment methods may require adaptation to meet the needs of protected areas used in climate adaptation strategies.

Focusing Some Management Specifically on Mitigation and Adaptation Needs

In addition to overall management effectiveness, some specific new management approaches may be needed, including special planning and assessment tools. Researchers face a number of important remaining questions about response strategies. More generally, within protected area agencies the implementation of such wide-ranging changes will require that a strategy be developed at both the protected-area-systems level and in management plans for individual protected areas. Capacity building will be needed to establish knowledge about emerging management challenges and opportunities. Other land managers will also require many of these skills, and protected area agencies may in some cases be a useful conduit for such information; this could, in effect, become another part of their functions.

Better forecasting is seen as important, to decide on the number and location of protected areas and on their relationship with the wider landscape or seascape, reflecting future ambient weather conditions, changes to biomes, refugia areas, etc. (Hannah et al. 2007). Protected area agencies will need to ensure that information to help manage rapidly changing environments is readily available to managers.

Introducing Restoration Strategies

In cases where protected areas have been established on previously degraded ecosystems, or where they have experienced illegal degradation following protection, some level of restoration may also be required (Dudley 2005), although restoration will need careful planning to ensure *additionality*, that is, the level of greenhouse gas emission reductions generated by a carbon offset project *over and above* what would have occurred in the absence of the project (Galatowitsch 2009). Restoration might be particularly important for mangrove ecosystems, where clearance has been commonplace, restoration relatively straightforward, and the ecosystem and carbon benefits quickly apparent (McKee and Faulkner 2000). In peat, drying out can release stored carbon, and several protected areas are now experimenting with restoration techniques to help reverse this process and restart sequestration.

USING AN ECOSYSTEM APPROACH TO ADDRESS ADAPTATION

Maintaining ecosystem function generally requires management of large areas, often larger than the boundaries of an individual protected area. The following questions need to be answered in developing an ecosystem-based adaptation strategy in a particular location, to ensure that it is assessed and developed as part of a comprehensive national adaptation strategy:

1. What ecosystem-based options are available, and what evidence (scientific or Traditional Ecological Knowledge) exists to show that the options are feasible?
2. What are the thresholds for failure in buffering risks (this question also applies to engineered solutions; a typical question would be, *what is the maximum rainfall that wetlands can absorb, without leading to catastrophic flooding?*).
3. What measures are needed to maintain resilience within the ecosystem in the face of climate change?
4. What ecosystem management options exist?
5. What other adaptation options exist? This would require that the feasibility, costs, and benefits of engineered solutions or behavior-based solutions be addressed and compared.
6. What other threats to the protected area exist?
7. Which option is most suitable given the local socioeconomic and ecological context?
8. What incentives are needed to sustain ecosystem-based adaptation? These may include, for instance, tax credits, payments for ecosystem services, and insurance schemes, but also incentives based on thorough understanding of costs and benefits among those involved.
9. What can existing protected areas do to contribute to ecosystem-based adaptation, and what new protected areas might need to be established to supply the necessary services?
10. How can these protected areas be mainstreamed into broader adaptation policies?
11. What other benefits (economic and noneconomic) might such protected areas supply to be included within cost comparisons?
12. How do local communities and other stakeholders view the various options?

FINANCING EFFECTIVE PROTECTED AREA NETWORKS

Since the CBD came into force in 1993, the world's protected areas have grown by almost 100% in number and 60% in area. Yet in the same period, international financing for biodiversity conservation has grown only 38% (Mulongoy et al 2008). Current financing for protected areas is generally judged to be inadequate; estimates of global shortfalls range from US$1.0 to $1.7 billion a year (Bruner et al. 2004), to US$23 billion a year (Quintela et al. 2004), up to US$45 billion per year (Balmford et al. 2002). A separate estimate suggested that funding a comprehensive marine protected-area system covering 20–30% of the seas and oceans would cost US$5–19 billion a year (Balmford et al. 2004). These shortfalls seem to represent massive amounts of money until they are compared with the annual value of total goods and services provided by protected areas, which are estimated to be between US$4,400 and US$5,200 billion (Balmford et al. 2002). This funding gap is currently not being addressed. An analysis of government funding of protected areas in over 50 countries suggested that financial support is generally declining (Mansourian and Dudley 2008).

Climate change incentive mechanisms open up a number of new opportunities for financing protected areas. Various climate-related initiatives, both market and nonmarket, could be considered as a way to support the creation and management of protected areas. There is a growing regulated international market for biocarbon offsets (e.g., the emerging REDD approach outlined in more detail later) alongside the existing voluntary international market for biocarbon offsets. In addition, voluntary payment for ecosystem service schemes for watershed protection, disaster mitigation, and other ecosystem functions have clear links to climate stabilization. Voluntary environmental offsets for households could also contribute. Funds from the Global Environment Facility for global biodiversity conservation could and should directly contribute to such efforts. There is also the potential for both voluntary and regulated international business biodiversity offsets (Mulongoy et al. 2008).

In addition to the development of funding relating to ecosystem services, economic measures could be put in place to remove environmentally perverse subsidies to sectors such as agriculture, fisheries, and energy that promote development without factoring in environmental externalities. Other economic and policy measures include implementing appropriate pricing policies for natural resources, establishing mechanisms to reduce nutrient releases and promote carbon uptake, and applying fees, taxes, levees, and tariffs to discourage activities that degrade ecosystem services (CBD 2009).

THE USE OF PROTECTED AREAS AS TOOLS TO STRENGTHEN REDD SCHEMES

One particular issue deserves further attention, given its current high profile. Forests and possibly other habitats contained within protected areas may offer important potential for funding under "reducing emissions from deforestation and forest degradation" (REDD) schemes, although the extent to which this will be possible still remains unclear. Methods to measure and verify reductions resulting from changes in land use and management are being developed under the UNFCCC. Many institutions already assume that protected areas will be a part of REDD and the need for a global network of forest protected areas has been identified under the CBD (Pistorius et al. 2008), which is also investigating the potential synergies between protected areas and carbon sequestration and storage. Most discussions about REDD focus on avoiding forest loss in multiple-use landscapes, but forests in protected areas also offer important options. It is possible that maintenance of carbon stored in protected areas in other ecosystems, such as grasslands, peat, and wetlands, could eventually become eligible for funding under REDD-type mechanisms.

Potential Advantages of Protected Areas within REDD Schemes

Well-managed protected areas offer advantages for carbon storage and sequestration and for REDD. Individual attributes occur in other management systems, but the combination provided by protected areas is unique and includes *management objectives*, *governance*, *legal constitution*, *management processes*, and *evidence of effectiveness*. Each is examined in the following:

Management objectives: Protected areas have objectives compatible with carbon storage. First, they aim to protect and restore natural ecosystems; such ecosystems—particularly forests and peat—provide good conditions for carbon sequestration. Conversely, while tree plantations accumulate biomass quickly, establishment can cause major carbon loss; one calculation found it would take 420 years of biofuel production on peat to replace carbon lost in establishment (Fargione et al. 2008). Protected areas are by definition set aside in perpetuity in that they are based around a commitment to permanence. Also, they are supported by government policy, which usually includes commitments under the CBD that focus attention on protected areas, adding to their protection.

Governance, legal constitution, and safeguards: As pre-existing management tools, protected areas have defined borders that are often legally defined and physically marked out, delineating precise areas that can be used to measure carbon potential. They operate under legal or other effective frameworks, usually including systems for establishing and codifying land tenure agreements already identified as REDD key requirements. Most protected areas also have an agreed governance structure. Although most protected areas remain in government control, a growing number are managed by private individuals, trusts, indigenous peoples, and communities. Protected area systems are also backed by a range of supportive conventions and agreements including the CBD's *Program of Work on Protected Areas*, UNESCO Man and the Biosphere and World Heritage, the Ramsar Convention on Wetlands, plus regional agreements like the European Union's Natura 2000. Many protected area managers have experience in implementing accessible, local approaches involving people in management.

Management processes in protected areas: These contain elements particularly useful in applying REDD, including that they have organized and populated data sources to set baselines and facilitate monitoring. Existing tools include the IUCN management categories, governance types and Red List, and the UNEP World Conservation Monitoring Center's World Database on Protected Areas. An increasing number of protected areas also have established systems for measuring effectiveness as discussed above. Techniques could be modified to include carbon accounting. Systems of certification are being discussed (Dudley 2004). In addition, protected area systems are backed up by networks of experts ready to provide advice and assistance including, in particular, the IUCN World Commission on Protected Areas, development agencies, and nongovernmental organizations (NGOs). Established protected areas have already invested in startup costs and are often supported by a management plan. Such areas also have staff and equipment that provide management expertise and capacity, so that REDD projects could use existing infrastructure. Importantly, most protected areas have a government budget and often also funding from the Global Environmental Facility, bilateral donors, and NGOs. Additional funding needed to help establishment or to increase effectiveness would build on existing resources.

The mechanism is not perfect, however, and protected areas sometimes fail to deliver expected benefits or raise important social concerns; some but not all of these challenges paradoxically add to their suitability as classic REDD projects with demonstrable additionality. Theory does not always match practice and protected areas are often under threat and losing values, including loss of vegetation and thus also carbon (Carey et al. 2000). Protected areas are also coming under increasing criticism for their social impacts, particularly loss of land and resources to local people (Dowie 2009). Although permanence is an explicit requirement, in practice, governments and others sometimes downgrade levels of protection, excise parts of a protected area, or completely remove legal protection and convert the land or water to other uses. REDD could provide resources to help address social challenges and threats to the area's integrity.

Developing Options for REDD

Negotiations currently under way envision the establishment of reference emissions levels and monitoring, reporting, and verification systems on a nationwide basis. Governments would therefore negotiate a scientifically defensible reference emission level from deforestation and forest degradation, and reduce emissions below that level in order to receive compensation through REDD mechanisms. Existing forest conservation, including protected areas and indigenous and community conserved areas, that have reduced reference deforestation levels need to be taken into account when establishing national REDD programs so as not to penalize these efforts. Compensation may occur within a system of nationally appropriate mitigation actions (NAMAs) with relatively flexible accounting standards and supported through fund-based mechanisms, or under a market-based approach that would be financed by private sector investors seeking more precisely measured emissions reductions.

Initial plans to make REDD incentives available only to high-emitting countries that have undertaken deep cuts seems to be giving way to the inclusion of a wider definition of REDD that includes conservation of standing forests and enhancement of carbon stocks (known as "REDD+"). There is also discussion about the need for

REDD to recognize the efforts and cater to the needs of countries that have already invested in conservation, either through the establishment and effective management of protected areas or other means, and historically have had low emissions levels from deforestation and forest degradation as a result (see discussion of stock-flow mechanisms in the following).

Strong international support exists for the development of social safeguard policies and other guidelines to ensure broad stakeholder consultations and program designs that avoid adverse effects, particularly on local and indigenous people. Depending on the national REDD implementation strategies adopted, project-based approaches may continue as a good way of addressing local drivers of forest loss and insuring accountability and equity of REDD strategies. National baselines will help insure against leakage that may occur in any individual project. Wide participation in REDD initiatives from among forested developing countries will guard against international displacement of deforestation (international leakage).

The resources needed to implement REDD effectively throughout the developing countries are substantial: figures of up to US$55 billion a year have been suggested (Saunders and Nussbaum 2008), although there are major differences in predictions about both the potential for reducing deforestation through financial incentives and the likely money available. The Stern report (Stern 2006) suggests that US$10 billion per year would be needed to implement REDD mechanisms. REDD has the potential to address several critical issues within a single mechanism: mitigation of climate change, reduced land degradation, improved biodiversity conservation, and increased human well-being and poverty alleviation. Institutions such as the World Bank and the United Nations are investing in REDD projects, which will require capacity building and continuous, predictable, and long-term funding.

Pros and Cons of REDD

However, there will be challenges for protected areas in implementing REDD in overall forest policy. REDD policies will, for example, take into account the need to fortify protected areas against potential incursions due to the implementation of REDD elsewhere. Similarly, REDD incentives focused on the most carbon-rich forest ecosystems may devalue other habitats, such as wetlands and grasslands, which are also crucial for biodiversity and other significant ecosystem services. REDD schemes need to be part of a broader national land-use planning process that takes into account the requirements of people and wildlife as it optimizes land-based carbon sequestration. At a subnational level, mechanisms are needed to account for accidental forest loss, for instance, through extreme wind, fire, or disease (that may, in itself, be caused indirectly by climate change); this might be achieved by "pooling" several areas. In general, these stochastic events would not be large enough to effect national targets for emissions reductions, but in some cases (such as El Niño–caused changes in fire regimes) they may be significant enough to warrant pooling among several countries.

More generally, some analysts fear that badly managed REDD projects will increase pressure on poor communities in terms of security of land tenure and access to resources (Mehta and Kill 2007). A substantial proportion of forest loss is due to the actions of poor farmers and subsistence gatherers who will be left with few other options if these resources are locked up. Some activist groups and indigenous peoples' organizations have already stated opposition to REDD on the basis that it will rely on sacrifices made by the poorest people rather than cutting energy and fossil fuel consumption by the world's rich. These challenges mean that strong social safeguards will be required (Smith and Scherr 2002; Peskett et al. 2006), coupled with a strong policy framework; indigenous peoples' organizations and communities are already investigating REDD schemes suggesting that many do not see these problems as intractable.

Stock-Flow Mechanisms

The most serious criticism of some approaches to REDD is that they simply reward (and encourage) bad management; only countries or areas that already have serious deforestation or threats to forests get money, while those that have already protected their forests or invested in protected areas are omitted from the scheme. There is a built-in "perverse incentive" for governments and others to put forests under pressure in order to access REDD funds (Woods Hole Research Center 2009).

A number of options have been suggested for addressing this paradox. One growing in popularity is the stock-flow approach, which is suitable where there is high forest cover and low deforestation (Cattaneo 2008). It proposes a "payment for performance" regime where payments are linked directly to reduction in deforestation or maintenance of certain levels of carbon stocks. This could be determined through either international accord or some kind of national arrangement. The approach separates incentives for lower deforestation from maintenance of existing carbon stocks and negotiates a common price to be paid to avoid emissions from deforestation. A proportion of REDD funds would be distributed as a dividend per ton of standing forest carbon stock. The system is not perfect and new iterations are still being proposed. Nonetheless, it remains one of the best options yet proposed to reward countries for good rather than bad management. It could help address, for example, options for REDD in a country like the Democratic Republic of Congo, where poverty goes hand in hand with high levels of forest cover and low carbon emissions (Laporte et al. 2009).

Additionality

There may also be a specific REDD-related question regarding additionality in protected areas. If protected areas are already in place, there may be little additional benefit in putting money into their protection. It is likely that REDD funding in protected areas will be limited to particular conditions. Perhaps the clearest case would be where the protected area is being newly created in areas at risk of forest loss and degradation, although it may also be applicable where an existing protected area is under-resourced and losing forest cover or quality. Other potential factors favoring REDD would be where there are no alternative, long-term funding sources and without REDD funding deforestation is likely to increase. Also, analysts believe situations where REDD payments can be used to support development and livelihoods in surrounding communities, in a way that will encourage long-term forest conservation, might also have potential. If proposals for REDD+ schemes as described previously come to fruition, the conditions may be less stringent. There is still considerable confusion about how these issues will be addressed in large-scale REDD implementation.

Ensuring Social Equity and Environmental Success

There are clearly some potential benefits for protected areas from a REDD mechanism, both for biodiversity conservation and for people living in natural forests, but only if there are sufficient social and environmental safeguards in place to ensure that REDD delivers real benefits within a framework that maximizes social benefits to those most in need.

There have been several attempts to address social issues through codes of practice. Proper application will likely be a prerequisite of success and for public acceptance of REDD offset schemes (Oestreicher et al. 2009). The Climate, Community, and Biodiversity Alliance has developed and tested detailed guidelines for such projects (CCBA 2008). WWF has identified critical steps needed to ensure that potential REDD projects are effective and socially equitable (Rietbergen-McCracken 2008). These steps can be considered in a framework based on broad issues (Table 11.2).

Making protected areas eligible for REDD funding would also help to increase synergy between Rio conventions and other international instruments (Kapos et al. 2007), by forming a direct link with existing conventions (e.g., the CBD *Program of Work on Protected Areas*). This will help nations faced with challenges of meeting multiple commitments regarding environmental protection and climate change.

SOME EXAMPLES OF UTILIZING CARBON STORAGE WITHIN PROTECTED AREAS

Potential gains in terms of climate change will vary depending on the type of forest, its age, and associated soils and vegetation. Forests that would be particularly valuable include those with the highest levels of biomass, such as the peat forests of Southeast Asia where carbon in living trees is dwarfed by carbon stored below ground (Swallow et al. 2007) and other forests of the tropics. Work commissioned by the government of Tanzania with funding from the United Nations Development Program Global Environment Facility (UNDP-GEF) has shown that the Eastern Arc mountains constitute an important carbon reservoir. The study (Burgess, personal communication, 2009) calculated that 149.3 million t C is stored in the mountains; 60% of this carbon is in existing forest reserves. Deforestation has resulted in the loss of around 33.5 million t C in the past 20 years—primarily in unprotected forests and woodlands. The study further calculated that disturbed forest stores around 84 t C per hectare (ha), whereas undisturbed forest stores between 100 and 400 t per ha (mean 306 t per ha). Similarly, around 6 million hectares of new protected areas are

TABLE 11.2
Comparison of Elements in the WWF Meta-Standard Framework for Carbon Projects with Likely Conditions in Protected Areas

Issue	Details	Protected Area Implications
Carbon accounting	Additionality	REDD funding should only usually be applicable to new protected areas in areas where forests are at risk or to protected areas where independent assessment shows clearly that vegetation is being lost or degraded and where additional resources could reduce this.
	Leakage	Analysis will be needed to ensure that establishment of a protected area does not simply move forest loss elsewhere, i.e., that any loss of resources to local communities is adequately compensated, e.g., by establishment of timber plantations or other renewable energy sources.
	Permanence	Protected areas aim to protect native vegetation in perpetuity. This could be complicated if vegetation removal is part of management, e.g., if fire control uses prescribed burns to reduce fuel. This will only apply to some places in some countries (and would be applicable in forest outside protected areas). Approaches exist for accounting for such losses.
Social and environmental impacts	Stakeholder consultation	Protected areas are increasingly required to have strong stakeholder processes; for example, this is a requirement for new protected areas established under the CBD *Program of Work on Protected Areas*. It is reflected in a growing number of self-declared protected areas by indigenous peoples' communities.
	Sustainable development	Protected areas increasingly adhere to rigorous social and environmental safeguards to ensure that they do not undermine livelihoods. Application of a range of management approaches and governance types can help; for example, IUCN Category VI extractive reserves facilitate sustainable collection of valuable products (such as nontimber forest products) while maintaining living trees: an ideal scenario for a REDD project if the forest would otherwise have been under threat.
	Identification of high conservation values	Protected areas are selected specifically for their value to conservation and an increasingly sophisticated set of tools are available to identify suitable sites.
	Assessment of environmental impacts	Similarly, there is now a range of methodologies for assessing the environmental benefits of protected areas in terms of, e.g., water supply, soil stabilization, or protection of communities from climatic extremes.

(*continued*)

TABLE 11.2 (continued)
Comparison of Elements in the WWF Meta-Standard Framework for Carbon Projects with Likely Conditions in Protected Areas

Issue	Details	Protected Area Implications
	Long-term viability	The IUCN definition of a protected area stresses the long-term nature of protection as a key feature that distinguishes protected areas from other forms of sustainable and nature-friendly land use.
Validation and certification	Validation	Methodologies for monitoring and assessing management effectiveness of protected areas have developed rapidly over the past decade. Some of these already address issues relating to carbon (for example, monitoring of forest cover through remote sensing) and it would be possible to integrate carbon accounting into existing assessments, although some development work would be needed.
	Certification	Some protected area certification schemes exist, e.g., the Pan Parks scheme in Europe and green ecotourism schemes; others are being developed. Some protected areas also use existing schemes, such as the Forest Stewardship Council, to certify forests in protected areas. Either approach could be applied to carbon accounting under REDD. There are also a growing number of certification schemes developing, especially for REDD projects.

Note: Some purely technical issues common to all carbon offset projects, such as avoidance of double counting, proper registration procedures, and issuance and tracking, are not discussed.

being created in Madagascar—which is responsible for 4 million t of avoided CO_2 a year. Protected areas are expected to provide triple benefits in the form of carbon storage and capture, provision of a range of ecosystem services, and biodiversity conservation (Hannah et al. 2008). For example, the Mantadia forest corridor restoration project is restoring 3,020 ha of forest linking the Antasibe and Mantidia protected areas (Pollini 2009). Habitat restoration and reforestation together are expected to sequester 111,215 t CO_2 equivalent by 2012 and 1.2 million t CO_2 equivalent over 30 years. Among other benefits, the current protected areas are important in ameliorating floods (Kramer et al. 1997). Options for exploring REDD in the context of the large protected areas program in the Amazon are also being actively pursued (Ricketts et al. 2010).

SUMMARY

The protected areas community has been abuzz with interest in the interface between protected areas and climate change for the past couple of years—a combination of fears about the implications of climate change within protected area conservation strategies and hopes (some perhaps unrealistic) about the potential funds that might flow into protected areas as a result of REDD. The extent to which either the hopes or fears will materialize remains uncertain at the present time.

Due to its complexity and the array of causes, impacts, and responses, climate change requires synergy between many international instruments (McNeely 2008), cooperation between different government departments within countries, and the involvement of different stakeholder groups including a wide range of industries. At present, this is frequently not happening. Governments are focusing on "brown solutions" (emissions reductions, etc.) and not always considering the knock-on effects to the "green" or "blue" solutions (carbon stored in terrestrial vegetation or in the seas and oceans). For example, a narrow focus on emissions reductions has encouraged biofuel production, which, if not properly planned, frequently results in additional carbon being lost from terrestrial systems.

More integrated approaches are urgently required, including recognition of the wider role of protected area systems in addressing both mitigation and adaptation. A report on climate change from the Economics of Ecosystems and Biodiversity study (TEEB 2009) outlines the options:

> Direct conservation, e.g. via protected areas, or sustainable use restrictions, are means of maintaining our ecological infrastructure healthy and productive, delivering ecosystem services. Very high benefit-cost ratios are observed, so long as we include amongst benefits a valuation of the public goods and services of ecosystems, and compute social returns on investment.

Policy decisions made in the next few years will determine the extent to which these benefits are fully recognized.

REFERENCES

Asner, G.P., D.E. Knapp, E.B. Broadbent, et al. 2005. Selective logging in the Amazon. *Science* 310: 480–482.

Aylward, B. 2000. *Economic Analysis of Land-Use Change in a Watershed Context.* Presented at a UNESCO Symposium/Workshop on Forest-Water-People in the Humid Tropics: Kuala Lumpur, Malaysia, July 31–August 4.

Baker, T.R., O.L. Phillips, Y. Malhi, S. Almeida, L. Arroyo, A. Di Fiore, T. Erwin, N. Higuchi, T.J. Killeen, S.G. Laurance1, W.F. Laurance, S.L. Lewis, A. Monteagudo, D.A. Neill, P. Núnez Vargas, N.C. A. Pitman, J.N.M. Silva, and R.V. Martínez. 2004. Increasing biomass in Amazon forest plots. *Philosophical Transactions of the Royal Society B* 359: 353–365.

Balmford, A., A. Bruner, P. Cooper, R. Costanza, S. Farber, R.E. Green, M. Jenkins, P. Jefferies, V. Jessamy, J. Madden, K. Munro, N. Myers, S. Naeem, J. Paavola, M. Rayment, S. Rosendo, J. Roughgarden, K. Trumper, and R.K., Turner. 2002. Economic reasons for conserving wild nature. *Science* 292: 950–953.

Balmford, A., P. Gravestock, N. Hockley, C.J. McClean, and C.M. Roberts. 2004. The worldwide costs of marine protected areas. *PNAS* 101(26): 9694–9697.

Barker, T., I. Bashmakov, A. Alharthi, M. Amann, L. Cifuentes, J. Drexhage, M. Duan, O. Edenhofer, B. Flannery, M. Grubb, M. Hoogwijk, F.I. Ibitoye, C.J. Jepma, W.A. Pizer, and K. Yamaji. 2007. Mitigation from a cross-sectoral perspective. In *Climate Change 2007: Mitigation. Contribution of Working Group III to the Fourth Assessment Report of the Intergovernmental Panel on Climate Change*, Eds. B. Metz, O.R. Davidson, P.R. Bosch, R. Dave, L.A. Meyer. United Kingdom: Cambridge University Press.

Borrini-Feyerabend, G., A. Kothari, and G. Oviedo. 2004. *Indigenous and Local Communities and Protected Areas: Towards equity and enhanced conservation.* IUCN/WCPA Best Practice Series no. 11. Cambridge, UK: IUCN.

Brändli, U.B., and A. Gerold. 2001. Protection against natural hazards. In *Swiss National Forest Inventory: Methods and Models of the Second Assessment,* Eds. P. Brassel and H. Lischke. Birmensdorf: WSL Swiss Federal Research Institute.

Bruner, A.G., R.E. Gullison, and A. Balmford. 2004. Financial costs and shortfalls of managing and expanding protected-area systems in developing countries. *BioScience* 54: 1119–1126.

Bruner, A.G., R.E. Gullison, R.E. Rice, and G.A.B. da Fonseca. 2001. *Science* 291: 125–129.

Campbell, A., V. Kapos, I. Lysenko, J. Scharlemann, B. Dickson, H. Gibbs, M. Hansen, and L. Miles. 2008. *Carbon Emissions from Forest Loss in Protected Area.* Cambridge, UK: UNEP WCMC.

Carey, C., N. Dudley, and S. Stolton. 2000. *Squandering Paradise?* Gland, Switzerland: WWF.

Cattaneo, A. 2008. Stock-flow: Balancing efficiency, effectiveness and country participation. Cape Cod, MA, and Arlington, VA: WHRC, IPAN, and TNC.

CBD (Convention on Biological Diversity). 2009. *Connecting Biodiversity and Climate Change Mitigation and Adaptation*. Report of the second ad hoc technical expert group on biodiversity and climate change, CBD Technical Series No. 41. Montreal: Convention on Biological Diversity.

Campbell et al. 2008. Effectiveness of parks in protecting tropical biodiversity.

CCBA. 2008. Climate, Community & Biodiversity Project Design Standards Second Edition. Arlington, VA: Climate Community and Biodiversity Alliance.

Chmura G.L., S.C. Anisfeld, D.R. Cahoon, and J.C. Lynch. 2003. Global carbon sequestration in tidal, saline wetland soils. *Global Biogeochemical Cycles* 17: 1111, DOI: 10.1029/2002GB001917.

Conant, R.T., K. Paustian, and E.T. Elliott. 2001. Grassland management and conversion into grassland: Effects on soil carbon. *Ecological Applications* 11: 343–355.

Damania, A. 1996. Biodiversity conservation: A review of options complementary to standard *ex situ* methods. *Plant Genetic Resources Newsletter* 107: 1–18.

Dowie, M. 2009. *Conservation Refugees*. Cambridge, MA: The MIT Press.

Dudley, N. 2004. Protected areas and certification. In *Environmental Law Paper 49,* Eds. J. Scanlon and F. Burhenne-Guilmin, 41–56. Gland: IUCN.

———. 2005. Restoration of protected area values. In *Beyond Planting Trees*, Eds. S. Mansourian, N. Dudley, and D. Vallauri, 208–212. New York: Springer.

———. Ed. 2008. *Guidelines for Applying Protected Area Management Categories*. Gland, Switzerland: IUCN.

Dudley, N., and J. Parrish. 2006. *Closing the Gap: Creating Ecologically Representative Protected Area Systems*. CBD Technical Series 24. Montreal: Convention on Biological Diversity.

Dudley, N., A. Belokurov, L. Higgins-Zogib, M. Hockings, S. Stolton, and N. Burgess. 2007. *Tracking Progress in Managing Protected Areas*. Gland, Switzerland: WWF.

Dudley, N., and S. Stolton. Eds. 2003. *Running Pure: The Importance of Forest Protected Areas to Drinking Water*. Gland, Switzerland, and Washington, DC: WWF International and the World Bank.

Dudley, N., S. Stolton, A. Belokurov, L. Krueger, N. Lopoukhine, K. MacKinnon, T. Sandwith, and N. Sekhran. 2009. *Natural Solutions: Protected Areas Helping People Cope with Climate Change.* Gland, Switzerland, Washington, DC, and New York: IUCN-WCPA, TNC, UNDP, WCS, the World Bank, and WWF.

Emerton, L., and L. Pabon-Zamora. 2009. *Valuing Nature: Why Protected Areas Matter for Economic and Human Wellbeing.* Arlington, VA: The Nature Conservancy.

Fargione, J., J. Hill, D. Tilman, S. Polasky, and P. Hawthorne. 2008. Land clearing and the biofuel carbon debt. *Science* 319: 1235–1238.

Galatowitsch S. M. 2009. Carbon offsets as ecological restorations. *Restoration Ecology* 17: 563–570.

Halpern, B.S. 2003. The impact of marine reserves: Do reserves work and does reserve size matter? *Ecological Applications* 13: 1, 117–137.

Hamilton, L.S, J.O. Juvik, and F.N. Scatena. 1994. *Tropical Montane Cloud Forests.* Ecological Studies Series Vol. 110. New York: Springer-Verlag.

Hannah, L., R. Dave, P.P. Lowry II, S. Andelman, M. Andrianarisata, L. Andriamaro, A. Cameron, R. Hijmans, C. Kremen, J. MacKinnon, H.H. Randrianasolo, S. Andriambololonera, A. Razafimpahanana, H. Randriamahazo, J. Randrianarisoa, P. Razafinjatovo, C. Raxworthy, G. Schatz, M. Tadross, and L. Wilme. 2008. Opinion piece: Climate change adaptation for conservation in Madagascar. *Biodiversity Letters* 4: 590–594.

Hannah, L., G.F. Midgley, S. Andelman, M Araújo, G. Hughes, E. Martinez-Meyer, R. Pearson, and P. Williams. 2007. Protected area needs in a changing climate. *Frontiers of Ecology and the Environment* 5: 131–138.

Hannah, L. G.F. Midgley, and D. Millar. 2002. Climate-change integrated conservation strategies. *Global Ecology and Biogeography* 11: 485–495.

Hockings, M., S. Stolton, F. Leverington, N. Dudley and J. Courrau. 2006. *Evaluating Effectiveness: A Framework for Assessing Management Effectiveness of Protected Areas*, 2nd edition. Gland, Switzerland: IUCN.

ISDR (International Strategy for Disaster Reduction). 2004. *Living with Risk: A Global Review of Disaster Reduction Initiatives.* Geneva: UN/ISDR.

Jeng, H., and Y.J. Hong. 2005. Assessment of a natural wetland for use in wastewater remediation. *Environmental Monitoring and Assessment* 111: 113–131.

Joppa, L.N., S.R. Loarie, and S.L. Pimm. 2008. *Proceedings of the National Academy of Sciences* 105: 6673–6678.

Kapos, V., P. Herkenrath, and L. Miles. 2007. *Reducing Emissions from Deforestation: A Key Opportunity for Attaining Multiple Benefits.* Cambridge, UK: UNEP World Conservation Monitoring Centre.

Kramer, R.A., D.D. Richter, S. Pattanayak, and N.P. Sharma. 1997. Ecological and economic analysis of watershed protection in eastern Madagascar. *Journal of Environmental Management* 49: 277–295.

Kuczera G. 1987. Prediction of water yield reductions following a bushfire in ash-mixed species eucalypt forest. *Journal of Hydrology* 94:215–236.

Kulshreshtha, S.N., S. Lac, M. Johnston, and C. Kinar. 2000. *Carbon Sequestration In Protected Areas Of Canada: An Economic Valuation.* Economic Framework Project, Report 549. Warsaw, Canada: Canadian Parks Council.

Laffoley, D., and G. Grimsditch. Eds. 2009. *The Management of Natural Coastal Carbon Sinks.* Gland, Switzerland: IUCN.

Lal, R. 2004a. Soil carbon sequestration impacts on global climate change and food security. *Science* 304: 1623–1627.

———. 2004b. Soil sequestration to mitigate climate change. *Geoderma* 123: 1–22.

Laporte, N., F. Merry, and A. Baccini. 2009. *Reducing CO_2 Emissions from Deforestation and Degradation in DRC.* Cape Cod, MA: Woods Hole Research Center.

Leverington, F., K. Lemos Costa, J. Courdu, H. Pavese, C. Nolte, M. Marr, L. Coad, N. Burgess, B. Bomhard, and M. Hockings. 2010. *Management Effectiveness Evaluation in Protected Areas.* A global study Second Edition. Brisbane: The University of Queensland.

Maginnis, S., W. Jackson, and N. Dudley. 2004. Conservation landscapes. Whose landscapes? Whose tradeoffs? In *Getting Biodiversity Projects to Work: Towards More Effective Conservation and Development*, Eds. T.O. McShane and M.P. Wells, 321–339. New York: Columbia University Press.

Mansourian, S., and N. Dudley. 2008. *Public Funds to Protected Areas.* Gland, Switzerland: WWF International.

McGray, H., A. Hammill, and R. Bradley. 2007. *Weathering the Storm: Options for Framing Adaptation and Development.* Washington, DC: World Resources Institute.

McKee, K.L., and P.L. Faulkner. 2000. Restoration of biogeochemical function in mangrove forests. *Restoration Ecology* 8: 247–259.

McNeely, J.A. 2008. Applying the diversity of international conventions to address the challenges of climate change. *Michigan State Journal of International Law* 17: 123–137.

Mehta, A., and J. Kill. 2007. *Seeing Red? "Avoided deforestation" and the Rights of Indigenous Peoples and Local Communities.* Brussels and Moreton-in-the-Marsh, UK: Fern.

Mulongoy, K.J., S.B. Gidda, L. Janishevski, and A. Cung. 2008. Current funding shortfalls and innovative funding mechanisms to implement the Programme of Work on Protected Areas. *Parks* 17: 31–36.

Nagendra, H. 2008. Do parks work? Impact of protected areas on land cover clearing. *Ambio* 37: 330–337.

Nelson, A., and K. Chomitz. 2009. *Protected Area Effectiveness in Reducing Tropical Deforestation.* Washington, DC: The World Bank.

Oestreicher, J.S., K. Benessaiah, M.C. Ruiz-Jaen, S. Sloan, K. Turner, J. Pelletier, B. Guay, K.E. Clark, D.G. Roche, M. Meiners, and C. Potvin. 2009. Avoiding deforestation in Panamanian protected areas: An analysis of protection effectiveness and implications for reducing emissions from deforestation and forest degradation. *Global Environmental Change* 19: 279–291.

Pattanayak, S.K., C.G. Corey, Y.F. Lau, and R.A. Kramer. 2003. Forest malaria: A microeconomic study of forest protection and child malaria in Flores, Indonesia, Duke University, USA. http://www.env.duke.edu/solutions/documents/forest-malaria.pdf (accessed July 1, 2009).

Pena, N. 2008. *Including Peatlands in Post-2012 Climate Agreements: Options and Rationales.* Wageningen, Netherlands: Wetlands International.

Peskett, L., C. Luttrell, and D. Brown. 2006. *Making Voluntary Carbon Markets Work Better for the Poor: The Case of Forestry Offsets.* ODI Forestry Briefing number 11. London: Overseas Development Institute.

Pistorius, T., C. Schmitt, and G. Winkel. 2008. *A Global Network of Forest Protected Areas under the CBD.* Germany: University of Freiburg.

Pollini, J. 2009. Carbon sequestration for linking conservation and rural development in Madagascar: The case of the Vohidrazana-Mantadia Corridor Restoration and Conservation Carbon Project. *Journal of Sustainable Forestry* 28: 322–342.

Quintela, C.E., L. Thomas, and S. Robin. 2004. *Proceedings of the Workshop Stream "Building a Secure Financial Future: Finance & Resources": 5th IUCN World Parks Congress.* Gland, Switzerland, and Cambridge, UK: IUCN.

Rakovich, V.A., and N.N. Bambalov. (in press) *Methodology for Measuring the Release and Sequestration of Carbon from Degraded Peatlands* (in Russian, Oct. 2009—being prepared for print).

Ricketts, T., B. Soares-Filho, G.A.B. da Fonseca, D. Nepstad, A. Pfaff, A. Petsonk, A. Anderson, D. Boucher, A. Cattane, M. Conte, K. Creighton, L. Linden, C. Maretti, P. Moutinho, R. Ullman, and R. Victurine. 2010. Indigenous lands, protected areas and carbon sequestration. *PLoS Biology* 8: E1000331.

Rietbergen-McCracken, J. Ed. 2008. *Green Carbon Guidebook*. Washington, DC: WWF US.

Roberts, C.M., and J.P. Hawkins. 2000. *Fully-Protected Marine Reserves: A Guide*. Washington, DC, and York: UK: WWF and University of York.

Sabine, C.L., M. Heimann, P. Artaxo, D.C.E. Bakker, C.T.A. Chen, C.B. Field, N. Gruber, C. Le Queré, R.G. Prinn, J.E. Richey, P. Romero Lankao, J.A. Sathaye, and R. Valentini. 2004. Current status and past trends of the global carbon cycle. In *The Global Carbon Cycle: Integrating Humans, Climate and the Natural World*, eds. C.B. Field and M.R. Raupach. 17–44. Washington, DC: Island Press.

Saunders, J., and R. Nussbaum. 2008. *Forest Governance and Reduced Emissions from Deforestation and Degradation (REDD)*. Briefing Paper EEDP LOG BP 08/01. London: Royal Institute of International Affairs.

SCBD (Secretariat of the Convention on Biological Diversity). 2004. *Decisions Adopted by the Conference of the Parties to the Convention on Biological Diversity at Its Seventh Meeting*. UNEP/CBD/COP/7/21. Montreal: SCBD. http://biodiv.org/decisions/?dec=VII/28

Schuman, G.E., H.H. Janzen, and J.E. Herrick. 2002. Soil carbon dynamics and potential carbon sequestration by rangelands. *Environmental Pollution* 116: 391–396.

Smith, J., and S.J. Scherr. 2002. *Forest Carbon and Local Livelihoods: Assessment of Opportunities and Policy Recommendations*. CIFOR Occasional Paper number 37. Bogor, Indonesia: Center for International Forestry Research.

Steffen, W., A. Burbidge, L. Hughes, R. Kitching, D. Lindenmayer, W. Musgrave, M. Stafford Smith, and P. Werner. 2009. *Australia's Biodiversity and Climate Change*. Canberra: Department of Climate Change.

Stern, N. 2006. *Stern Review on the Economics of Climate Change*. London: HM Treasury.

Stolton, S., T. Boucher, N. Dudley, J. Hoekstra, N. Maxted, and S. Kell. 2008a. Ecoregions with crop wild relatives are less well protected. *Biodiversity* 9: 52–55.

Stolton, S., and N. Dudley. 2010. *Vital Sites: The Contribution of Protected Areas to Human Health*. Gland, Switzerland, and Bristol, UK: WWF and Equilibrium Research.

Stolton, S., N. Dudley, and J. Randall. 2008b. *Natural Security: Protected Areas and Hazard Mitigation*. Gland, Switzerland: WWF.

Swallow, B., M. van Noordwijk, S. Dewi, D. Murdiyarso, D. White, J. Gockowski, G. Hyman, S. Budidarsono, V. Robiglio, V. Meadu, A. Ekadinata, F. Agus, K. Hairiah, P.N. Mbile, D.J. Sonwa, and S. Weise. 2007. *Opportunities for Avoided Deforestation with Sustainable Benefits: An Interim Report by the ASB Partnership for the Tropical Forest Margins*. Nairobi, Kenya: ASB Partnership for the Tropical Forest Margins.

TEEB. 2009. *TEEB Climate Issues Update*. Bonn: The Economics of Ecosystems and Biodiversity.

UNEP-WCMC. 2008. *State of the World's Protected Areas: An Annual Review of Global Conservation Progress*. Cambridge, UK: UNEP World Conservation Monitoring Centre.

Vittor, A.Y., R.H. Gilman, J. Tielsch, G. Glass, T. Shields, W.S. Lozano, V. Pinedo-Cancino, and J.A. Patz. 2006. The effect of deforestation on the human-biting rate of Anopheles darling, the primary vector of falciparum malaria in the Peruvian Amazon. *American Journal of Tropical Medicine and Hygiene* 74: 3–11.

Woods Hole Research Center. 2009. *Making REDD a Success: Readiness and Beyond*. Cape Code, MA: WHRC.

Worboys, G.L., W.L. Francis, and M. Lockwood. Eds. 2010. *Connectivity Conservation Management: A Global Guide*. London: Earthscan.

World Bank. 2009. *Convenient Solutions to an Inconvenient Truth: Ecosystem-Based Approaches to Climate Change*. Washington, DC: World Bank.

12 Resource Managers Rise to the Challenge of Climate Change

Melinda G. Knutson and Patricia J. Heglund

CONTENTS

INTRODUCTION

Climate change presents unprecedented challenges and is likely to stimulate profound changes in the practice of natural resource management. Many management practices that were routine in the past will need to be reevaluated as ecosystems adjust to changes in temperature, precipitation, storm events, and sea level rise (Palmer et al. 2009; Sutherland et al. 2004). Uncertainties will increase regarding the appropriate conservation targets, the outcomes of alternative management practices, and how to evaluate trade–offs among competing conservation objectives. New tools will be needed to cope with this uncertainty. Resource managers will need to work closely with species and ecosystem scientists, as well as modelers, decision analysts, statisticians, information technologists, and other technical experts to efficiently monitor and improve the outcomes of their management practices.

Challenges are also opportunities. In the past, protected lands were often designated through opportunity rather than need; lands that were not in demand for agriculture or urban development became protected (Scott et al. 2001). Through better cooperative planning, resource managers can seek opportunities across the landscape to meet the habitat needs of priority species. In this chapter, we highlight some of the challenges that climate change presents to the resource manager, suggest new conservation opportunities, and describe how a structured approach to management decision making will be useful. We use examples from the National Wildlife Refuge System (Refuge System), a US land–management agency, to illustrate our points.

The US Fish and Wildlife Service (FWS) manages the Refuge System, covering more than 37 million hectares and comprising 635 units and 37 Wetland Management Districts in the United States and its territories (Griffith et al. 2009; www.fws.gov/refuges/). The mission of the Refuge System is to "administer a national network of lands and waters for the conservation, management, and, where appropriate, restoration of the fish, wildlife, and plant resources and their habitats within the United States for the benefit of present and future generations of Americans" (US Congress 1997, Sec. 4). The Refuge System is home to more than 700 species of birds, 220 mammal species, 250 reptile and amphibian species, and 200 fish species. About 23% of the more than 1200 federally listed threatened or endangered species in the United States are found on Refuge System lands. In addition to its primary task of conservation, the Refuge System also supports six wildlife–dependent recreational uses enjoyed by nearly 40 million people annually: hunting, fishing, wildlife observation, photography, environmental education, and interpretation.

THE THREAT OF CLIMATE CHANGE

Climate change poses unprecedented challenges to the conservation community, primarily because the predicted and the observed rates of change will exceed anything modern society has experienced (IPCC 2007). Climate change is predicted to magnify the threat and greatly increase species extinction rates across the globe (Folke et al. 2004). Scientists estimate that globally as many as 20–30% of species will be at risk of extinction by 2100 as global mean temperatures exceed 2–3°C above pre-industrial levels (Thomas et al. 2004; Fischlin et al. 2007).

What will the proximate drivers of these extinctions be and how will resource managers address them? The following are some of the primary threats to the integrity of North American ecosystems posed by climate change, along with some potential management responses (Griffith et al. 2009; US Department of the Interior 2009b; Karl et al. 2009; Scott et al. 2009):

- Changes in water quality and quantity will occur. Climate change has the potential to decrease supply and increase demand for water, creating new water scarcity problems and exacerbating existing ones. Increasing water temperatures will impact riparian habitats and the fish and wildlife dependent on them as well as effecting ecosystem–wide changes in flora. This threat is a major challenge for the Refuge System because much of its land base is composed of wetland habitats that may dry out. Land acquisition

and conservation partnerships may need to expand from a focus on existing protected areas to include new areas with enough water to sustain wetland habitats, upon which many species of conservation concern depend.
- Large–scale shifts in species ranges and timing of the seasons and animal migration will occur (Folke et al. 2004). Phenological mismatches may occur between migrating or hibernating species and the food resources available upon their arrival or emergence. Lands that were initially protected for a specific species may no longer support that species in the future. Resource managers may need to consider strategies such as creating migration corridors, translocating species, or food supplementation to address these threats (Scott et al. 2009).
- The numbers of threatened and endangered species will increase as a result of climate change. In addition, the extinction risk will increase for those species already designated as threatened and endangered. Species with limited dispersal abilities and specialized habitat requirements are likely to be most at risk (Thomas et al. 2004). Addressing the needs of threatened and endangered species will tax the resources of agency staff charged with protecting and restoring these populations. Difficult decisions will arise regarding which species should receive scarce conservation resources (Scott et al. 2005; McCarthy et al. 2008).
- Dry conditions and changes in forest health will change the duration, frequency, intensity, and extent of wildland fires. Deserts and dry lands will likely become hotter and drier, exacerbating threats from drought, invasive plants, fire, and erosion. Some forested areas may transition to savannah or grasslands. Fire risks across the landscape will need to be reevaluated and resource managers will need to invest more resources in addressing fire risks at the wildland–urban interface. Resource managers will need to advocate for the needs of fish and wildlife in the face of growing human demands for water.
- Coastal and marine, as well as arctic and subarctic ecosystems are threatened by multiple stressors; climate change will exacerbate these stresses. Management agencies can adjust their land acquisition policies to acquire higher elevation coastal land and take other measures to mitigate some of these stressors. However, it is doubtful whether management actions can be implemented at an intensity or spatial scale sufficient to reverse many climate–related ecological effects in these vast ecosystems (Munday et al. 2008; Greene et al. 2008).

A STRUCTURED APPROACH TO DECISION MAKING

How should the practice of resource management evolve to meet the challenge of climate change, as well as a host of other threats to conservation lands, including the widespread conversion of diverse native ecosystems to agriculture, forest monocultures, and urban expansion? The United Nations Convention on Biological Diversity (http://www.biodiv.org) advocates an "ecosystem approach," which is a strategy for the integrated management of land, water, and living resources that promotes conservation and sustainable use in an equitable way. How can resource managers promote conservation and sustainable use under the threat of climate change? The diverse threats associated with climate change, not to mention threats due to widespread land–use changes and increases in the "human footprint" on the planet, seem overwhelming.

A structured approach that begins with clarifying what is desired from the land, both short term and long term, is useful and sets the stage for selecting the appropriate conservation practices. Resource management is primarily an exercise in making decisions that best achieve the manager's objectives, i.e., making "good" decisions. Management decisions fall along a wide spectrum, from doing nothing to taking major actions, such as creating or removing a dam on a river—actions that may be costly, risky, and irreversible from a practical standpoint. Doing nothing may also involve cost (lost opportunities) and risk. Decision making is about deciding what is valued (i.e., forming objectives) (Keeney 1992).

Fortunately, decision making is a process that has been studied extensively. There is a strong body of literature that describes the theory and practice of decision making; this information is potentially very useful to resource managers who are seeking tools for dealing with rapid environmental change (Keeney 1992; Walters and Holling 1990). The terms "structured decision making (SDM)" and "decision analysis" are applied to a process of breaking down a problem into component parts for purposes of improving decision making (Table 12.1) (Gregory and Keeney 2002; Martin et al. 2009). When decision making is iterative and focused on learning for the purpose of improving future decisions, the process is called adaptive management (AM) (Williams et al. 2007).

Adaptive management is a structured process that can help managers address the uncertainty associated with climate change by learning from past strategies and adjusting future management accordingly (Lyons et al. 2008; McCarthy and Possingham 2007). Biome shifts predicted under climate change may lead to

TABLE 12.1
Elements of Structured Decision Making (Making Smarter Decisions)

Element	Description
Problem	Define your decision problem to solve the right problem.
Objectives	Clarify what you're really trying to achieve with your decision.
Alternatives	Create better alternatives from which to choose.
Consequences	Describe how well each alternative meets your objectives.
Trade-offs	Equate the value of different levels of achievement on different objectives.
Uncertainty	Identify and quantify the major uncertainties affecting your decision.
Risk tolerance	Account for your willingness to take risks.
Linked decisions	Plan ahead by coordinating current and future decisions.

Source: Adapted from Gregory, R. S., and R. L. Keeney. 2002. Making Smarter Environmental Management Decisions *Journal of American Water Resources Association* 38:1603. With permission.

difficulties in sustaining both species and ecosystems, because most existing natural and conservation lands were acquired with specific conservation targets (rare or otherwise valued elements of biodiversity, ecosystems, or populations of species) in mind (Griffith et al. 2009; Scott et al. 2001; Noss 2004). Under a climate–change scenario, managers are concerned with both conservation targets and conservation opportunities. It is unlikely that biome shifts that are under way can be halted directly, so adaptive strategies must be sought. Resource managers will need to address the following questions: Where will species occur? How can we preserve self–sustaining populations? Where can we focus our conservation actions most cost–effectively? They will also need to make decisions about what species to save, and when to control invasive species and when to accept them as new members of a changing assemblage (Millar et al. 2007). Resource managers also face institutional obstacles; agency leaders will need to consider organizational changes (Scott et al. 2009; Meretsky et al. 2006; Griffith et al. 2009). For example, resource–management agencies will need to hire staff with specialized skills such as modeling, decision analysis, and data management.

Structured decision making is useful in a climate–change context because no assumptions are made about the scale or scope of the problem, what the management objectives are, or what the range of management options are. By reviewing basic questions such as "what are we trying to accomplish?," SDM allows the managers or groups of stakeholders to revisit the fundamental aspects of the problem before making a decision or implementing a solution. Structured decision making addresses both policy (values) and science (analysis) (Gregory and Wellman 2001). Climate change demands that we think about our values and engage in creative thinking about management options; SDM is a framework that integrates both activities (Gregory, Ohlson, et al. 2006; Gregory and Keeney 2002).

How does AM differ from scenario planning? Scenario planning has been used by groups of stakeholders to envision alternative futures in the face of large and irreducible uncertainty (Peterson et al. 2003). Scenarios are useful for capturing a wide range of variables and integrating them into alternative "stories" that capture the main uncertainties about the future (Brooke 2008). These alternative futures are used to evaluate public policies such as zoning laws, energy consumption, conservation incentives, and other broad public concerns. Scenario planning and AM are related; both tools address uncertainty about the future. Scenario planning is most useful when uncertainty is high and system manipulations are difficult or impossible (Peterson et al. 2003); AM is useful for evaluating the relative effectiveness of alternative management strategies. For example, scenario planning was used to evaluate climate models for the midcontinental area of North America (Minnesota) and to recommend specific management and restoration strategies, based on those models (Galatowitsch et al. 2009). Adaptive management is used to reduce uncertainty about the outcomes of management strategies when they are applied in specific locations. We focus this chapter on the problem of resource management and the use of AM as a tool for coping with and reducing the management uncertainties associated with climate change.

DEFINING THE PROBLEM, SETTING OBJECTIVES

The first steps in a structured approach to management are defining the problem and setting objectives. These steps are strongly linked. Defining the problem involves defining the geographic, temporal, and administrative scope of the problem. Are we concerned with management at the local, regional, or national scale? What are we trying to accomplish by managing land? Are we most concerned with managing species, habitats, ecosystems, or something else (e.g., recreational opportunities)? In this context, setting objectives involves defining what is valued by the landowner (Keeney 1992). In the case of the Refuge System and other federal land, this means defining what the American people value about the Refuge System and the specific locale that is under consideration. Objectives are an explicit statement of values that are informed by science; therefore, science alone cannot define resource management objectives.

Most land–management agencies (governmental and nongovernmental) have policies that provide agency–wide guidance about what the landowner values and what the priorities are. These conservation priorities are a product of the history of the agency and its past conservation challenges. The challenge for managers today is to take this general guidance and translate it into specific objectives and priorities for the land unit(s) they manage (US Fish and Wildlife Service 2009). There are often differences of opinion within the agency or among stakeholders about what the highest priority biological objectives for a given unit should be. Given all of these issues, setting management objectives is generally the most difficult step in the management process (Johnson 2000). Climate change will make setting objectives even harder. Differences of opinion about biological objectives will escalate, given the uncertainty surrounding climate forecasts for specific geographic locations. The numbers of species at risk of population decline or extinction will increase, making it even more difficult to set priorities. The number and extent of geographic locations where conservation planning and action are indicated will expand, again challenging a conservation community struggling to set priorities and allocate scarce resources.

The Refuge System derives its agency–wide objectives from establishing (organic) legislation and a set of associated policies (Keiter 2004; Fischman 2003, 2007). The National Wildlife Refuge System Improvement Act (RIA) directs FWS to provide for the conservation of fish, wildlife, and plants, and their habitats throughout the Refuge System (US Congress 1997). The RIA defines the terms "conserving," "conservation," "manage," "managing," and "management" as meaning to sustain and, where appropriate, restore and enhance healthy populations of fish, wildlife, and plants. The terms "fish," "wildlife," and "fish and wildlife" were defined as any wild member of the animal kingdom. The RIA further requires FWS to "ensure that the biological integrity, diversity, and environmental health of the System are maintained for the benefit of present and future generations of Americans" (US Fish and Wildlife Service 2001, Sec. 5). According to Meretsky et al. (2006, p. 136), the provision related to biological integrity, diversity, and environmental health is "one of the most emphatic ecosystem conservation directives ever written by Congress." Biological management objectives for units within the Refuge System are derived from this legislation and the associated policies (US Fish and Wildlife Service 2009).

The 1997 Refuge Improvement Act mandates the development of a Comprehensive Conservation Plan (CCP) for each refuge; this plan addresses a 15–year time period (Schroeder 2006, 2008; US Fish and Wildlife Service 2000). Goals, objectives, and strategies for management of each refuge are defined in the CCP; public input into the process is required. Each CCP needs to define specifically the goals, objectives, and strategies planned for that land unit, based upon these broad agency policies. However, there is considerable variation in the specificity of these plans (Schroeder 2006). Only a few of the highest–priority strategies associated with the objectives can be implemented because resources are limited.

What can be done to help managers clarify and refine their objectives? One technique for clarifying management objectives is to prioritize objectives based on the decisions that must be made by a manager this week, this month, or this year regarding the management unit(s) in question (Lyons et al. 2008). A focus on management actions or decisions narrows the priorities to those objectives that the manager can directly address through management. Such decisions encompass a wide range of possibilities, including implementing some action on the ground, acquiring land, repairing roads and physical structures, working with local or state agencies on cooperative conservation or preservation, setting hunting regulations or bag limits, and broadening or limiting public access to certain management units. Questions such as "why is that important?" will eventually uncover the manager's fundamental objective, i.e., the primary reason the manager acts/decides in the first place (Gregory and Keeney 2002; Lyons et al. 2008).

In the Refuge System, fundamental objectives are generally focused on sustaining populations of high–priority species (e.g., migratory birds, threatened or endangered species, etc.) (Johnson 2000). However, some attributes of high–priority ecosystems like forests, salt marshes, and tall grass prairies (e.g., ecological integrity, ecosystem function, patterns of disturbance) can also be fundamental management objectives. Once the fundamental objective is identified, it can be linked to the management decisions or actions. Why are you burning that grassland every three years? Are you trying to restore the native plant community, attract specific species of nesting birds, or reduce the threat of wild fires? This seems elementary, but in our experience, defining specific and measurable objectives is a difficult process (Fischman 2007; Schroeder 2008). No real conservation progress can be made without answering these questions.

IS HISTORY A BENCHMARK FOR THE FUTURE?

Climate change will force ecologists and resource managers to reconsider their conservation targets. In the past, managers have looked to history to give them a benchmark for restoration objectives (Brook and Bowman 2006; Schroeder 2008). Restoration targets are usually plant communities that were present historically at that location. In the United States, this usually means communities that were there prior to 1800. Explorer journals, land–surveyor records, and paleoecology have allowed geographers to reconstruct historic vegetation for many locations (Rentch and Hicks 2005; Nielsen and Odgaard 2005). If climate change prompts each species to follow its own individualistic adaptation trajectory, as predicted, we will see plant communities developing that have no analog today (Williams and Jackson 2007). For example, we know from the paleoecological record that contemporary tree species were once found in assemblages that are different from any that we see today. If we consider the huge number of invasive and non–native species that have been translocated globally, we already have many no–analog communities.

Restoration benchmarks like "historical range of variation" become obsolete if past climate is not like the present or the future (Mitchell and Duncan 2009; Scott et al. 2009; Griffith et al. 2009). This is a problem, in that our knowledge of past and present native plant communities is the foundation of restoration ecology. If no–analog communities start to replace native species assemblages across the landscape, it will be more difficult to define what the appropriate management objectives at a management unit should be. What constitutes biological integrity in a no–analog plant community? Even ecological processes such as stream flow, fire frequency, and nutrient cycling will change as temperature, precipitation, and the biota change.

In the Refuge System, climate change has the potential to require adjustments in CCP objectives. The Refuge System Policy on Biological Integrity, Diversity, and Environmental Health directs stations to "assess historic conditions and compare them to current conditions" (US Fish and Wildlife Service 2001, Sec. 3.9 C). This provides a benchmark of comparison for the relative intactness of ecosystems' current functions and processes. With contemporary climate change, this historic benchmark may not be an appropriate conservation target.

Ecosystem services are much discussed in the ecological literature, but they are rarely explicitly represented in agency or government resource management policies (Daily et al. 2009). However, it is likely that the public will demand new policies that require resource managers to account for ecosystem services derived from public lands. They may need to adjust their management objectives and work to optimize ecosystem services as well as more traditional conservation values. For example, conservation lands filter and deliver clean water; also, carbon is stored in old–growth and long–rotation forests, perennial grassland soils, and wetlands. Landscapes with a high interspersion of native plants support native bee pollinators as well as honey bees and thus provide "free" ecosystem services to adjacent fruit and nut farmers (Winfree 2008). Carbon credits provide a currency for calculating the "carbon" costs and benefits involved in alternative land–uses; these credits may provide a source of revenue to both private and public resource managers in the future (Nelson et al. 2008).

Given that the area of land dominated by native plants and animals is shrinking worldwide due to competing human land uses, public–resource managers may be called upon to work with private landowners to modify private land–management objectives to include conservation of species, ecosystems, and ecosystem services (Polasky et al. 2008; Swift et al. 2004; Herzon and Helenius 2008). This will include both lands managed for agriculture as well as urban and suburban landscapes. There are many potential ways to support native species assemblages as well as important ecological processes in both agricultural and urban landscapes, but it is likely that private landowners will need financial incentives to fully realize this potential

(Nelson et al. 2008). Private landowners will need to consider multiple, perhaps competing management objectives, given that most private landowners need to reap some economic benefit from owning land (Firbank 2005; Juutinen 2008).

Setting management objectives is the first and most difficult step in conservation delivery. Climate change will add another layer of complexity to an already difficult endeavor. The conservation targets with which we have scientific and managerial experience are changing; this will have ripple effects on all aspects of resource management. For example, many resources have been devoted to planning for the management of the existing protected areas, regardless of ownership. In the United States, federal agencies must follow the National Environmental Policy Act (NEPA) and consider the environmental impacts of their actions and reasonable alternatives to those actions—a fairly lengthy process. If climate change brings rapid ecological change, planning documents will need to be revisited more frequently than most agencies are prepared to do. Managers will also be challenged by the need to engage in multiple management strategies for which the conservation community has had little experience or success (creation of landscape–scale movement and dispersal corridors, assisted dispersal of native species to avoid extinctions, and species triage) (Millar et al. 2007).

The challenge is greater than it has ever been to envision a desired (valued) future condition that is feasible, ecologically sustainable, and consistent with broad agency directives and public values. Those who manage land all or in part for conservation purposes are facing a major challenge when the very objectives (conservation targets) for which we manage are in flux. We need tools for working with multistakeholder groups to redefine management objectives and for managing in a flexible way that accommodates changes in objectives over time (Brooke 2008).

MODELING THE FUTURE

Modeling will help managers envision new land–management objectives that are congruent with a changing climate. Models integrate data from multiple sources to forecast future ecosystem condition and how that may translate into new ecosystem and community assemblages at specific locations (Galatowitsch et al. 2009; Nemani et al. 2009). These forecasts will be used to assess the vulnerability of species and ecosystems for the purpose of setting management priorities. As temperature increases, species are predicted to respond by moving toward the poles or upslope in mountainous terrain (Weiss et al. 1993; Zuckerberg et al. 2009). As precipitation patterns change, some species will find their only available habitat too dry during critical life stages to support their populations. Habitat conditions are predicted to change too rapidly for many species to move, even if suitable habitat is available somewhere else. For many rare species, we do not know precisely what constitutes suitable habitat, so it will be difficult to predict whether any will remain (Scott et al. 2002). As populations adjust to the changing climate, some species will benefit and other populations will decline; conservation priorities will need to be updated at a rate that matches these changes. The threat of climate change may elevate the conservation priority of the most–vulnerable species and ecosystems and lower the priority of the least vulnerable. The most vulnerable may demand more attention and cooperation among a wider set of agencies than in the past.

Models of predicted changes in temperature and precipitation will need to be translated into predicted changes in species assemblages at the management–unit level in order to guide the setting of objectives (Magness et al. 2008); however, there is great uncertainty associated with these projections. Such models will help managers think about possible future conditions and consider what the appropriate management responses might be. Managers will need some idea of how climate change may affect their conservation targets, and they need to be prepared to support species' movements across the landscape as species track favorable habitats. For example, the Refuge System has 161 coastal refuges with approximately 0.4 million ha in the contiguous 48 states (Scott et al. 2009). Managers need to identify where the lowest–elevation saltmarshes are located and begin immediately to plan for their migration to higher ground as sea level rises.

There are many kinds of models. Finding the most efficient, useful modeling approach for any particular conservation problem combines art and science, in much the same way that the practice of medicine provides diagnoses and treatments for a wide array of human maladies. Most importantly, modeling in a conservation context requires modelers and managers to work together, each contributing their own expertise to the modeling problem. These modeler–manager partnerships are uncommon today but will be required in the future as climate change puts stronger pressure on already fragmented and degraded ecosystems. For example, the need for triage to set conservation priorities for limited resources will demand a modeling approach. Modeling will play an important role in helping clarify where scarce conservation resources should be spent outside of the existing conservation estate. This involves a host of challenges, including collaborative work with federal, state, and local governments, nongovernmental organizations, and private landowners. Any collaborative project that involves many people is a sociological challenge that requires special skills to achieve efficiency. The conservation community will need to improve communication to accommodate these new needs (Knight et al. 2008).

The FWS recognizes that meeting its conservation mission in the face of climate change will require a coordinated effort involving all stakeholders with an interest in our natural heritage of fish and wildlife. The FWS has issued "Rising to the Challenge: Strategic Plan for Responding to Accelerating Climate Change" and a 5–Year Action Plan to implement the Strategic Plan (US Fish and Wildlife Service 2009). The plans call for organizing scientific units called Landscape Conservation Cooperatives (LCCs) to provide scientific leadership at broad geographic scales. Staff of the LCCs will work with other agency and conservation partners to plan, design, and evaluate landscape–scale conservation, in the face of climate change and other stressors. The goals of the FWS Strategic and Action climate change plans were reinforced under Secretarial Order 3289, issued on September 14, 2009, tasking the newly proposed DOI Regional Climate Change Response Centers (USGS) and the LCCs with developing adaptation strategies to respond to climate change (US Department of the Interior 2009a).

CREATIVE ALTERNATIVES

Creativity is essential when considering alternative management strategies designed to achieve management objectives. Climate change will stimulate a search for creative alternatives, if only because past methods may fail to meet objectives. A structured approach to management decision making provides a framework for evaluating several alternative management strategies simultaneously. In this way, the manager learns as quickly as possible what works and what does not, from the process of management itself. This field–based learning is not simply trial and error, because alternative strategies are devised a priori and a framework for evaluating these alternative strategies is designed. The monitoring program is designed specifically to reduce uncertainty about the outcomes of the alternative strategies.

A common pitfall in any kind of planning effort is the failure to consider multiple management options (Gregory and Keeney 2002). A recent review summarized some of the strategies available for sustaining wildlife and biodiversity in the face of climate change (Mawdsley et al. 2009) (Table 12.2). Scott et al. (2009) describe a wide range of specific strategies that the Refuge System and other land–management agencies may need to undertake to adapt to climate change, including assisted dispersal for species with limited dispersal abilities, use of plant species in restoration projects that are better adapted to future climate conditions, identifying and conserving climate refugia, and supplying interim food for mis–timed migrants. Other suggested strategies are more familiar to managers, such as prescribed burning, restoration of historical hydrologic or fire regimes, and reducing nonclimate stressors. The conservation community has experience with some of the strategies; others will require new tools and approaches; none are simple.

Millar et al. (2007) identified three general types of adaptation strategies—resistance, resilience, and facilitation—scaled relative to the magnitude of climate change. Management options that focus on helping ecosystems resist the influence of climate change are likely to be most successful if the magnitude of change is small. However, if changes in temperature or precipitation are large, a strategy of resistance to change is likely to fail or be very costly. Instead, facilitation strategies designed to mimic ecological processes such as disturbance, dispersal, and colonization may be more effective. Resource managers will need to evaluate model predictions and select strategies that are compatible with the rate and magnitude of change predicted to occur (Galatowitsch et al. 2009).

CONSIDERING CONSEQUENCES, EMBRACING UNCERTAINTY

Monitoring and Adaptive Management

After setting objectives and selecting alternative management strategies, a structured approach calls for envisioning the consequences (outcomes) of each set of management actions (Gregory, Failing, et al. 2006). If this is a one–time decision, monitoring is planned to evaluate the outcome. If it is an iterative decision, then AM is indicated (Figure 12.2). Monitoring is a required component of AM, but too often monitoring

TABLE 12.2
Climate Change Adaptation Strategies Available to Resource Managers

Type	Strategy
Land and water protection	1. Increase extent of protected areas.
	2. Improve representation and replication within protected-area networks.
	3. Improve management and restoration of existing protected areas to facilitate resilience.
	4. Design new natural areas and restoration sites to maximize resilience.
	5. Protect movement corridors, stepping-stones, and refugia.
	6. Manage and restore ecosystem function rather than focusing on specific components (species or assemblages).
	7. Improve the matrix by increasing landscape permeability to species movement.
Direct species management	8. Focus conservation resources on species that might become extinct.
	9. Translocate species at risk of extinction.
	10. Establish captive populations of species that would otherwise go extinct.
	11. Reduce pressures on species from sources other than climate change.
Monitoring and planning	12. Evaluate and enhance monitoring programs for wildlife and ecosystems.
	13. Incorporate predicted climate-change impacts into species and land-management plans, programs, and activities.
	14. Develop dynamic landscape conservation plans.
	15. Ensure that wildlife and biodiversity needs are considered as part of the broader societal adaptation process.
Law and policy	16. Review and modify existing laws, regulations, and policies regarding wildlife and natural resource management.

Source: Mawdsley, J. R., R. O'Malley, and D. S. Ojima. 2009. A review of climate-change adaptation strategies for wildlife management and biodiversity conservation. *Conservation Biology* 23:1080-1089.

is considered an end in itself (Nichols and Williams 2006). In a resource–limited world, there will be pressing need for tools that help managers and policy makers learn quickly what actions or decisions are most effective and efficient. Fortunately, AM provides a framework for deciding what to monitor, how intensively to monitor, how to analyze the monitoring data, and when to stop monitoring. In the absence of a clear link between monitoring and management decisions, these questions are difficult to answer.

Nearly all forms of adaptation to climate change involve setting management objectives, delivering management on the ground, and evaluating progress toward meeting those objectives, i.e., AM (Figure 12.1). The monitoring component of AM,

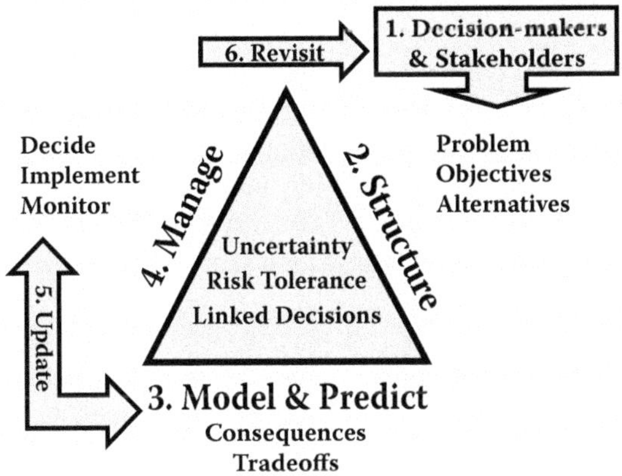

FIGURE 12.1 Diagram of the adaptive management process. 1. Assemble decision makers and key stakeholders. 2. Structure the problem by defining the scope of the problem, clarifying objectives, and generating alternatives. Consider the sources of management uncertainty and risk tolerance of the decision maker, and identify linked decisions. 3. Model and predict the consequences of management alternatives and associated trade-offs. 4. Manage by making a decision, implementing that decision, and monitoring the outcomes. 5. Update the models with monitoring information; repeat steps 3, 4, and 5, reducing management uncertainty with each iteration. 6. Periodically revisit the structuring phase of the process by assembling key stakeholders.

usually referred to as targeted monitoring, supports management decisions, helps managers to evaluate the achievement of management objectives at multiple spatial scales, and helps to evaluate the effects of climate change (Holthausen et al. 2005). Targeted monitoring under AM is flexible in terms of the resources and expertise needed to implement it. Targeted monitoring of species and ecosystems can contribute to vulnerability assessments and provide an early warning of impending ecosystem changes and effects on high–priority resources (Lawler et al. 2010).

Targeted monitoring is perhaps the type of monitoring most frequently conducted by Refuge System field stations and is the most resource–intensive component of the AM process (Holthausen et al. 2005; Nichols and Williams 2006; Sauer and Knutson 2008). All but the simplest forms of monitoring require protocols, sampling designs, documentation, data management, and reporting. If widely employed across the Refuge System, targeted monitoring under AM has the potential to improve the quality, efficiency, and transparency of management decisions. The Refuge System is implementing a new Inventory and Monitoring Program that will support targeted monitoring under AM as one element of the program.

VALUES AND TRADE–OFFS

Regardless of what management decisions are made, there are necessary trade–offs among objectives. For example, should we manage for grassland or shrub–associated bird species? Fire is an important tool for inducing the disturbances needed to keep some North American ecosystems healthy, but fire releases carbon into the atmosphere. These trade–offs link back to the values associated with the management objectives. We need a process for evaluating these trade–offs. Evaluating risks and trade–offs involves some form of modeling, followed by review and consideration by the stakeholders (Williams et al. 2007; Gregory 2000; Gregory and Wellman 2001; Tompkins et al. 2008). Scientific information supports the process, but the process is fundamentally focused on values.

Competing models of how the system is expected to respond to management are used to address both uncertainty and trade–offs among alternatives. A wide range of models are used to support AM, including system models, population models, decision trees, Bayesian updating, and stochastic dynamic programming, to name a few (Walters 1986; Teeter et al. 1993; Kendall 2001; Nichols 2001; Nyberg et al. 2006; Starfield and Bleloch 1991). The primary purpose for a model in a decision–making context is to capture key relationships between alternative management decisions and possible outcomes (ecosystem responses), which are tied directly to the management objectives. Decision analysts begin with simple conceptual models and add complexity as needed to address the decision problem—a process referred to as rapid prototyping (Starfield et al. 1994). The model plays a central role in both SDM and AM by uniting all of the key elements of the problem into a unified framework. This modeling framework supports clear thinking, transparency, and can be used to conduct sensitivity analyses to assess which elements of the problem may have the largest effect on the management outcomes (Starfield and Bleloch 1991).

EXAMPLE: COASTAL WETLAND IMPOUNDMENTS AND SEA–LEVEL RISE

We present a case study from the Refuge System that illustrates several components of the AM process described previously, as applied to a climate change–related problem. National Wildlife Refuges in the northeastern United States are using a structured process to help them make decisions about whether to maintain, restore, or abandon freshwater coastal impoundments. Generally, these freshwater units were created by impounding a low–lying saltmarsh area. Some impoundments may have a small freshwater stream as a water source, while others are entirely dependent upon precipitation. Freshwater impoundments were originally created to support waterfowl and other waterbirds—species identified as conservation targets by the refuges. As sea level rises and the dikes that maintain these impoundments age, managers face a costly decision about whether to maintain the dike, remove it, or abandon it. Maintaining the dikes is an example of a "resistance" strategy aimed at maintaining the status quo (Galatowitsch et al. 2009; Millar et al. 2007). Removing the dikes and allowing the impoundment to revert to saltmarsh is a strategy likely to increase ecosystem resilience and facilitate ecosystem adaptation to future sea–level rises. The "right" strategy depends, in part, on the rate and magnitude of sea–level rise. Workshops, attended by refuge staff, partners, and species and ecosystem experts, were held to review the problem, revisit management objectives, and consider creative solutions. The scope of the problem was defined to include refuges in the northeastern United States that are maintaining dike systems that support coastal freshwater impoundments.

The fundamental objective is to effectively and efficiently support conservation targets (as defined by the refuges' Comprehensive Conservation Plans or Habitat Management Plans); means objectives include supporting target species' use of the impoundments and ecological integrity. These objectives will be translated into attributes (monitoring metrics) that represent forage quality, level of disturbance, and vegetative cover (target species use) and desirable ecological processes, an index of floristic quality, and presence/absence of contaminants (ecological integrity). The refuge also has objectives for maintaining visitor safety and supporting visitor use and education that are related to the issue of maintaining the dikes; refuge visitors enjoy using the dikes as access to wetland areas for bird watching. These objectives were considered constraints for the purposes of this problem.

Linked decisions are involved in that first, a manager needs to consider a large set of factors that affect the decision to maintain, remove, or abandon the dike. For example, the status of the freshwater source, annual cost of dike repairs, public safety considerations, elevation of the dike and the projected rate of sea level rise, presence of invasive species or disease agents, and stakeholder concerns are all important factors. Once a decision has been made, periodic monitoring of specific ecological attributes of the impoundment (ecological integrity) and the target species using the impoundment will be useful in informing future decisions. A model is under development that will support the initial decision to remove, repair, or rebuild the dike as well as incorporate the monitoring data to help inform future decisions. The decision to remove the dike is a "one–time" decision (because it is unlikely that it will be rebuilt); the decision to repair or rebuild the dike can be revisited at a later time. By working together and sharing their monitoring information, managers can learn much more from a series of these "one–time" decisions made in different places than any individual manager could learn by monitoring independently.

A central role for AM is reducing the uncertainty associated with alternative management actions (Lyons et al. 2008). In this example, uncertainty about the rate of sea–level rise is a major issue, but the outcomes of any management actions will also be influenced by drought and precipitation patterns (environmental variation). The primary uncertainties are the outcomes of management actions that are planned (structural uncertainty). Few coastal freshwater impoundments have been restored to saltmarsh (dike removed), so little information is available about the outcomes of such actions. Management alternatives were discussed in the workshops that might not have been considered by a single manager working alone. For example, relocating an impoundment, if the benefits outweigh the costs, or "mothballing" the impoundment by maintaining the dike with minimal effort while more information is gathered were options considered. Creating fish passage around or through the dike was also mentioned as a possibility. Creativity is encouraged by the structured process.

Trade–offs are an important consideration in the impoundment problem. Freshwater impoundments maintained by a dike are not a "natural" component of the coastal environment; if the dike is removed, either a saltmarsh or a mudflat will likely replace the impoundment. These habitats support different sets of species; any decision will trade off the needs of one set of species for another. Different decisions carry different costs; decisions will be made based upon whether the value of the impoundments to the target species outweighs the costs, among other criteria. The model will provide the accounting system for calculating the cost–benefit ratio. However, the manager retains the decision–making authority; the model simply informs that decision by summarizing and reporting the available information.

The structured process helped this group of managers think through a complex, shared problem and facilitated input from managers higher in the organization as well as partners outside of the Refuge System. The workshops allowed the managers to discuss their management objectives for these impoundments in depth and consider how to measure the achievement of their management objectives. Workshop minutes have been translated into a project record that documents decisions made by the team and details of the model. The project record is updated as the project evolves. This project is in the initial, structuring phase of AM. The managers are committed to developing a monitoring plan that will help inform future decisions.

SUMMARY

Climate change brings with it greater uncertainty about what public resource managers should be managing for (conservation targets) and what strategies are likely to meet their objectives. We suggest that a structured approach to management decision making fosters high–quality communication among managers, scientists, and other stakeholders and broad stakeholder input into the value–based components, and promotes the application of the best available science in the modeling and monitoring components. The structured approach breaks down the problem into various components that can be considered individually and in context with the other components. It can be scaled to the complexity and time frame of the problem; a relatively simple approach is applied to simple problems or short–term decisions and a more intensive, model–based approach can be used for more complex problems and long–term decisions.

The focus of decision analysis is on resolving a management problem and reducing uncertainty regarding management decisions. Information about how ecosystems are structured or how they function is relevant in this context only if it helps reduce management uncertainty. Thus, applying SDM and AM on the ground requires recognizing that the practice of resource management is different than the practice of scientific inquiry and requires a set of tools designed specifically to address the needs of resource managers. This is a paradigm shift for both ecological scientists and for resource managers.

ACKNOWLEDGMENTS

We thank our colleagues for sharing ideas about how climate change will affect the practice of resource management, especially Hal Laskowski, Jennifer Casey, Andy Loranger, John Morton, Todd Sutherland, Socheata Lor, Pauline Drobney, Clinton Moore, and Eric Lonsdorf. The findings and conclusions of this article are those of the authors and do not necessarily reflect the view of the US Fish and Wildlife Service.

REFERENCES

Brook, B. W., and D. M. J. S. Bowman. 2006. Postcards from the past: charting the landscape–scale conversion of tropical Australian savanna to closed forest during the 20th century. *Landscape Ecology* 21:1253–1266.

Brooke, C. 2008. Conservation and adaptation to climate change. *Conservation Biology* 22:1471–1476.

Daily, G. C., S. Polasky, J. Goldstein, et al. 2009. Ecosystem services in decision making: time to deliver. *Frontiers in Ecology and the Environment* 7:21–28.

Firbank, L. G. 2005. Striking a new balance between agricultural production and biodiversity. *Annals of Applied Biology* 146:163–175.

Fischlin, A., G. F. Midgley, J. T. Price, et al. 2007. Ecosystems, their properties, goods, and services. In *Climate change 2007: impacts, adaptation and vulnerability. Contribution of Working Group II to the Fourth Assessment Report of the Intergovernmental Panel on Climate Change*, Eds. M. L. Parry, O. F. Canziani, J. P. Palutikof, P. J. van der Linden and C. E. Hanson, 211–272. United Kingdom: Cambridge University Press.

Fischman, R. L. 2003. *The National Wildlife Refuges: coordinating a conservation system through law*. Washington, DC: Island Press.

———. 2007. From words to action: the impact and legal status of the 2006 National Wildlife Refuge System management policies. *Stanford Environmental Law Journal* 26:77.

Folke, C., S. Carpenter, B. Walker, et al. 2004. Regime shifts, resilience, and biodiversity in ecosystem management. *Annual Review of Ecology Evolution and Systematics* 35:557–581.

Galatowitsch, S., L. Frelich, and L. Phillips–Mao. 2009. Regional climate change adaptation strategies for biodiversity conservation in a midcontinental region of North America. *Biological Conservation* 142:2012–2022.

Greene, C. H., A. J. Pershing, T. M. Cronin, and N. Ceci. 2008. Arctic climate change and its impacts on the ecology of the North Atlantic. *Ecology* 89:S24–S38.

Gregory, R. 2000. Using stakeholder values to make smarter environmental decisions. *Environment* 42:34–44.

Gregory, R., L. Failing, and P. Higgins. 2006. Adaptive management and environmental decision making: a case study application to water use planning. *Ecological Economics* 58:434–447.

Gregory, R., D. Ohlson, and J. Arvai. 2006. Deconstructing adaptive management: criteria for applications to environmental management. *Ecological Applications* 16:2411–2425.

Gregory, R., and K. Wellman. 2001. Bringing stakeholder values into environmental policy choices: a community–based estuary case study. *Ecological Economics* 39:37–52.

Gregory, R. S., and R. L. Keeney. 2002. Making smarter environmental management decisions. *Journal of the American Water Resources Association* 38:1601–1612.

Griffith, B., J. M. Scott, R. Adamcik, et al. 2009. Climate change adaptation for the US National Wildlife Refuge System. *Environmental Management* 44:1043–1052.

Herzon, I., and J. Helenius. 2008. Agricultural drainage ditches, their biological importance and functioning. *Biological Conservation* 141:1171–1183.

Holthausen, R., R. L. Czaplewski, D. DeLorenzo, et al. 2005. Strategies for monitoring terrestrial animals and habitats, Gen. Tech. Rep. RMRS–GTR–161. Fort Collins, CO: US Department of Agriculture, Forest Service, Rocky Mountain Research Station.

IPCC. 2007. Climate change 2007: *Impacts, adaptation and vulnerability. Contribution of Working Group II to the Fourth Assessment Report of the Intergovernmental Panel on Climate Change*, Eds. M. L. Parry, O. F. Canziani, J. P. Palutikof, P. J. van der Linden and C. E. Hanson, 976. United Kingdom: Cambridge University Press.

Johnson, D. H. 2000. Statistical considerations in monitoring birds over large areas. In *Strategies of bird conservation: the Partners in Flight planning process*, Eds. R. Bonney, D. N. Pashley, R. J. Cooper, and L. J. Niles. USDA Forest Service Proceedings RMRS–P–16.

Juutinen, A. 2008. Cost–effective forest conservation and criteria for potential conservation targets: a Finnish case study. *Environmental Science & Policy* 11:613–626.

Karl, T. R., J. M. Melillo, and T. C. Peterson, Eds. 2009. *Global climate change impacts in the United States.* New York: Cambridge University Press.

Keeney, R. L. 1992. *Value focused thinking: a path to creative decision making.* Cambridge, MA: Harvard University Press.

Keiter, R. B. 2004. Ecological concepts, legal standards, and public land law: an analysis and assessment. *Natural Resources Journal* 44:943–988.

Kendall, W. L. 2001. Using models to facilitate complex decisions. In *Modeling in natural resource management*, Eds. T. M. Shenk and A. B. Franklin, 147–170. Washington, DC: Island Press.

Knight, A. T., R. M. Cowling, M. Rouget, et al. 2008. Knowing but not doing: selecting priority conservation areas and the research–implementation gap. *Conservation Biology* 22:610–617.

Lawler, J. J., T. H. Tear, C. Pyke, et al. 2010. Resource management in a changing and uncertain climate. *Frontiers in Ecology and the Environment* 8:35–43.

Lyons, J. E., M. C. Runge, H. P. Laskowski, and W. L. Kendall. 2008. Monitoring in the context of structured decision making and adaptive management. *Journal of Wildlife Management* 72:1683–1692.

Magness, D. R., F. Huettmann, and J. M. Morton. 2008. Using random forests to provide predicted species distribution maps as a metric for ecological inventory and monitoring programs. In *Applications of computational intelligence in biology: current trends and open problems. Studies in computational intelligence, vol. 122*, Eds. T. G. Smolinski, M. G. Milanova, and A.–E. Hassanien, 209–229. Heidelberg, Berlin: Springer-Verlag.

Martin, J., M. C. Runge, J. D. Nichols, B. C. Lubow, and W. L. Kendall. 2009. Structured decision making as a conceptual framework to identify thresholds for conservation and management. *Ecological Applications* 19:1079–1090.

Mawdsley, J. R., R. O'Malley, and D. S. Ojima. 2009. A review of climate–change adaptation strategies for wildlife management and biodiversity conservation. *Conservation Biology* 23:1080–1089.

McCarthy, M. A., and H. P. Possingham. 2007. Active adaptive management for conservation. *Conservation Biology* 21:956–963.

McCarthy, M. A., C. J. Thompson, and S. T. Garnett. 2008. Optimal investment in conservation of species. *Journal of Applied Ecology* 45:1428–1435.

Meretsky, V. J., R. L. Fischman, J. R. Karr, et al. 2006. New directions in conservation for the National Wildlife Refuge System. *Bioscience* 56:135–143.

Millar, C. I., N. L. Stephenson, and S. L. Stephens. 2007. Climate change and forests of the future: managing in the face of uncertainty. *Ecological Applications* 17:2145–2151.

Mitchell, R. J., and S. L. Duncan. 2009. Range of variability in southern coastal plain forests: its historical, contemporary, and future role in sustaining biodiversity. *Ecology and Society* 14:17–17.

Munday, P. L., G. P. Jones, M. S. Pratchett, and A. J. Williams. 2008. Climate change and the future for coral reef fishes. *Fish and Fisheries* 9:261–285.

Nelson, E., S. Polasky, D. Lewis, et al. 2008. Efficiency of incentives to jointy increase carbon sequestration and species conservation on a landscape. *Proceedings of the National Academy of Science* 105:9471–9476.

Nemani, R., H. Hashimoto, P. Votava, et al. 2009. Monitoring and forecasting ecosystem dynamics using the Terrestrial Observation and Prediction System (TOPS). *Remote Sensing of Environment* 113:1497–1509.

Nichols, J. D. 2001. Using models in the conduct of science and management of natural resources. In *Modeling in natural resource management*, Eds. T. M. Shenk and A. B. Franklin, 11–34. Washington, DC: Island Press.

Nichols, J. D., and B. K. Williams. 2006. Monitoring for conservation. *Trends in Ecology and Evolution* 21:668–673.

Nielsen, A. B., and B. V. Odgaard. 2005. Reconstructing land cover from pollen assemblages from small lakes in Denmark. *Review of Palaeobotany and Palynology* 133:1–21.

Noss, R. F. 2004. Conservation targets and information needs for regional conservation planning. *Natural Areas Journal* 24:223–231.

Nyberg, J., B. G. Marcot, and R. Sulyma. 2006. Using Bayesian belief networks in adaptive management. *Canadian Journal of Forest Research/Revue Canadienne de Recherche Forestiere* 36:3104–3116.

Palmer, M. A., D. P. Lettenmaier, N. L. Poff, et al. 2009. Climate change and river ecosystems: protection and adaptation options. *Environmental Management* 44:1053–1068.

Peterson, G. D., G. S. Cumming, and S. R. Carpenter. 2003. Scenario planning: a tool for conservation in an uncertain world. *Conservation Biology* 17:358–366.

Polasky, S., E. Nelson, J. Camm, et al. 2008. Where to put things? Spatial land management to sustain biodiversity and economic returns. *Biological Conservation* 141:1505–1524.

Rentch, J. S., and R. R. Hicks. 2005. Changes in presettlement forest composition for five areas in the central hardwood forest, 1784–1990. *Natural Areas Journal* 25:228–238.

Sauer, J. R., and M. G. Knutson. 2008. Objectives and metrics for wildlife monitoring. *Journal of Wildlife Management* 72:1663–1664.

Schroeder, R. L. 2006. A system to evaluate the scientific quality of biological and restoration objectives using National Wildlife Refuge Comprehensive Conservation Plans as a case study. *Journal for Nature Conservation* 14:200–206.

———. 2008. Comprehensive conservation planning and ecological sustainability within the United States National Wildlife Refuge System. *Sustainability: Science, Practice, & Policy* 4:1–7.

Scott, J. M., F. W. Davis, R. G. McGhie, et al. 2001. Nature reserves: do they capture the full range of America's biological diversity? *Ecological Applications* 11:999–1007.

Scott, J. M., D. D. Goble, J. A. Wiens, et al. 2005. Recovery of imperiled species under the Endangered Species Act: the need for a new approach. *Frontiers in Ecology and the Environment* 3:383–389.

Scott, J. M., B. Griffith, R. S. Adamcik, et al. 2009. Adaptation options for climate–sensitive ecosystems and resources, National Wildlife Refuges. In *Preliminary review of adaptation options for climate–sensitive ecosystems and resources, final report, Synthesis and Assessment Product 4.4. A report by the U.S. Climate Change Science Program and the Subcommittee on Global Change Research*, Eds. S. H. Julius, J. M. West, J. S. Baron, L. A. Joyce, P. Karieva, B. D. Keller, M. A. Palmer, C. H. Peterson, and J. M. Scott, 55. Washington, DC: US Environmental Protection Agency, US Climate Change Science Program.

Scott, J. M., P. J. Heglund, M. L. Morrison, et al. 2002. *Predicting species occurrences: issues of accuracy and scale*. Covelo, CA: Island Press.

Starfield, A. M., and A. L. Bleloch. 1991. *Building models for conservation and wildlife management*, 2nd edition. Edina, MN: Burgess International Group, Inc.

Starfield, A. M., K. A. Smith, and A. L. Bleloch. 1994. *How to model it: problem solving for the computer age*. Edina, MN: Interaction Book Company.

Sutherland, W. J., A. S. Pullin, P. M. Dolman, and T. M. Knight. 2004. The need for evidence–based conservation. *Trends in Ecology & Evolution* 19:305–308.

Swift, M. J., A. M. N. Izac, and M. vanNoordwijk. 2004. Biodiversity and ecosystem services in agricultural landscapes—are we asking the right questions? *Agriculture Ecosystems & Environment* 104:113–134.

Teeter, L., G. Somers, and J. Sullivan. 1993. Optimal forest harvest decisions: a stochastic dynamic programming approach. *Agricultural Systems* 42:73–84.

Thomas, C. D., A. Cameron, R. E. Green, et al. 2004. Extinction risk from climate change. *Nature* 427:145–148.

Tompkins, E. L., R. Few, and K. Brown. 2008. Scenario–based stakeholder engagement: incorporating stakeholders preferences into coastal planning for climate change. *Journal of Environmental Management* 88:1580–1592.

US Congress. 1997. National Wildlife Refuge System Improvement Act of 1997. PL 105–57, Statute 1252.

US Department of the Interior. 2009a. Addressing the impacts of climate change on America's water, land, and other natural and cultural resources (Secretarial Order 3289). Edited by Interior Secretary Ken Salazar. Washington, DC: US Department of the Interior.

———. 2009b. An analysis of climate change impacts and options relevant to the Department of the Interior's managed lands and waters. Edited by Subcommittee on Land and Water Management Task Force on Climate Change. Washington, DC: US Department of the Interior.

US Fish and Wildlife Service. 2000. Comprehensive Conservation Planning Process. 602 FW 3. Edited by Division of Refuges. Washington, DC: US Fish and Wildlife Service.

———. 2001. Biological Integrity, Diversity, and Environmental Health. 601 FW 3. Edited by Division of Natural Resources. Washington, DC: US Fish and Wildlife Service.

———. 2009. Identifying refuge resources of concern and management priorities: a handbook (Draft). Edited by National Wildlife Refuge System. Arlington, VA: US Fish and Wildlife Service.

Walters, C. J. 1986. *Adaptive management of renewable resources*. Caldwell, NJ: Blackburn Press.

Walters, C. J., and C. S. Holling. 1990. Large–scale management experiments and learning by doing. *Ecology* 71:2060–2068.

Weiss, S. B., D. D. Murphy, P. R. Ehrlich, and C. F. Metzler. 1993. Adult emergence phenology in checkerspot butterflies: the effects of macroclimate, topoclimate, and population history. *Oecologia* 96:261–270.

Williams, B. K., R. C. Szaro, and C. D. Shapiro. 2007. Adaptive management: the US Department of the Interior technical guide. Washington, DC: Adaptive Management Working Group, US Department of the Interior.

Williams, J. W., and S. T. Jackson. 2007. Novel climates, no–analog communities, and ecological surprises. *Frontiers in Ecology* 5:475–482.

Winfree, R. 2008. Wild bee pollinators provide the majority of crop visitation across land–use gradients in New Jersey and Pennsylvania, USA. *Journal of Applied Ecology* 45:793–802.

Zuckerberg, B., A. M. Woods, and W. F. Porter. 2009. Poleward shifts in breeding bird distributions in New York State. *Global Change Biology* 15:1866–1883.

Section VI

Conclusions and Future Research Needs

13 Ecological Consequences of Climate Change: *Synthesis and Research Needs*

Erik A. Beever and Jerrold L. Belant

CONTENTS

As has been evidenced throughout the chapters, characterizations of biotic response to recent climate change continue to occur, and with a quickening pace compared to earlier research on the topic. However, as also evidenced throughout this book, the complete story involved in most responses is invariably more nuanced. Some species will exhibit dramatic response to a particular aspect of altered climate, others may exhibit a minimal or no detectable response, while a few exhibit a response counter to what would be predicted by a tight relationship to climate. For example, under warmer climates in a portion of a mountain range, some species' elevational limits will retract upslope at the lower edge, other species' limits at the upper edge, while still others may shift their lower- or upper-elevational boundary (or both) downslope (Moritz et al. 2008; Lenoir et al. 2010; Wilson and Gutiérrez *this volume*). Similarly, the same species may exhibit dramatic reductions in distribution in some regions, yet experience few or no apparent losses in other portions of the species' geographic range (e.g. Walther et al. 2002; Bruggerman 2009; Beever et al. 2010). Alternatively, the same species may have its distribution controlled by one factor in a portion of its range, and by another factor at different elevations, latitudes, years, seasons, or even weeks (Stenseth et al. 2002; Hallett et al. 2004; Murphy and Lovett-Doust 2007). However, just as a city's environment is better described by its climate than by its forecast for the next day's weather, the complete story of responses to recent climate change is most accurately told by synthesizing our understanding of multiple evidences of biotic response.

We have made an intentional effort to highlight investigations that involve diverse taxa from many corners of the globe. Species within clades as diverse as butterflies (Fleishman and Murphy *this volume*), coastal plants (Jones *this volume*), montane and arctic mammals (Guralnick et al. *this volume*), wetland-breeding amphibians (Blaustein et al. *this volume*), and boreal-forest inhabitants (McLennan *this volume*) all possess unique combinations of life-history traits that may make them particularly responsive to contemporary changes to climate in some contexts, relatively resilient to altered climates in other contexts, and particularly sensitive to mitigation and conservation-adaptation efforts in still other contexts. As the various chapters have illustrated, understanding of the biotic responses to climate is only achieved when an investigator has intimate familiarity with the ecosystem and its drivers, the natural history of the focal species and the species with which it interacts, and the other factors acting on the focal species. Furthermore, given that climate will typically interact with other drivers and stressors (Root and Schneider 2006; Thomas et al. 2006; Parmesan et al. 2011), alternative explanations that involve no influence of climate or a mix of climatic and nonclimatic factors should be retained explicitly as competing hypotheses (e.g., Kelly and Goulden 2008).

As reflected in the organization of section themes of the book, science related to contemporary climate change begins with understanding the climatology of past, current, and future regimes as it relates to biotas. It is incumbent upon biologists to remain vigilant for new ways in which aspects of micro-, meso, macro-, and global climates may influence one or more focal ecosystem components. For example, currently in most contexts it is relatively poorly understood how relative humidity affects short- and long-term reproductive success of many animals, although it is possible to conceive how experiments could be devised to clarify the variable's role. Similarly, although several kilometers above the earth's surface, changes in the earth's troposphere may directly or indirectly affect demographic trends of wildlife species or freshwater distribution, independently of the well-known greenhouse effect. Beyond climatology, biotic responses to climatic changes can occur at an autecological level and at a community-wide, synecological level (Jones *this volume*; Wilson and Gutiérrez *this volume*).

Given this body of primary research, long-term monitoring is continuing and new efforts are being initiated at numerous spatial and temporal scales (Peterson et al. *this volume*; Goodin *this volume*; McLennan *this volume*). These monitoring efforts seek to detect some meaningful subset of the changes caused by recent climate change at places and times other than those associated with the original research that demonstrated those changes. In some cases, monitoring has provided some very novel insights about the what, where, and when of biotic responses, and occasionally even some how's (Peterson et al. *this volume*). In turn, both investigative research and monitoring will be used to make conservation decisions and strategies in the face of contemporary climate change and other contemporary ecosystem drivers (Dudley et al. *this volume*; Knutson *this volume*).

IMPORTANCE OF MECHANISTIC UNDERSTANDING

Throughout the book, authors consistently made an effort to discuss, propose, and highlight known or suspected mechanisms that account for biotic responses to climate change. Although great rigor in experimental and monitoring design is needed to acquire a mechanistic understanding, it is only with such understanding that conservation and management actions will be best informed (Hallett et al. 2004; Beever et al. 2010). The longer that one ponders a particular type of wildlife–climate relationship, the more demanding that a *full* mechanistic understanding seems. That is, complete understanding of the mechanisms by which climate affects an ecological phenomenon would ideally include relatively long-term data on (a) macroclimate and microclimates relevant for the target species or process; (b) distribution of the target species through time, at fine spatial resolutions and broad extent; (c) demography, including reproductive success or knowing which genders and age classes are being positively or negatively affected; (d) physiology of each individual being affected; (e) behavior of affected and unaffected individuals or populations (e.g., how much plasticity exists in diel behavior patterns?); and (f) sympatric species or other alternative sources of influence on the target species. Given that the full complement of these data sources are rarely present in an entire research program (let alone in any individual investigation), understanding will usually have to be constructed piecemeal, and at least occasionally via proxies of the variables truly affecting biotas.

Even in cases where an aspect of climate is suspected to affect an ecosystem component or ecological process, our understanding of biotic responses will be much improved if greater specificity is achieved. For example, consider the hypothetical assertion that "increasing temperatures" are reducing survivorship of juveniles of a particular species of plethodontid salamander. First, are temperature maxima, minima, or means increasing? Second, in which season(s) of the year are temperatures rising in a manner that is biologically relevant? Alternatively, is season even the right time period to investigate? In some cases, month, week, day, or even single extreme events may be more informative (Gutschick and BassiriRad 2010). Third, over what spatial extents and resolutions are temperatures increasing? Fourth, can you distinguish whether the effects are not from increasing temperatures, but are instead from increased variability in temperatures? Careful and thoughtful study of those portions of a species' distribution in which it is being lost can provide additional insight (*sensu* Wilson et al. 2005). Such focused study may alternatively be performed across important biophysical gradients within a species' distribution (Breshears et al. 2008).

SOME THINGS WE KNOW, AND THINGS WE HAVE LEARNED

As evidenced in this book, work on biotic responses to contemporary climate change is burgeoning, and scientists are achieving new milestones of understanding every month. As also seen in this book, biotic responses reflect influence not only of climate, but also of other factors—interacting in nuanced and sometimes nonintuitive ways (Post and Stenseth 1999; Jones *this volume*; Peterson et al. *this volume*; Blaustein et al. *this volume*; Fleishman and Murphy *this volume*; Wilson and Gutiérrez *this volume*). Furthermore, even in situations where anthropogenic drivers are contributing minimally, most commonly, effects of several aspects of climate will be acting simultaneously (Peterson et al. *this volume*; Jones *this volume*; Martínez-Meyer *this volume*). Climate will act on biotas in both direct and indirect ways (Jones *this volume*; Fleishman and Murphy *this volume*), and pathways of climatic influence may involve as intermediates (a) only abiotic ecosystem components, (b) both biotic and abiotic components, or (c) (climate acting directly on) a suite of only biotic components.

Studies of all kinds of taxa cement the observation that species within clades can vary dramatically in the magnitude of their response to climatic influence, as a function of life-history characteristics, synecology, and innumerable other factors (Blaustein et al. *this volume*; Peterson et al. *this volume*; Fleishman and Murphy *this volume*; Walther et al. 2002; Root et al. 2003; Parmesan 2006; Lawler et al. 2010). We also know that populations from different portions of a species' geographic range may exhibit apparently different responses to climatic influence (Murphy and Lovett-Doust 2007; Beever et al. 2010; Blaustein et al. *this volume*; Fleishman and Murphy *this volume*). Similarly, different models and algorithms that predict future distributions for the same species and time period can vary widely among models; combining predictions into an ensemble forecast will be useful in some contexts, but understanding the reasons behind that variability may be important in other contexts (Martínez-Meyer *this volume*).

Particular insights can be gained by empirically comparing responses of species that might be predicted to respond very differently to climate, across a gradient of important contexts (e.g., latitude, elevation, precipitation, distance to water) (Fleishman and Murphy *this volume*). Robust data will need to be combined with solid biological understanding to disentangle effects of climate from other factors that can influence the measured response variables (Fleishman and Murphy *this volume*). We know that some aspects of climate such as precipitation may respond to one climate-forcing phenomenon (e.g., Southern Oscillation Index) in some years, but to others (e.g., North Atlantic Oscillation) in other years (Goodin *this volume*). Analogously, parameters of biotic response, even within the same population, can vary across weeks, months, seasons, years, and sex and age classes (Hallett et al. 2004). The various chapters have also explicitly illustrated the types of caveats that will be necessary to increase the pace of our learning through time about biotic responses to climate.

REMAINING GAPS IN OUR UNDERSTANDING—SOME EXAMPLES

As evidenced in the chapters, great strides have been achieved, but many biotic responses to climate change are poorly understood. Here we outline a few areas that may be considered as relative gaps in our understanding, and consequently could serve as avenues for further research—across a range of taxa, ecosystems, and temporal resolutions.

On initial inspection, it is remarkable how much uncertainty remains in our understanding of the climate system itself. Although there are general predictions that the frequency of anomalous events (i.e., strong departures from the mean, in a given climate variable) will generally increase under continued climate change, it is not clear for most aspects of climate (e.g., wind speed, temperature, frequency of drought) what the magnitude of departure or the spatial pattern in such deviations will be. Additionally, for both hindcasts and forecasts, much higher levels of uncertainty accompany predictions of precipitation than predictions of temperature (Hulme 2005; IPCC 2007). Also, much remains to be learned of how climatic variables (even ones as fundamental as temperature) vary in both their status and trends at multiple scales (Pepin and Lundquist 2008), how any trends will interact with trends in other climatic parameters, and how these dynamics will affect ecological systems. Importantly for managers of individual land-management units, modelers are still currently struggling with statistical and dynamical downscaling of regional-climate models that will produce climate forecasts at (finer) resolutions that are meaningful to manageable portions of those units and to biological entities within the units (McLennan *this volume*).

Correspondingly, ecological uncertainties in response to climate are compounded by these uncertainties in predicting climate itself. For example, for many taxa, it remains unknown how species (e.g., with different life-history characteristics, such as *r* vs. *K* selection) will respond to more frequent extreme-weather events (McLennan *this volume*; Wilson and Gutiérrez *this volume*); it is even less well known how such altered frequencies will affect species interactions and ecological processes. Past research shows that at least for populations living in topographically heterogeneous systems, responses to weather are mediated by the topography of the habitat patch (Singer 1972; Singer and Ehrlich 1979; Weiss et al. 1987; 1988; Murphy and Weiss 1988). To illustrate the complexity of alterations to multiple climate factors, there are many possible ways in which species may respond to not only altered precipitation regimes, but also to the interaction of altered precipitation and temperature. For example, although some areas of Alaska are predicted to receive increased amounts of precipitation in the coming decades, the rate of potential evapotranspiration (or water stress) at those areas may actually increase, if rising temperatures effectively outstrip the higher precipitation (O'Brien et al. in preparation). As evidenced throughout the book, nearly all climate–biota interactions will be more fully understood with better understanding of the spatial and temporal scales of climate (and other stressors) that are important for various ecosystem components and processes. Temporally, for example, the time lags after which climatic influence will be exhibited in species can vary across taxa and years, but are generally unknown (Hallett et al. 2004; Thomas et al. 2004; Cole 2010). In low-productivity, event-driven systems such as the Grand Canyon, however, Cole (2010), using paleo-reconstruction of 81 assemblages, found that multiple centuries were required for plant communities to re-establish and stabilize after periods of climate warming in the early Holocene.

The understanding of more-complex dynamics that is beginning to emerge from research and monitoring suggests a further suite of questions. For example, as of now we only know particular case examples of how synergies of climate with other ecosystem drivers, disturbances, and stressors (both natural and anthropogenic) affect biotic response. Furthermore, although it has been modeled for some ecosystems (e.g., Luo et al. 2008), it is largely unknown how historically unprecedented levels of atmospheric CO_2 around the globe (IPCC 2007) will affect species' physiology as they respond to changes in climate (but see, e.g., Smith et al. 2000; Russell et al. 2009). In some instances, one or more aspects of a species' synecology will be the means by which climatic influence may either expand or contract a species' local (or regional) distribution, depending on whether the changed climate ultimately weakens or strengthens the species in its interactions. Examples of synecological interactions include disease or parasitism, predation pressure, and competition from other species that may also be undergoing changes in abundance or distribution in response to climate change (LaVal 2004; Pounds et al. 2006). In other instances, synecological factors will interact in concert with direct (e.g., physiological) effects of climate on organisms. For many species, it remains unknown what combinations of physiology, diet, water stress, symbiotic relationships, and other stressors (e.g., fire, invasive species, harvest) produce losses in distribution. More rare yet is quantification of the relative contributions of these various factors in various geographic and topographic contexts. As with many biological phenomena, synergy with other factors appears likely to be the norm (Murphy and Weiss 1992; Thomas et al. 2004; Hulme 2005; Root and Schneider 2006; Parmesan et al. 2011). Changes in land use and resultant landscape fragmentation, as well as invasive species appear poised to be factors that commonly interact with climate (Noss 2001; Thomas et al. 2004; Wilson et al. 2005).

Although there have been many predictions of which life-history traits may make individual species more or less vulnerable to relatively rapid changes in climate, there remains a need to broadly and rigorously test the generalities that have been postulated (Keith et al. 2008). For example, there still remains a need to find traits, markers, and even proxies for predicting not only which species, but also which populations within a species may be most vulnerable to climate-induced losses (Beever et al. 2012). The domain of vulnerability assessment has already spawned an ever-enlarging literature on species, guilds, and regions predicted to be at greater risk; however, relatively few such assessments have been quantitatively validated or verified. Although some traits are intuitively obvious predictors of risk (e.g., limited dispersal capacity and low reproductive capacity [Thomas et al. 2004]), other relationships may yield more surprises. For mountain-dwelling vertebrates, for example, persistence within a particular mountain range might reflect any combination of habitat-diversity and climatic surrogates: total area of habitat present, topographic diversity (e.g., availability of different aspects), presence of cold-air pooling, elevation of the highest peak, extent of perennial flowing water, abundance of rock-ice features, number of species that provide habitat structure (e.g., conifers), or number of major habitat types (Johnson 1978; Brown 1978; Murphy and Weiss 1992; Millar and Westfall 2010). Future challenges will include not only understanding why patterns of loss of some species unexpectedly depart from predictions of ecological theory (e.g., "holes" in nestedness analyses), but also why other species are surprisingly resilient. Even for demonstrated relationships between organisms and climate, we generally do not know where the "tipping points," thresholds, and nonlinearities lie, in terms of climatic tolerance—especially when multiple factors are involved (even if all are related to climate). From the opposite side of the coin, it is virtually untested which species (if any) will be most effectively aided by assisted migration, assuming that all of the attendant concerns are handled satisfactorily (Richardson et al. 2009). In most cases, models of biota–climate relationships that have been postulated or demonstrated need to be further corroborated, validated, and confirmed (Hulme 2005).

Although considerable insights and knowledge of the effects of contemporary climate change have been gained, much remains unknown. As is the case for vulnerability of various species or taxa under altered climates, it remains to be confirmed which communities, ecosystems, and ecoregions may be least and most affected by contemporary climate change. As with most species (Beever et al. 2010), we do not know whether it is the status of potentially stressful aspects of climate or the magnitude of change in those aspects that determines the vulnerability of communities and ecosystems. At these broader scales, although multi- and cross-scale hierarchies and dynamics have been described in a few systems (e.g., Allen 2007; Peters et al. 2007), these are the exceptions. Often, data with high resolution from broad spatial and temporal extents will be necessary to illuminate such relationships. We also suspect more surprises and novel insights in community dynamics, wherein an ecosystem component or process is affected in a secondary or tertiary manner, as a function of ecosystems' organization and inter-connectedness.

LOOKING FORWARD TO THE FUTURE

Nearly all indications suggest that the climate changes the earth is currently experiencing will not only *not* subside in the foreseeable future, but will likely become more pronounced in the coming decades (IPCC 2007; evidence presented in Wilson and Gutiérrez *this volume*). Consequently, in spite of the uncertainty and gaps in our understanding detailed previously, conservation planners, land managers, adaptation and mitigation specialists, and other conservation practitioners must continue to make decisions regarding how to manage and conserve ecosystems and their components in the face of contemporary climate change. Given the substance of the previous chapters, we argue that these efforts must be prepared for greater uncertainty than has been evident in past decades. To best inform understanding of this variability, we urge research and monitoring efforts to regularly employ existing and emerging tools to quantify uncertainty in displays of their results. Such tools include the classic confidence intervals (e.g., 95%), standard errors (e.g., ±1 SE), and standard deviations; CVs (coefficients of variation); a posteriori distributions for Bayesian analyses; quantitative mention of dispersion and skew, when relevant; probability density functions; simulation (e.g., Monte Carlo) modeling; fuzzy logic; and explicit quantification of process variance. Fleishman and Murphy (*this volume*) provide numerous approaches that may minimize at least some of the types of uncertainty attendant with wildlife–climate investigations. Furthermore, adaptive management can help land managers address the uncertainties associated with climate change by iteratively learning from past strategies and adjusting future management accordingly, rather than acting on assumptions (Knutson and Heglund from Chapter 12; McLennan *this volume*).

In addition, given the seeming magnitude of the challenge, we encourage researchers, land managers, and wildlife and conservation practitioners who may have already been wrestling with analogous issues to seek out other comparable efforts, and thereby avoid "reinventing the wheel." Similarly, merging the time and talents of multiple team members (McLennan *this volume*) can bring surprisingly helpful insights and improvements to most projects. Such collaboration may involve only similar disciplines or may merge expertise from highly divergent disciplines. For example, wildlife ecologists may find synergy when working with climatologists, economists, ecosystems-simulation modelers, ecological-niche modelers (Martínez-Meyer *this volume*), soil scientists, botanists, hydrologists, physiologists, epidemiologists and veterinarians, and sociologists. The right team will ideally reflect the system and hypotheses or questions being addressed, and less strongly reflect personalities and political allegiances. We agree with Jones (*this volume*) that seeking to sustain, conserve, and promote resilience will serve as a valuable long-term strategy, and with Wilson and Gutiérrez (*this volume*) that reducing nonclimate threats to species and ecosystems that appear particularly vulnerable to rapid change (especially decline) may render a more-palatable palette of management and conservation options for the future.

SUMMARY

Finally, we cannot stress enough how important a mechanism-focused approach will be to clarify and unravel the nuances that often typify wildlife–climate relationships. As illustrated in this book, such approaches may involve exploratory modeling and analyses; development and use of novel field instruments, tools, and analytical approaches (e.g., Wilson et al. 2005); and well-reasoned thinking from design through interpretation stages of research or monitoring. In some cases, parallel experiments may be needed to remove alternative explanations, one at a time; in others, factorial and other more complex designs may be required to tease out how various factors are combining synergistically in biotic responses to changes in climate. Explicit testing of our assumptions, both new and long held, may yield additional insight (Fleishman and Murphy *this volume*; Guralnick et al. *this volume*). Such understanding from these and other approaches will be the energy to fuel the adaptation, mitigation, and restoration engines, as biotas respond at various spatial and temporal scales. The results of these and other efforts may well be some of the more enduring legacies by which the current generation may be remembered decades into the future.

REFERENCES

Allen, C. D. 2007. Interactions across spatial scales among forest dieback, fire, and erosion in northern New Mexico landscapes. *Ecosystems* 10:797–808.

Beever, E. A., C. Ray, P. W. Mote, and J. L. Wilkening. 2010. Testing alternative models of climate-mediated extirpations. *Ecological Applications* 20:164–178.

Beever, E. A., C. Ray, J. L. Wilkening, P. F. Brussard, and P. W. Mote. 2011. Contemporary climate change hastens extinctions and alters the factors governing extinction. *Global Change Biology* 17(6):2054–2070.

Breshears, D. D., T. E. Huxman, H. D. Adams, C. B. Zou, and J. E. Davison. 2008. Vegetation synchronously leans upslope as climate warms. *Proceedings of the National Academy of Sciences of the United States of America* 105:11591–11592.

Brown, J. H. 1978. The theory of insular biogeography and the distribution of boreal birds and mammals. *Great Basin Naturalist Memoirs* 2:209–227.

Cole, K. L. 2010. Vegetation response to Early Holocene warming as an analog for current and future changes. *Conservation Biology* 24:29–37.

Gutschick, V. P., and H. BassiriRad. 2010. Biological extreme events: a research framework. *Eos* 91:85–86.

Hallett, T. B., T. Coulson, J. G. Pilkington, T. H. Clutton-Brock, J. M. Permberton, and B. T. Grenfell. 2004. Why large scale climate indices seem to predict ecological processes better than local weather. *Nature* 430:71–75.

Hulme, P. E. 2005. Adapting to climate change: is there scope for ecological management in the face of a global threat? *Journal of Applied Ecology* 42:784–794.

IPCC (Intergovernmental Panel on Climate Change). 2007. *Climate change 2007: the physical science basis*. New York: Cambridge University Press.

Johnson, N, K. 1975. Controls of number of bird species on Montane Islands in the Great Basin. *Evolution* 29:545–567.

Keith, D. A., H. R. Akçakaya, W. Thuiller, G. F. Midgley, R. G. Pearson, S. J. Phillips, H. M. Regan, M. B. Araújo, and T. G. Rebelo. 2008. Predicting extinction risks under climate change: coupling stochastic population models with dynamic bioclimatic habitat models. *Biology Letters* 4:560–563.

Kelly, A. E., and M. L. Goulden. 2008. Rapid shifts in plant distribution with recent climate change. *Proceedings of the National Academy of Sciences of the United States of America* 105:11823–11826.

LaVal, R. K. 2004. Impact of global warming and locally changing climate on tropical cloud forest bats. *Journal of Mammalogy* 85:237–244.

Lawler, J. J., S. L. Shafer, and A. R. Blaustein. 2010. Projected climate impacts for the amphibians of the western hemisphere. *Conservation Biology* 24:38–50.

Lenoir, J., J. C. Gegout, A. Guisan, P. Vittoz, T. Wohlgemuth, N. E. Zimmermann, S. Dullinger, H. Pauli, W. Willner, and J. C. Svenning. 2010. Going against the flow: potential mechanisms for unexpected downslope range shifts in a warming climate. *Ecography* 33:295–303.

Luo, Y. Q., D. Gerten, G. Le Maire, W. J. Parton, E. S. Weng, X. H. Zhou, C. Keough, C. Beier, P. Ciais, W. Cramer, J. S. Dukes, B. Emmett, P. J. Hanson, A. Knapp, S. Linder, D. Nepstad, and L. Rustad. 2008. Modeled interactive effects of precipitation, temperature, and [CO_2] on ecosystem carbon and water dynamics in different climatic zones. *Global Change Biology* 14:1986–1999.

Millar, C. I., and R. D. Westfall. 2010. Distribution and climatic relationships of the American Pika (*Ochotona princeps*) in the Sierra Nevada and Eastern Great Basin. USA Periglacial landforms as srefugia in warming climates. *Arctic, Antartic, and Alpine Research* 42:76–88.

Moritz, C., J. L. Patton, C. J. Conroy, J. L. Parra, G. C. White, and S. R. Beissinger. 2008. Impact of a century of climate change on small-mammal communities in Yosemite National Park, USA. *Science* 322:261–264.

Murphy, H. T. and J. Lovett-Doust. 2007. Accounting for regional niche variation in habitat suitability models. *Oikos* 116:99–110.

Noss, R. F. 2001. Beyond Kyoto: forest management in a time of rapid climate change. *Conservation Biology* 15:578–590.

O'Brien, B. J., A. L. Springsteen, T. S. Rupp, and W. M. Loya. in preparation. Future climate-change impacts on growing-season water availability in Alaska. *Climatic Change*.

Parmesan, C., C. Duarte, E. Poloczanka, A.J. Richardson, and M.C. Singer. 2011. Overstretching Attribution. *Nature Climate Change* 1:2–4.

Parmesan, C. 2006. Ecological and evolutionary responses to recent climate change. *Annual Review of Ecology, Evolution, and Systematics* 37:637–669.

Pepin, N. C. and J. D. Lundquist. 2008. Temperature trends at high elevations: patterns across the globe. *Geophysical Research Letters* 35:1–6.

Peters, D. P. C., O. E. Sala, C. D. Allen, A. Covich, and M. Brunson. 2007. Cascading events in linked ecological and socioeconomic systems. *Frontiers in Ecology and the Environment* 5:221–224.

Post, E. and N. C. Stenseth. 1999. Climatic variability, plant phenology, and northern ungulates. *Ecology* 80:1322–1339.

Pounds, A. J., M. R. Bustamante, L. A. Coloma, J. A. Consuegra, M. P. L. Fogden, P. N. Foster, E. La Marca, K. L. Masters, A. Merino-Viteri, R. Puschendorf, S. R. Ron, G. A. Sanchez-Azofeifa, C. J. Still, and B. E. Young. 2006. Widespread amphibian extinctions from epidemic disease driven by global warming. *Nature* 439:161–167.

Richardson, D. M., J. J. Hellmann, J. S. McLachlan, D. F. Sax, M. W. Schwartz, P. Gonzalez, E. J. Brennan, A. Camacho, T. L. Root, O. E. Sala, S. H. Schneider, D. M. Ashe, J. R. Clark, R. Early, J. R. Etterson, E. D. Fielder, J. L. Gill, B. A. Minteer, S. Polasky, H. D. Safford, A. R. Thompson, and M. Vellend. 2009. Multidimensional evaluation of managed relocation. *Proceedings of the National Academy of Sciences of the United States of America* 106:9721–9724.

Root, T. L., J. T. Price, K. R. Hall, S. H. Schneider, C. Rosenzweig, and A. J. Pounds. 2003. Fingerprints of global warming on wild animals and plants. *Nature* 421:57–60.

Root, T. L. and S. H. Schneider. 2006. Conservation and climate change: the challenges ahead. *Conservation Biology* 20:706–708.

Russell, B. D., J. A. I. Thompson, L. J. Falkenberg, and S. D. Connell. 2009. Synergistic effects of climate change and local stressors: CO_2 and nutrient-driven change in subtidal rocky habitats. *Global Change Biology* 15:2153–2162.

Singer, M.C. 1972. Complex components of habitat suitability within a butterfly colony. *Science* 173:75–77.

Singer, M. C., and P. R. Ehrlic. 1979. Population dynamics of the Checkerspot butterfly.

Smith, S. D., T. E. Huzman, S. F. Zitzer, T. N. Charlet, D. C. Housman, J. S. Coleman, L. K. Fenstermaker, J. R. Seemann, and R. S. Nowak. 2000. Elevated CO_2 increases productivity and invasive species success in an arid ecosystems. *Nature* 408:79–82.

Stenseth, N. C., A. Mysterud, G. Otterson, J. W. Hurrel, K.-S. Chan, and M. Lima. 2002. Ecological effects of climate fluctuations. *Science* 297:1292–1296.

Thomas, C. D., A. Cameron, R. E. Green, M. Bakkenes, L. J. Beaumont, Y. C. Collingham, B. F. N. Erasmus, M. Ferreira de Siqueira, A. Grainger, L. Hannah, L. Hughes, B. Huntley, A. S. van Jaarsveld, G. F. Midgley, L. Miles, M. A. Ortega-Huerta, A. T. Peterson, O. L. Phillips, and S. Williams. 2004. Extinction risk from climate change. *Nature* 427:145–148.

Thomas, C. D., A. M. A. Franco, and J. K. Hill. 2006. Range retractions and extinction in the face of climate warming. *Trends in Ecology & Evolution* 21:415–416.

Walther, G.-R., E. Post, P. Convey, A. Menzel, C. Parmesan, T. J. C. Beebee, J.-M. Fromentin, O. Hoegh-Guldberg, and F. Bairlein. 2002. Ecological responses to recent climate change. *Nature* 416:389.

Weiss, S. B., R. R. White, D. D. Murphyk, and P. R. Ehrlich. 1987. Growth and dispersal of larvae of the Checkerspot butterfly, *Euphydras edithaoikos* 50:161–166.

Weiss, S. B., D. D. Murphy, and R. R. White. 1988. Sun, slope and butterflies: Topographic determinants of habitat quality for *Euphydras editha bayensis*. *Ecology* 69:1486–1496.

Wilson, R. J., D. Gutiérrez, J. Gutiérrez, D. Martínez, R. Agundo, and V. J. Monserrat. 2005. Changes to the elevational limits and extent of species ranges associated with climate change. *Ecology Letters* 8:1138–1146.

Index

D

F

N

O

P

Q

R

S

U

V

W

Y

Z

For Product Safety Concerns and Information please contact our EU representative GPSR@taylorandfrancis.com
Taylor & Francis Verlag GmbH, Kaufingerstraße 24, 80331 München, Germany

www.ingramcontent.com/pod-product-compliance
Ingram Content Group UK Ltd.
Pitfield, Milton Keynes, MK11 3LW, UK
UKHW022122190926
475702UK00013B/1585
9781138114692